Frontiers in Space

Robot Spacecraft

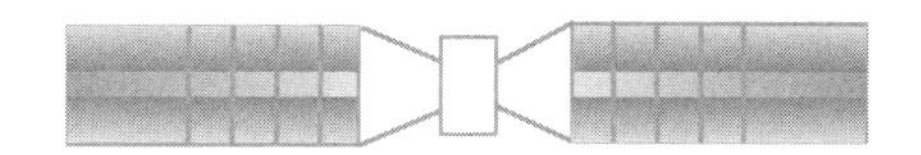

Joseph A. Angelo, Jr.

To the memory of my paternal (Italian) grandparents, Antonio and Nina, who had the great personal courage to leave Europe early in the 20th century and embrace the United States as their new home. Through good fortune they met, married, and raised a family. Their simple, hardworking lives taught me what is most important in life.

This book also carries a special dedication to Mugsy-the-Pug (February 23, 1999, to January 2, 2006)—my faithful canine companion—who provided so much joy and relaxation during the preparation of this book and many other works.

ROBOT SPACECRAFT

Facts On File, Inc.
An imprint of Infobase Publishing
132 West 31st Street
New York NY 10001

Library of Congress Cataloging-in-Publication Data
Angelo, Joseph A.
Robot spacecraft / Joseph A. Angelo, Jr.
p. cm.— (Frontiers in space)
Includes bibliographical references and index.
ISBN 0-8160-5773-7
1. Space robotics—Juvenile literature. 2. Space probes—Juvenile literature. 3. Roving vehicles (Astronautics)—Juvenile literature. I. Title. II. Series.
TL1097.A54 2007
629.47—dc22 2006001118

Text design by Erika K. Arroyo
Cover design by Salvatore Luongo
Illustrations by Sholto Ainslie

Printed in the United States of America

VB FOF 10 9 8 7 6 5 4 3 2 1

This book is printed on acid-free paper.

Contents

Preface

It is difficult to say what is impossible, for the dream of yesterday is the hope of today and the reality of tomorrow.

—Robert Hutchings Goddard

Frontiers in Space is a comprehensive multivolume set that explores the scientific principles, technical applications, and impacts of space technology on modern society. Space technology is a multidisciplinary endeavor, which involves the launch vehicles that harness the principles of rocket propulsion and provide access to outer space, the spacecraft that operate in space or on a variety of interesting new worlds, and many different types of payloads (including human crews) that perform various functions and objectives in support of a wide variety of missions. This set presents the people, events, discoveries, collaborations, and important experiments that made the rocket the enabling technology of the space age. The set also describes how rocket propulsion systems support a variety of fascinating space exploration and application missions—missions that have changed and continue to change the trajectory of human civilization.

The story of space technology is interwoven with the history of astronomy and humankind's interest in flight and space travel. Many ancient peoples developed enduring myths about the curious lights in the night sky. The ancient Greek legend of Icarus and Daedalus, for example, portrays the age-old human desire to fly and to be free from the gravitational bonds of Earth. Since the dawn of civilization, early peoples, including the Babylonians, Mayans, Chinese, and Egyptians, have studied the sky and recorded the motions of the Sun, the Moon, the observable planets, and the so-called fixed stars. Transient celestial phenomena, such as a passing comet, a solar eclipse, or a supernova explosion, would often cause a great deal of social commotion—if not out right panic and fear—because these events were unpredictable, unexplainable, and appeared threatening.

It was the ancient Greeks and their geocentric (Earth-centered) cosmology that had the largest impact on early astronomy and the emergence of Western Civilization. Beginning in about the fourth century B.C.E., Greek philosophers, mathematicians, and astronomers articulated a geocentric model of the universe that placed Earth at its center with everything else revolving about it. This model of cosmology, polished and refined in about 150 C.E. by Ptolemy (the last of the great early Greek astronomers), shaped and molded Western thinking for hundreds of years until displaced in the 16th century by Nicholas Copernicus and a heliocentric (sun-centered) model of the solar system. In the early 17th century, Galileo Galilei and Johannes Kepler used astronomical observations to validate heliocentric cosmology and, in the process, laid the foundations of the Scientific Revolution. Later that century, the incomparable Sir Isaac Newton completed this revolution when he codified the fundamental principles that explained how objects moved in the "mechanical" universe in his great work *Principia Mathematica.*

The continued growth of science over the 18th and 19th centuries set the stage for the arrival of space technology in the middle of the 20th century. As discussed in this multivolume set, the advent of space technology dramatically altered the course of human history. On the one hand, modern military rockets with their nuclear warheads redefined the nature of strategic warfare. For the first time in history, the human race developed a weapon system with which it could actually commit suicide. On the other hand, modern rockets and space technology allowed scientists to send smart robot exploring machines to all the major planets in the solar system (including tiny Pluto), making those previously distant and unknown worlds almost as familiar as the surface of the Moon. Space technology also supported the greatest technical accomplishment of the human race, the Apollo Project lunar landing missions. Early in the 20th century, the Russian space travel visionary Konstantin E. Tsiolkovsky boldly predicted that humankind would not remain tied to Earth forever. When astronauts Neil Armstrong and Edwin (Buzz) Aldrin stepped on the Moon's surface on July 20, 1969, they left human footprints on another world. After millions of years of patient evolution, intelligent life was able to migrate from one world to another. Was this the first time such an event has happened in the history of the 14-billion-year-old universe? Or, as some exobiologists now suggest, perhaps the spread of intelligent life from one world to another is a rather common occurrence within the galaxy. At present, most scientists are simply not sure. But, space technology is now helping them search for life beyond Earth. Most exciting of all, space technology offers the universe as both a destination and a destiny to the human race.

Each volume within the Frontiers in Space set includes an index, a chronology of notable events, a glossary of significant terms and concepts, a helpful list of Internet resources, and an array of historical and current print sources for further research. Based upon the current principles and standards in teaching mathematics and science, the Frontiers in Space set is essential for young readers who require information on relevant topics in space technology, modern astronomy, and space exploration.

Acknowledgments

I wish to thank the public information specialists at the National Aeronautics and Space Administration (NASA), the National Oceanic and Atmospheric Administration (NOAA), the United States Air Force (USAF), the Department of Defense (DOD), the Department of Energy (DOE), the National Reconnaissance Office (NRO), the European Space Agency (ESA), and the Japanese Aerospace Exploration Agency (JAXA), who generously provided much of the technical material used in the preparation of this series. Acknowledgment is made here for the efforts of Frank Darmstadt and other members of the editorial staff at Facts On File, whose diligent attention to detail helped transform an interesting concept into a series of publishable works. The support of two other special people merits public recognition here. The first individual is my physician, Dr. Charles S. Stewart III, M.D., whose medical skills allowed me to successfully complete the series. The second individual is my wife, Joan, who, as she has for the past 40 years, provided the loving spiritual and emotional environment so essential in the successful completion of any undertaking in life, including the production of this series.

Introduction

Modern space robots are sophisticated machines that have visited all the major worlds of the solar system, including (soon) tiny Pluto. *Robot Spacecraft* examines the evolution of these fascinating, far-traveling spacecraft—from the relatively unsophisticated planetary probes flown at the dawn of the Space Age to the incredibly powerful exploring machines that now allow scientists to conduct detailed, firsthand investigations of alien worlds within this solar system. Emerging out of the space race of the cold war, modern robot spacecraft have dramatically changed what we know about the solar system.

In this century, an armada of ever more sophisticated machine explorers will continue this legacy of exploration as they travel to the farthest reaches of the solar system and beyond. Robot spacecraft have formed a special intellectual partnership with their human creators by allowing us to explore more "new worlds" in one human lifetime than in the entire history of the human race. This unprecedented wave of discovery and the continued acquisition of vast quantities of new scientific knowledge—perhaps even the first definitive evidence of whether alien life exists—will transform how human beings view themselves and their role in the universe.

Robot Spacecraft describes the historic events, scientific principles, and technical breakthroughs that now allow complex exploring machines to orbit around, or even land upon, mysterious worlds in our solar system. The book's special collection of illustrations presents historic, contemporary, and future robot spacecraft—allowing readers to appreciate the tremendous aerospace engineering progress that has occurred since the dawn of the space age. A generous number of sidebars are strategically positioned throughout the book to provide expanded discussions of fundamental scientific concepts and robot-spacecraft engineering techniques. There are also capsule biographies of several space exploration visionaries and scientists, to allow the reader to appreciate the human dimension in the development and operation of robot spacecraft.

It is especially important to recognize that, throughout the 20th and 21st centuries and beyond, sophisticated robot spacecraft represent the enabling technology for many exciting scientific discoveries for the human race. Awareness of these technical pathways should prove career-inspiring to those students now in high school and college who will become the scientists, aerospace engineers, and robot designers of tomorrow. Why are such career choices important? Future advances in robot spacecraft for space exploration no longer represent a simple societal option that can be pursued or not, depending upon political circumstances. Rather, continued advances in the exploration of the solar system and beyond form a technical, social, and psychological imperative for the human race. We can decide to use our mechanical partners and become a spacefaring species as part of our overall sense of being and purpose; or we can ignore the challenge and opportunity before us and turn our collective backs on the universe. The latter choice would confine future generations to life on just one planet around an average star in the outer regions of the Milky Way Galaxy. The former choice makes the human race a spacefaring species with all the exciting social and technical impacts that decision includes.

Robot Spacecraft examines the role the modern space robot has played in human development since the middle of the 20th century and then projects the expanded role space robots will play throughout the remainder of this century and beyond. Who can now predict the incredible societal impact of very smart machines capable of visiting alien worlds around other suns? One very exciting option on the space-robot technology horizon is that of the self-replicating system—a robot system so smart it can make copies of itself out of the raw materials found on other worlds. Later in this century, as a wave of such smart robots start to travel through interstellar space, people here on Earth might be able to answer the age-old philosophical question: Are we alone in this vast universe?

Robot Spacecraft also shows that the development of modern space robots did not occur without problems, issues, and major financial commitments. Selected sidebars within the book address some of the most pressing contemporary issues associated with the application of modern robot technology in space exploration—including the long-standing space-program debate concerning the role of human explorers (i.e., astronauts and cosmonauts) versus machine explorers. For some managers within the American space program, this debate takes on an "either/or" conflict; for others, the debate suggests the need for a more readily embraced human-machine partnership. *Robot Spacecraft* also describes how future advances in robot technology will exert interesting social, political, and technical influences upon our global civilization. The technology inherent in very smart space robots will exert a tremendous influence upon the trajectory of human civilization that extends well beyond this century.

Some interesting impacts of smart space robots include their use in the development of permanent human settlements on the Moon and Mars, in exploration of the outermost regions of the solar system, as interstellar emissaries of the human race, and in operation of a robot-spacecraft-enabled planetary defense system against killer asteroids or rogue comets. Sophisticated space robots also have a major role to play in the discovery of life (extinct or existing) beyond Earth and in the emergence of a successfully functioning solar-system civilization. Advanced space-robot systems, endowed with high levels of machine intelligence by their human creators, are unquestionably the underlying and enabling technology for many interesting future developments.

Robot Spacecraft has been carefully designed to help any student or teacher who has an interest in robots discover what space robots are, where they came from, how they work, and why they are so important. The back matter contains a chronology, glossary, and an array of historical and current sources for further research. These should prove especially helpful for readers who need additional information on specific terms, topics, and events in space-robot technology.

From Pioneer Lunar Probes to Interstellar Messengers

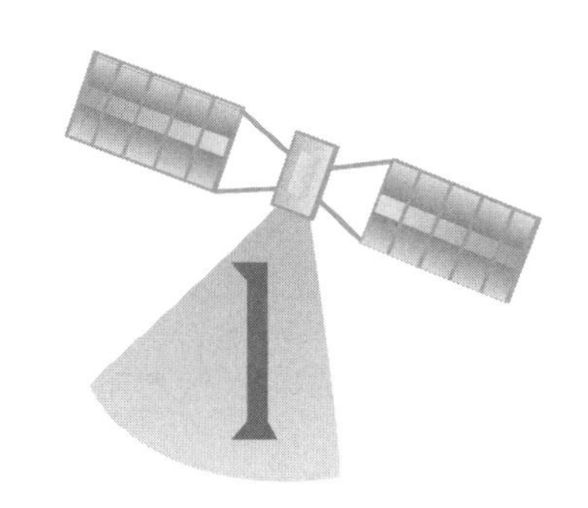

Robot spacecraft have opened up the universe to exploration. Modern space robots are sophisticated exploring machines that have, or will have, visited all the major worlds of the solar system, including tiny Pluto. Emerging out of the politically charged space race of the cold war, a progressively more capable family of robot spacecraft have dramatically changed what scientists know about the alien worlds that journey together with Earth around a star called the Sun. In a little more than four decades, scientists have learned a greater amount about these wandering lights, called πλαυετες (planets) by the ancient Greek astronomers, than in the previous history of astronomy. Thanks to space robots, every major planetary body—and (where appropriate) its collection of companion moons—has now become a much more familiar world. Similarly, sophisticated robot astronomical observatories placed on strategically located platforms in space have allowed astronomers and astrophysicists to meet the universe face-to-face, across all portions of the electromagnetic spectrum. No longer is the human view of the universe limited to a few narrow bands of radiation that trickle down to Earth's surface through an intervening atmosphere that is often murky and turbulent.

This chapter introduces the basic principles of robotics. Space robots share certain common features with their terrestrial counterparts. They also involve, however, a blending of aerospace and computer technologies that is far more demanding, unusual, and sophisticated than that generally needed for robots operating here on Earth. Space robots have to work in the harsh environment of outer space and sometimes up on strange alien worlds about which little is previously known.

Under certain circumstances, telepresence and virtual reality technologies will allow a human being to form a real-time, interactive partnership with an advanced space robot, which serves as a dextrous mechanical surrogate capable of operating in a hazardous, alien world environment. For

Artificial Intelligence

The term *artificial intelligence (AI)* is commonly taken to mean the study of thinking and perceiving as general information-processing functions—or the science of machine intelligence (MI). In the past few decades, computer systems have been programmed to diagnose diseases; prove theorems; analyze electronic circuits; play complex games such as chess, poker and backgammon; solve differential equations; assemble mechanical equipment using robotic manipulator arms and end effectors (the "hands" at the end of the manipulator arms); pilot uncrewed vehicles across complex terrestrial terrain, as well as through the vast reaches of interplanetary space; analyze the structure of complex organic molecules; understand human speech patterns; and even write other computer programs.

All of these computer-accomplished functions require a degree of intelligence similar to mental activities performed by the human brain. Someday, a general theory of intelligence may emerge from the current efforts of scientists and engineers who are now engaged in the field of artificial intelligence. This general theory would help guide the design and development of even smarter robot spacecraft and exploratory probes.

Artificial intelligence generally includes a number of elements or subdisciplines. Some of the more significant of these elements or subdisciplines are: planning and problem solving, perception, natural language, expert systems, automation, teleoperation and robotics, distributed data management, and cognition and learning.

All artificial intelligence involves elements of planning and problem solving. The problem-solving function implies a wide range of tasks, including decision making, optimization, dynamic-resource allocation, and many other calculations or logical operations.

Perception is the process of obtaining data from one or more sensors and processing or analyzing these data to assist in making some subsequent decision or taking some subsequent action. The basic problem in perception is to extract from a large amount of (remotely) sensed data some feature or characteristic that then permits object identification.

One of the most challenging problems in the evolution of the digital computer has been the communications that must occur between the human operator and the machine. The human operator would like to use an everyday, or natural, language to gain access to the computer system. The process of communication between machines and people is very complex and frequently requires sophisticated computer hardware and software.

An expert system permits the scientific or technical expertise of a particular human being to be stored in a computer for subsequent use by other human beings who have not had the

example, an advanced future space robot might explore remote regions of the Moon, while its human controller, working inside a permanent lunar surface base or even back on Earth, uses virtual reality-technologies to make important new discoveries.

As a space robot operates farther away, the round-trip communications distance with human controllers back on Earth must soon be measured not in thousands of miles (or kilometers), but rather in light

equivalent professional or technical experience. Expert systems have been developed for use in such diverse fields as medical diagnosis, mineral exploration, and mathematical problem solving. To create such an expert system, a team of software specialists will collaborate with a scientific expert to construct a computer-based interactive dialogue system that is capable, at least to some extent, of making the expert's professional knowledge and experience available to other individuals. In this case, the computer, or thinking machine, not only stores the scientific (or professional) expertise of one human being, but also uses its artificial intelligence to permit ready access to this valuable knowledge base by other human users.

Automatic devices are those that operate without direct human control. NASA has used many such automated smart machines to explore alien worlds. For example, the *Viking 1* and *2* lander spacecraft placed on the Martian surface in 1976 represent one of the great early triumphs of robotic space exploration. After separation from the Viking orbiter spacecraft, the lander (protected by an aeroshell) descended into the thin Martian atmosphere at speeds of approximately 9,940 miles per hour (16,000 km per hour). The descending lander was slowed down by aerodynamic drag until its aeroshell was discarded. Each robot lander spacecraft slowed down further by releasing a parachute and then achieved a gentle landing by automatically firing retrorockets. Both Viking landers successfully accomplished the entire soft landing sequence automatically, without any direct human intervention or guidance.

Teleoperation implies that a human operator is in remote control of a mechanical system. Control signals can be sent by means of hardwire (if the device under control is nearby) or in a wireless mode via transmitted electromagnetic signals—for example, laser or radio frequency—(if the robot system is some distance away and operates within line-of-sight of the transmitter). NASA's Pathfinder mission to the surface of Mars in 1997 successfully demonstrated teleoperation of a mini-robot rover at interplanetary distances. The highly successful Mars Pathfinder mission consisted of a stationary lander spacecraft and a small surface rover. NASA named the lander spacecraft the *Carl Sagan Memorial Station* in honor of the American astronomer Carl Sagan (1934–96), who popularized astronomy and the search for extraterrestrial life. The mini-rover was called *Sojourner,* after the American civil rights crusader Sojourner Truth.

The six-wheeled mini-robot rover vehicle was actually controlled (or teleoperated) by the Earth-based flight team at the Jet Propulsion Laboratory (JPL) in Pasadena, California. The human operators used images of the Martian surface obtained by both the rover and the lander systems. These interplanetary teleoperations required that the rover be capable of some semi-autonomous operation, since there was a time delay of signals that averaged between 10 and 15 minutes in duration—

(continues)

minutes. The great distances associated with deep-space exploration make the real-time control of a robot spacecraft by human managers impractical, if not altogether impossible. So, in order to survive and function around or on distant worlds, space robots need to be smart—that is, they need to contain various levels of machine intelligence, or artificial intelligence (AI). As levels of machine intelligence continue to improve in this century, truly autonomous space robots will become a reality. Someday,

(continued)

depending upon the relative positions of Earth and Mars over the course of the mission. This rover had a hazard avoidance system and surface movement was performed very slowly.

Starting in 2004, NASA's Mars Exploration Rovers, *Spirit* and *Opportunity,* provided even more sophisticated and rewarding teleoperation experiences at interplanetary distances, as they rolled across different portions of the Red Planet.

Of course, in dealing with the great distances in interplanetary exploration, a situation eventually arises in which electromagnetic wave transmission cannot accommodate any type of effective "real-time control." When the device to be controlled on an alien world is many light-minutes or even light-hours away, and when actions or discoveries require split-second decisions, teleoperation must yield to increasing levels of autonomous, machine-intelligence-dependent robotic operation.

Robot devices are computer-controlled mechanical systems that are capable of manipulating or controlling other machine devices, such as end effectors. Robots may be mobile or fixed in place and either fully automatic or teleoperated. The more AI a robot has, the less dependent it is upon human supervision.

Large quantities of data are frequently involved in the operation of automatic robotic devices. The field of distributed data management is concerned with ways of organizing cooperation among independent, but mutually interacting, databases. Instead of transmitting enormous quantities of data back to Earth, an advanced robot explorer will use AI to selectively sort and send only the most interesting data.

In AI, the concept of cognition and learning refers to the development of a level of machine intelligence that can deal with new facts, unexpected events, and even contradictory information. Today's smart machines handle new data by means of preprogrammed methods or logical steps. Tomorrow's smarter machines will need the ability to learn, possibly even to understand, as they encounter new situations and are forced to change their mode of operation.

Perhaps late in this century, as the field of artificial intelligence sufficiently matures, scientists can send fully automatic robot probes on interstellar voyages. Each interstellar probe must be capable of independently searching a candidate star system for suitable extrasolar planets that might support extraterrestrial life.

human engineers will construct an especially intelligent robot that exhibits a cognitive "machine mind" of its own. Artificial intelligence experts suggest that smart exploring machines of the future will have (machine) intelligence capabilities sufficient to repair themselves, to avoid hazardous circumstances on alien worlds, and to recognize and report all of the interesting objects or phenomena they encounter.

Starting in the late 1950s—at about the same time that the space race of the cold war began—robots (terrestrial and extraterrestrial) became more practical and versatile. One of the reasons for this important transformation was the vast improvement in computer technology and electronics (especially the invention of the transistor) that took place during this same period. The information-processing-and-storage revolution continues. As tomorrow's computer chips and microprocessors pack more information-technology punch, future space robots will enjoy far more sophisticated levels of artificial intelligence than those existing today. Over the next four decades, robotic spacecraft will accomplish ever more exciting exploration missions throughout the solar system and beyond. Several of these very exciting missions are discussed in the latter portions of this book.

At this point, it is important simply to recognize that sophisticated robot spacecraft represent the enabling technology for many of the most important scientific discoveries that await the human race in the remainder of this century. Space robots are the mechanical partners that enable the human race to fulfill its destiny as an intelligent, spacefaring species. Failure to fully appreciate or to capitalize upon the opportunity offered by the space robot will confine future generations of human beings to life on just one planet around an average star in the outer regions of the Milky Way Galaxy. By recognizing the value of and vigorously using the space robot, the human race will, however, emerge within the galaxy as an active, spacefaring species. By initially reaching for the stars with very smart machines, future generations of human beings will experience all of the exciting social and technical impacts involved in becoming an interstellar spacefaring species.

Robot spacecraft have revolutionized knowledge about the solar system and visited all the major planets. This is a montage of planetary images taken by NASA spacecraft. Included are (from top to bottom) Mercury, Venus, Earth (and Moon), Mars, Jupiter, Saturn, Uranus, and Neptune. The inner planetary bodies (Mercury, Venus, Earth, Moon, and Mars) are roughly to scale with each other; the outer planets (Jupiter, Saturn, Uranus, and Neptune) are roughly to scale with each other. *(NASA/JPL)*

There is an interesting correlation between progress in space exploration by robots and parallel progress in computer technology and aerospace technology. To emphasize the connection, this chapter provides a brief look at some of the most interesting American space robots, as found in the Pioneer, Ranger, Mariner, Viking, and Voyager programs. Subsequent chapters provide more detailed insights into the technical features of these marvelous machines and many of the important scientific discoveries that they brought about. The main objective in this chapter is to provide a historic snapshot of how space robots emerged from simple, often unreliable, electromechanical exploring devices into sophisticated scientific platforms that now extend human consciousness and intelligent inquiry to the edges of the solar system and beyond.

✧ The Basic Principles of Robotics

Robotics is the science and technology of designing, building, and programming robots. Robotic devices, or robots as they are usually called, are primarily smart machines with manipulators that can be programmed to do a variety of manual or human labor tasks automatically, and with sensors that explore the surrounding environment, including the landscape of interesting alien worlds. A robot, therefore, is simply a machine that does mechanical, routine tasks on human command. The expression *robot* is attributed to Czech writer Karel Capek, who wrote the play *R.U.R. (Rossum's Universal Robots).* This play first appeared in English in 1923 and is a satire on the mechanization of civilization. The word *robot* is derived from *robata,* a Czech word meaning compulsory labor or servitude.

Here on Earth, a typical robot normally consists of one or more manipulators (arms), end effectors (hands), a controller, a power supply, and possibly an array of sensors to provide information about the environment in which the robot must operate. Because most modern robots are used in industrial applications, their classification is traditionally based on these industrial functions. So terrestrial robots frequently are divided into the following classes: *nonservo* (that is, pick-and-place), *servo,* programmable, computerized, sensory, and assembly robots.

The nonservo robot is the simplest type. It picks up an object and places it at another location. The robot's freedom of movement usually is limited to two or three directions.

The servo robot represents several categories of industrial robots. This type of robot has servomechanisms for the manipulator and end effector, enabling the device to change direction in midair (or midstroke) without having to trip or trigger a mechanical limit switch. Five to seven directions of motion are common, depending on the number of joints in the manipulator.

The programmable robot is essentially a servo robot that is driven by a programmable controller. This controller memorizes (stores) a sequence of movements and then repeats these movements and actions continuously. Often, engineers program this type of robot by "walking" the manipulator and end effector through the desired movement.

The computerized robot is simply a servo robot run by computer. This kind of robot is programmed by instructions fed into the controller electronically. These smart robots may even have the ability to improve upon their basic work instructions.

The sensory robot is a computerized robot with one or more artificial senses to observe and record its environment and to feed information back to the controller. The artificial senses most frequently employed are sight (robot or computer vision) and touch. Finally, the assembly robot is a computerized robot, generally with sensors, that is designed for assembly line and manufacturing tasks, both on Earth and eventually in space.

In industry, robots are designed mainly for manipulation purposes. The actions that can be produced by the end effector or hand include: (1) motion (from point to point, along a desired trajectory or along a contoured surface); (2) a change in orientation; and (3) rotation.

Nonservo robots are capable of point-to-point motions. For each desired motion, the manipulator moves at full speed until the limits of its travel are reached. As a result, nonservo robots often are called limit-sequence, bang-bang, or pick-and-place robots. When nonservo robots reach the end of a particular motion, a mechanical stop or limit switch is tripped, stopping the particular movement.

Servo robots are also capable of point-to-point motions; but their manipulators move with controlled variable velocities and trajectories. Servo robot motions are controlled without the use of stop or limit switches.

Four different types of manipulator arms have been developed to accomplish robot motions. These are the rectangular, cylindrical, spherical, and anthropomorphic (articulated or jointed) arms. Each of these manipulator arm designs features two or more degrees of freedom (DOF)—a term that refers to the direction a robot's manipulator arm is able to move. For example, simple straight-line or linear movement represents one DOF. If the manipulator arm is to follow a two-dimensional curved path, it needs two degrees of freedom: up and down and right and left. Of course, more complicated motions will require many degrees of freedom. To locate an end effector at any point and to orient this effector to a particular work volume requires six DOF. If the manipulator arm needs to avoid obstacles or other equipment, even more degrees of freedom are required. For each DOF, one linear or rotary joint is needed. Robot designers sometimes combine two or more of these four basic manipulator arm configurations to increase the versatility of a particular robot's manipulator.

Actuators are used to move a robot's manipulator joints. Three basic types of actuators are currently used in contemporary robots: pneumatic, hydraulic, and electrical. Pneumatic actuators employ a pressurized gas to move the manipulator joint. When the gas is propelled by a pump through a tube to a particular joint, it triggers or actuates movement. Pneumatic actuators are inexpensive and simple, but their movement is not precise. Therefore, this kind of actuator usually is found in nonservo, or pick-and-place, robots. Hydraulic actuators are quite common and are capable of producing a large amount of power. The main disadvantages of hydraulic actuators are their accompanying apparatuses (pumps and storage tanks) and problems with fluid leaks. Electrical actuators provide smoother movements, can be controlled very accurately, and are very reliable; however, these actuators cannot deliver as much power as hydraulic actuators of comparable mass. Nevertheless, for modest power actuator functions, electrical actuators often are preferred.

Many industrial robots are fixed in place or move along rails and guideways. Some terrestrial robots are built into wheeled carts, while others use their end effectors to grasp handholds and pull themselves along. Advanced robots use articulated manipulators as legs to achieve a walking motion.

A robot's end effector (hand or gripping device) generally is attached to the end of the manipulator arm. Typical functions of this end effector include grasping, pushing and pulling, twisting, using tools, performing insertions, and various types of assembly activities. End effectors can be mechanical, vacuum or magnetically operated; can use a snare device; or can have some other unusual design feature. The shapes of the objects that the robot must grasp determine the final design of the end effector. Usually most end effectors are some type of gripping or clamping device.

Robots can be controlled in a wide variety of ways, from simple limit switches tripped by the manipulator arm to sophisticated computerized remote-sensing systems that provide machine vision, touch, and hearing. In the case of a computer-controlled robot, the motions of its manipulator and end effector are programmed: that is, the robot memorizes what it is supposed to do. Sensor devices on the manipulator help to establish the proximity of the end effector to the object to be manipulated and then feed information back to the computer controller concerning any modifications needed in the manipulator's trajectory.

Another interesting type of terrestrial robot system, the field robot, has become practical recently. A field robot is a robot that operates in unpredictable, unstructured environments, typically outdoors (on Earth) and often operates autonomously or by teleoperation over a large workspace (typically a square mile [square kilometer] or more). For example, in surveying a potentially dangerous site, the human operator will stay at a safe distance away in a protected work environment and control (by cable or radio frequency link) the field robot, which then actually operates in the hazardous environment. The United States Air Force's *Predator* aerial surveillance robot and various bomb-sniffing, explosive-ordnance disposal (EOD) robots are examples of some of the most advanced field robots. These terrestrial field robots are technical first cousins to the more sophisticated, teleoperated robot planetary rovers that have roamed on the Moon and Mars. Most of the space robots mentioned in this book draw a portion of their design heritage from terrestrial robots.

The need to survive in outer space or on an unknown alien world has imposed much more stringent design requirements upon even the simplest of the space robots. When a factory robot has a part fail or a terrestrial field robot loses a wheel, human technicians are normally available to fix the problem quickly and efficiently. When a space robot that is millions of miles from Earth has a malfunction, it is on its own, and the difficulty can lead to catastrophic failure of an entire exploration mission. A simple example will illustrate this important point. When a mobile rover on Earth gets some dust or soil on the lenses of its machine vision system, a human technician is available to gently remove the troublesome material. When a sudden wind gust coats a surface rover with Martian soil, there is no person available to "dust it off." The rover either has to be able to clean itself or else function with reduced machine vision and possibly reduced electric power, if the troublesome red-colored dust has also coated its solar cells. Because of this and similar mission-threatening "simple problems," some aerospace engineers have suggested operating smart planetary rovers in teams. A team of advanced mechanical critters could be designed to help each other, whenever one runs into difficulty. In the dust-coating example, a second rover might come by, scan its dust-coated companion, and then use a special brush tool (grasped by its manipulator arm) to remedy the situation.

The operative concept here is to design future space robots that are robust with in-depth design redundancy. In that way, the smart machine, perhaps with a little coaxing from human controllers on Earth, can fix itself or at least implement appropriate "workarounds," and thus keep the exploration mission going. Another important design strategy is to engineer space robots so that they can work in teams. That way, one or more functional robots can assist and/or repair their companion robot in distress.

✧ Pioneer to the Moon and Beyond

The dictionary defines a pioneer as a person who ventures into the unknown. That definition proved very appropriate for the first family of American deep space robots, which were given the name *Pioneer.* The initial spacecraft to be launched and the first space missions to actually be carried out by the United States Air Force were the Pioneer lunar probes of 1958. Now just a frequently overlooked page in aerospace history, these early Pioneer lunar probes were the world's first attempted deep-space missions.

The first series of Pioneer spacecraft was flown between 1958 and 1960. *Pioneer 1, 2,* and *5* were developed by Space Technology Laboratories, Inc. and were launched for NASA by the Air Force Ballistic Missile Division (AFBMD). *Pioneer 3* and *4* were developed by the Jet Propulsion Laboratory (JPL) and launched for NASA by the U.S. Army Ballistic Missile Agency (ABMA) at Redstone Arsenal, Alabama—the technical team also responsible for the launch of *Explorer 1,* the first American satellite, on January 31, 1958.

In January 1958, the Air Force Ballistic Missile Division (AFBMD) and its technical advisory contractor, Space Technology Laboratories (STL) proposed using the newly developed Thor missile with the second stage of the Vanguard rocket to launch the first missions to the Moon. The new launch vehicle configuration was named the Thor Able. The stated purpose of these early lunar-probe missions were to gather scientific data from space and to gain international prestige for the United States by doing so before the former Soviet Union. During the cold war, both superpowers were bitter political rivals, and space exploration provided each country with a convenient showcase in which to display national superiority on a global basis.

After President Dwight Eisenhower's administration activated the Advanced Research Projects Agency (ARPA) on February 7, 1958, the new agency's first directives to the military services dealt with lunar probes. AFBMD was to launch three lunar probes using the Thor Able configuration; ABMA was to launch two lunar probes using its Juno II vehicle; and the Naval Ordnance Test Station (NOTS) at China Lake was to provide a miniature imaging system to be carried on the lunar probes.

Space Technology Laboratories (STL) designed and assembled the lunar probes known as *Pioneer 0, Pioneer 1,* and *Pioneer 2. Pioneer 0* was the first United States attempt at a lunar mission and the first attempt by any country to send a space probe beyond Earth orbit. The *Pioneer 0* robot probe was designed to go into orbit around the Moon and carried a television (TV) camera and other instruments as part of the first International Geophysical Year (IGY) science payload. Unfortunately, the 84-pound (38-kg) robot probe was lost when the Thor rocket vehicle exploded 77 seconds after launch from Cape Canaveral. The Thor rocket blew up at an altitude of 10 miles (16 km), when the launch vehicle and its payload were about 10 miles downrange over the Atlantic Ocean. Erratic telemetry signals were received from the *Pioneer 0* payload and upper rocket stages for 123 seconds after the explosion. Range safety officials tracked the upper stages and payload until they impacted in the Atlantic Ocean.

The original plan was for the *Pioneer 0* spacecraft to travel for 62 hours to the Moon, at which time a solid propellant rocket motor would fire to put the spacecraft into a 18,000-mile (28,960-km) lunar orbit that would last for about two weeks. *Pioneer 0*'s scientific instrument package had a mass of 25 pounds (11.3 kg). The package consisted of an image-scanning infrared television system to study the Moon's surface, a micrometeorite detector, a magnetometer, and temperature-variable resistors to record internal thermal conditions of the spacecraft. Batteries provided electric power. Finally, *Pioneer 0* was to be spin-stabilized at a rate of 1.8 revolutions per second.

Pioneer 1 was the second and most successful of the early American space-probe efforts, as well as the first spacecraft launched by the newly created civilian space agency, NASA. Similar in design to *Pioneer 0*, the 75-pound (34.2-kg) mass *Pioneer 1* was launched from Cape Canaveral on October 11, 1958, by a Thor Able rocket vehicle. Due to a launch vehicle malfunction, *Pioneer 1* only attained a ballistic trajectory and never reached the Moon as planned. The spacecraft's ballistic trajectory had a peak altitude of 70,730 miles (113,800 km). On October 13, after about 43 hours of flight, the spacecraft ended data transmission when it reentered Earth's atmosphere over the South Pacific Ocean. Despite the spacecraft's failure to reach the Moon because its launch vehicle did not provide sufficient velocity to escape Earth's gravity, *Pioneer 1*'s instruments did return some useful scientific data about the extent of Earth's trapped radiation belts. *Pioneer 1*'s scientific instrument package had a mass of 39 pounds (17.8 kg), making it slightly heavier than the scientific payload carried by *Pioneer 0*. *Pioneer 1* contained an image-scanning infrared television system to study the Moon's surface, an ionization chamber to measure radiation levels in space, a micrometeorite detector, a magnetometer, and temperature-variable sensors to record thermal conditions in the interior of the spacecraft. *Pioneer 1* was spin-stabilized at 1.8 revolutions per second and received its electric power from limited lifetime batteries.

Pioneer 2 was the last of the Thor Able space probes, which were designed to orbit the Moon and make measurements in interplanetary space between Earth and the Moon—a region called cislunar space. This spacecraft was nearly identical to *Pioneer 1*. Launched from Cape Canaveral on November 8, 1958, the space probe never achieved its intended lunar orbit. Instead, shortly after launch the third stage of the Thor Able rocket separated but failed to ignite. Given an inadequate velocity, *Pioneer 2* only attained an altitude of 963 miles (1,550 km) before reentering Earth's atmosphere over northwest Africa. Due to its short flight, *Pioneer 2* collected only a small amount of useful scientific data about near-Earth space.

Following the unsuccessful U.S. Air Force/NASA *Pioneer 0, 1,* and *2* lunar probe missions in 1958, the U.S. Army and NASA collaborated in launching two additional probe missions. Smaller than the previous Pioneer spacecraft, *Pioneer 3* and *4* each carried only a single experiment to detect cosmic radiation. It was the intention of the mission planners in both the U.S. Army and NASA that the two space probes would perform a flyby of the Moon and return data about the radiation environment in cislunar space. The Jet Propulsion Laboratory (JPL) constructed the *Pioneer 3* and *4* spacecraft, which were nearly identical in mass, shape, size, and functions.

Pioneer 3—a 12.9-pound (5.9-kg), spin-stabilized, cone-shaped spacecraft—was launched on December 6, 1958, from Cape Canaveral by the U.S. Army Ballistic Missile Agency (ABMA), using a Juno II rocket. Developed in conjunction with NASA, *Pioneer 3* was designed to pass close to the Moon some 34 hours after launch and then go into orbit around the Sun. Propellant depletion, however, caused the first-stage rocket engine to shut down 3.7 seconds early. This premature termination of thrust prevented *Pioneer 3* from reaching escape velocity. Instead, the spacecraft always remained a captive of Earth's gravity field and traveled on an enormously high ballistic trajectory, reaching a maximum altitude of 63,615 miles (102,360 km) before falling back to Earth. On December 7, *Pioneer 3* reentered Earth's atmosphere and burned up over Africa.

The *Pioneer 4* spacecraft being installed on top of its Juno II launch vehicle at Cape Canaveral in February 1959. *Pioneer 4* was the first United States spacecraft to orbit the Sun. *(NASA/MSFC)*

This planned lunar probe returned telemetry for about 25 hours of its approximately 38-hour journey. The other 13 hours (of missing telemetry) corresponded to communications-blackout periods owing to the location of the two tracking stations. Mercury batteries provided *Pioneer 3* with its electric power. The spacecraft's scientific payload included Geiger-Mueller tube radiation detectors, which provided data indicating the existence of two distinct trapped radiation belt regions around Earth.

Pioneer 4, launched on March 3, 1959, by a Juno II rocket, was the first U.S. spacecraft to escape Earth's gravity and also the first to go into orbit around the Sun. Like *Pioneer 3,* its technical sibling,

Pioneer 4 was a cone-shaped, spin-stabilized spacecraft built by the Jet Propulsion Laboratory and launched by the U.S. Army Ballistic Missile Agency in conjunction with NASA. The main scientific payloads of this 13.4-pound (6.1-kg) mass spacecraft were a lunar radiation environment experiment (using a Geiger-Mueller tube detector) and a lunar photography experiment.

The cone-shaped *Pioneer 4* probe was 20 inches (51 cm) high and 9.1 inches (23 cm) in diameter at its base. The cone itself was made of a thin fiberglass shell coated with a gold wash to make it an electrical conductor and painted with white stripes to assist in thermal control of the spacecraft's interior. A ring of mercury batteries at the base of the cone provided electric power.

After a successful launch, *Pioneer 4* achieved its primary objective (an Earth-Moon trajectory), returned radiation data, and served as a valuable space-probe-tracking exercise. The robot probe passed within 37,290 miles (60,000 km) of the Moon's surface on March 4, 1959, at a speed of 4,490 miles per hour (7,230 km/h). The lunar encounter distance was about twice the planned flyby altitude, so the spacecraft's photoelectric sensor for the lunar photography experiment did not trigger. Although *Pioneer 4* did indeed fly past the Moon, the Soviet Union's *Luna 1* spacecraft had passed by the Moon several weeks earlier (on January 4, 1959) and laid claim to the distinction of being the first human-made object to escape Earth's gravity and to fly past another celestial body. A Russian space robot, not an American robot, had won the first lap in the cold war's hotly contested, but officially undeclared, race to the Moon.

This politically uncomfortable "second-place" trend would continue for much of the 1960s, until that fateful day at the end of the decade (July 20, 1969), when American astronauts Neil Armstrong and Edwin "Buzz" Aldrin claimed the victory lap by leaving human footprints on the Moon's surface for the first time. The glare of this magnificent human spaceflight accomplishment often obscures the fact that the pathway to the Moon was paved by a family of American space robots named Ranger, Surveyor, and Lunar Orbiter.

After several early attempts to reach the Moon, the U.S. Air Force and NASA sent the spin-stabilized *Pioneer 5* spacecraft on a mission to investigate interplanetary space between Earth and Venus. The 95-pound (43-kg) robot space probe was successfully launched from Cape Canaveral on March 11, 1960, by a Thor Able rocket vehicle. Instrumentation onboard *Pioneer 5* measured magnetic field phenomena, solar flare particles, and ionization. On June 26, 1960, which was the spacecraft's last day of transmission, *Pioneer 5* established a communications link with Earth from a record distance of 22.5 million miles (36.2 million km). Among its scientific contributions, *Pioneer 5* confirmed the existence of interplanetary

Early Soviet Luna Missions

The name *Luna* was given to a series of robot spacecraft successfully sent to the Moon in the 1960s and 1970s by the former Soviet Union. Between 1958 and 1959, there were also several "unannounced" Luna launch failures, as the Soviet Union attempted to reach the Moon with a robot probe before the United States. Aerospace mission failures were not officially acknowledged by the Soviet Union during the cold war. However, post–cold war cooperation in space exploration has allowed Western analysts to assemble and reconstruct some details about these unsuccessful early lunar probe missions. Tentatively identified failed Luna launches include *Luna 1958A* (September 23, 1958), *Luna 1958B* (October 12, 1958), *Luna 1958C* (December 4, 1958), and *Luna 1959A* (June 18, 1959).

Luna 1 was the first robot spacecraft of any country to reach the Moon and the first in a series of Soviet automatic interplanetary stations successfully launched in the direction of the Moon. The 794-pound (361-kg) sphere-shaped *Luna 1* was also called *Mechta* (Dream). The robot probe was launched by a modified intercontinental ballistic missile from the Baikonur Cosmodrome (Tyuratam) on January 2, 1959. The Soviets sent *Luna 1* directly toward the Moon from the launch site, using a trajectory that suggested the spacecraft was most likely intended to crash-land on the Moon. After 34 hours of flight, however, *Luna 1* missed the Moon, passing within 3,725 miles (6,000 km) of the lunar surface on January 4. Following its close encounter with the Moon, *Luna 1* went into orbit around the Sun between the orbits of Earth and Mars. So *Luna 1* also became the first human-made object to escape from Earth's gravitational field and go into orbit around the Sun.

Luna 1 was a sphere-shaped spacecraft with five antennae extending from one hemisphere. The robot probe had no onboard propulsion system and relatively short-lived batteries provided all its electric power. The spacecraft contained radio equipment, a tracking transmitter, a telemetry system, and scientific instruments for examining interplanetary space. Measurements made by *Luna 1* provided scientists with new data about Earth's trapped radiation belts, as well as the important discovery that the Moon has no measurable magnetic field. Instruments on *Luna 1* also indicated the presence of the solar wind (ionized plasma emanating from the Sun), which streams through interplanetary space. Data transmissions from *Luna 1* ceased about three days after launch, when the spacecraft's batteries ran down. Because of its high velocity and its prominent package of various metallic emblems with the Soviet coat of arms, Western aerospace analysts concluded that *Luna 1* was primarily intended to crash on the Moon and (in a manner of speaking) to "plant the Soviet flag."

Luna 2 was the second of a series of early Soviet spacecraft launched in the direction of the Moon. *Luna 2* had the distinction of being the first human-made object to land on the Moon. The Soviet space probe made impact on the lunar surface east of Mare Serenitatis near the Archimedes, Aristides, and Autolycus craters. The 858-pound (390-kg) spacecraft was similar in design to *Luna 1*. This means that *Luna 2* was shaped like a sphere with protruding antennae and instrument ports. The science payload included radiation detectors, a magnetometer, and micrometeorite detectors. The spacecraft also carried a political payload, namely Soviet emblems and pennants.

The *Luna 2* space probe was launched on September 12, 1959, from the Baikonur Cosmodrome. On September 14, after almost 34 hours of spaceflight, radio signals from spacecraft abruptly ceased, indicating that *Luna 2* had made impact (crash landed) on the Moon. The robot space probe confirmed that the Moon has no appreciable magnetic field and also discovered no evidence that the Moon has trapped radiation belts.

Luna 3 was the third robot spacecraft successfully launched to the Moon by the former Soviet Union and the first spacecraft of any country to return photographic images of the lunar farside. The spacecraft's relatively coarse images showed that the Moon's farside was mountainous and quite different from the nearside, which always faces Earth. *Luna 3*'s images caused excitement among astronomers around the world, because these pictures (no matter how crude by today's space mission standards) allowed them to make the first tentative atlas of the lunar farside. *Luna 3* was spin-stabilized and radio-controlled from Earth.

The 613-pound (279-kg) spacecraft was a cylindrically shaped canister with hemispherical ends and a wide flange near the top end. The *Luna 3* robot spacecraft (sometimes called an automatic interplanetary station in the Russian aerospace literature) was 51 inches (130 cm) long and 47 inches (120 cm) wide at its maximum diameter (that is, at the flange). Soviet engineers mounted solar cells along the outside of the cylinder in order to recharge the chemical batteries within the spacecraft. The interior also contained a dual-lens camera, an automatic film processing system, a scanner, radio equipment, and gyroscopes for attitude control. When the film was processed, commands from Earth activated a sequence of automated actions that moved the film from the processor to the scanner. Each photograph was scanned and converted into electrical signals, which were then transmitted back to Earth.

The mission profile for *Luna 3* involved a loop around the Moon that allowed the robot spacecraft to automatically photograph the unknown farside. After launch from the Baikonur Cosmodrome on October 4, 1959, *Luna 3* departed Earth on an interplanetary trajectory to the Moon. About 40,400 miles (65,000 km) from the Moon, the attitude control system was activated and the spacecraft stopped spinning. The lower end of the spacecraft was oriented toward the Sun, which was shining on the lunar farside. On October 6, *Luna 3* passed within 3,850 miles (6,200 km) (at closest approach) of the Moon near its south pole and then continued on to the farside. On October 7, the photocell on the upper end of the spacecraft detected the sunlit farside and started the photography sequence. The first image was taken at a distance of 39,500 miles (63,500 km). *Luna 3* took its last photograph about 40 minutes later, when the spacecraft was at a distance of 41,500 miles (66,700 km) from the surface of the Moon. During this trail-blazing mission, a total of 29 photographs were taken, covering approximately 70 percent of the previously unseen and unknown farside. After the photography portion of its mission was completed, *Luna 3* resumed spinning, passed over the north pole of the Moon, and returned toward Earth. As *Luna 3* approached Earth, a total of 17 resolvable (but noisy and grainy) photographs were transmitted by October 18 to Soviet spacecraft controllers. Then, on October 22, they lost contact with the probe. Western analysts believe *Luna 3* remained in orbit until about April, 1960, at which point it reentered Earth's atmosphere and burned up.

magnetic fields and helped explain how solar flares trigger magnetic storms and the northern and southern lights (auroras) on Earth.

With the launch of *Pioneer 6* (also called *Pioneer A* in the new series of robot spacecraft) in December 1965, NASA resumed using these space probes to complement interplanetary data acquired by the Mariner spacecraft. Over the years, NASA's solar-orbiting Pioneer spacecraft have contributed an enormous amount of data concerning the solar wind, solar magnetic field, cosmic radiation, micrometeoroids, and other phenomena of interplanetary space.

Pioneers 7, 8, and *9* (second-generation robot spacecraft) were launched between August 1966 and November 1968 and continued NASA's investigation of the interplanetary medium. These spacecraft provided large quantities of valuable data concerning the solar wind, magnetic and electrical fields, and cosmic rays in interplanetary space. Data from second-generation Pioneer spacecraft helped space scientists draw a new picture of the Sun as the dominant phenomenon of interplanetary space.

The *Pioneer 10* and *11* spacecraft were designed as true deep-space robot explorers—the first human-made objects to navigate the main asteroid belt, the first spacecraft to encounter Jupiter and its fierce radiation belts, the first to encounter Saturn, and the first spacecraft to leave the solar system. This far-traveling pair of robot spacecraft also investigated magnetic fields, cosmic rays, the solar wind, and the interplanetary dust concentrations as they flew through interplanetary space.

The Pioneer Venus mission consisted of two separate spacecraft launched by the United States to the planet Venus in 1978. The *Pioneer Venus Orbiter* (also called *Pioneer 12*) was a 1,173-pound (553-kg) spacecraft that contained a 100-pound (45-kg) payload of scientific instruments. *Pioneer 12* was launched on May 20, 1978, and placed into a highly eccentric orbit around Venus on December 4, 1978. For 14 years (from 1978 to 1992) the *Pioneer Venus Orbiter* gathered a wealth of scientific data about the atmosphere and ionosphere of Venus and their interactions with the solar wind, as well as details about the planet's surface. Then, in October 1992, this spacecraft made an intended final entry into the Venusian atmosphere, collecting data up to its final fiery plunge and dramatically ending the operations portion of the Pioneer Venus mission. Data analysis and scientific discovery would continue for years afterward.

The *Pioneer Venus Multiprobe* (also called *Pioneer 13*) consisted of a basic bus spacecraft, a large probe, and three identical small probes. The *Pioneer Venus Multiprobe* was launched on August 8, 1978, and separated about three weeks before entry into the Venusian atmosphere. The four (now-separated) probes and their (spacecraft) bus successfully entered the Venusian atmosphere at widely dispersed locations on December 9, 1978, and returned important scientific data as they plunged toward the planet's surface. Although the probes were not designed to survive landing, one hardy probe did and transmitted data for about an hour after impact.

✧ Jet Propulsion Laboratory (JPL)—America's Premier Space Robot Factory

The American space age began on January 31, 1958, with the launch of the first U.S. satellite, *Explorer 1*—an Earth-orbiting spacecraft built and controlled by the Jet Propulsion Laboratory (JPL). For almost five decades, JPL has led the world in exploring the solar system with robot spacecraft.

The Jet Propulsion Laboratory (JPL) is a federally funded research and development facility managed by the California Institute of Technology for the National Aeronautics and Space Administration (NASA). The Laboratory is located in Pasadena, California approximately 20 miles (32 km) northeast of Los Angeles. In addition to the Pasadena site, JPL operates the worldwide Deep Space Network (DSN), including a DSN station, at Goldstone, California.

JPL's origin dates back to the 1930s, when Caltech professor Theodor von Kármán (1881–1963) supervised pioneering work in rocket propulsion for the U.S. Army—including the use of strap-on rockets for "jet-assisted take-off" of aircraft with extra heavy cargoes. At the time, von Kármán was head of Caltech's Guggenheim Aeronautical Laboratory. On December 3, 1958, two months after the U.S. Congress created NASA, JPL was transferred from the U.S. Army's jurisdiction to that of the new civilian space agency. The Laboratory now covers 177 acres (72 hectares) adjacent to the site of von Kármán's early rocket experiments in a dry riverbed wilderness area of Arroyo Seco.

In the 1960s, JPL began to conceive, design, and operate robot spacecraft to explore other worlds. This effort initially focused on NASA's Ranger and Surveyor missions to the Moon—robot spacecraft that paved the way for successful human landings by the Apollo Project astronauts. The Ranger spacecraft were the first U.S. robot spacecraft sent toward the Moon in the early 1960s to prepare the way for the Apollo Project's human landings at the end of that decade. The Rangers were a series of fully attitude-controlled robot spacecraft designed to photograph the lunar surface at close range before making impact. *Ranger 1* was launched on August 23, 1961, from Cape Canaveral Air Force Station and set the stage for the rest of the Ranger missions by testing spacecraft navigational performance. The *Ranger 2* through *9* spacecraft were launched from November 1961 through March 1965. All of the early Ranger missions (namely, *Ranger 1* through *6*) suffered setbacks of one type or another. Finally, *Ranger 7, 8,* and *9* succeeded, with flights that returned many

NASA's Ranger spacecraft were sent to the Moon in the early to mid-1960s to pave the way for the Apollo Project's human landings at the end of that decade. These attitude-controlled robot spacecraft were designed to photograph the lunar surface at close range before impacting. *(NASA/JPL)*

Minutes before impact on March 24, 1965, NASA's *Ranger 9* robot spacecraft took this close-up television picture of the lunar surface. *(NASA)*

thousands of lunar surface images (before impact) and greatly advanced scientific knowledge about the Moon.

NASA's highly successful Surveyor Project began in 1960. It consisted of seven robot lander spacecraft that were launched between May 1966 and January 1968, as an immediate precursor to the human expeditions to the lunar surface in the Apollo Project. These versatile space robots were used to develop soft-landing techniques, to survey potential Apollo mission landing sites, and to improve scientific understanding of the Moon.

The *Surveyor 1* spacecraft was launched on May 30, 1966, and soft-landed in the Ocean of Storms region of the Moon. The space robot discovered that the bearing strength of the lunar soil was more than adequate to support the Apollo Project's human-crewed lander spacecraft (called the lunar module, or LM). This finding contradicted the then-prevalent hypothesis that a heavy spacecraft like the LM might sink out of sight in the anticipated talcum-powder-like, ultra-fine lunar dust particles. The *Surveyor 1* spacecraft also telecast many images from the lunar surface.

Surveyor 2 was the second in this series of soft-landing robots. Successfully launched on September 20, 1966, by an Atlas-Centaur rocket from Cape Canaveral, this robot lander experienced a vernier engine failure during a midcourse maneuver while en route to the Moon. The failure of one vernier engine to fire resulted in an unbalanced thrust that caused *Surveyor 2* to tumble. Attempts by NASA engineers to salvage this mission failed.

Things went much better for NASA's next robot lander mission to the Moon. The *Surveyor 3* spacecraft was launched on April 17, 1967, and soft-landed on the side of a small crater in another region of the Ocean of Storms. The perky space robot used the shovel attached to its mechanical arm to dig a trench and thus it was discovered that the load-bearing strength of the lunar soil increased with depth. *Surveyor 3* also transmitted many images from the lunar surface.

At the same time that JPL engineers were busy with the Ranger and Surveyor missions, they also conducted Mariner spacecraft missions to Mercury, Venus, and Mars. The Mariner missions were true trail-blazing efforts that continued through the early 1970s and greatly revised scientific understanding of the terrestrial planets and the inner solar system. The first Mariner mission, called *Mariner 1,* was intended to perform a Venus flyby. (Chapter 3 presents the different types of robot spacecraft and their characteristic missions.) NASA and JPL engineers based the design of this spacecraft on the Ranger lunar spacecraft. A successful liftoff of *Mariner 1*'s Atlas-Agena B launch vehicle on July 22, 1962, soon turned tragic. When the rocket vehicle veered off course, the range safety officer at Cape Canaveral Air Force Station was forced to destroy it some 293 seconds after launch. Because of faulty guidance commands, the rocket

This photograph, taken during the *Apollo 12* lunar landing mission (November 1969), shows astronaut Charles Conrad, Jr., examining the *Surveyor 3* robot spacecraft. Between 1967 and 1968, NASA used several Surveyor lander spacecraft to carefully examine the lunar surface before sending humans to the Moon. *Surveyor 3* was launched from Cape Canaveral on April 17, 1967, and successfully soft-landed on the side of a small crater in the Ocean of Storms region on April 19, 1967. (The lunar module [*Intrepid*] used by the Moon-walking astronauts Conrad and Alan L. Bean appears in the background.) *(NASA)*

vehicle's steering was very erratic and the *Mariner 1* spacecraft was going to crash somewhere on Earth, possibly in the North Atlantic shipping lanes or in an inhabited area. Undaunted by the heartbreaking loss of the *Mariner 1* spacecraft, which was never given a chance to demonstrate its capabilities, the NASA/JPL engineering team quickly prepared its identical twin, named *Mariner 2,* to pinch-hit and perform the world's first interplanetary flyby mission.

Following a successful launch from Cape Canaveral on August 27, 1962, *Mariner 2* cruised through interplanetary space, and then became the first robot spacecraft to fly past another planet (in this case, Venus). *Mariner 2* encountered Venus at a distance of about 25,500 miles (41,000 km) on December 14, 1962. Following the flyby of Venus, *Mariner 2* went into orbit around the Sun. The scientific discoveries made by *Mariner 2* included a slow retrograde rotation rate for Venus, hot surface temperatures and high surface pressures, a predominantly carbon-dioxide atmosphere, continuous cloud cover with highest altitude of about 37 miles (60 km), and no detectable magnetic field. Data collected by *Mariner 2* during its interplanetary journey to Venus showed that the solar wind streams continuously in interplanetary space and that the cosmic dust density is much lower than in the region of space near Earth.

The *Mariner 2* encounter helped scientists dispel many pre-space age romantic fantasies about Venus, including the widely held speculation

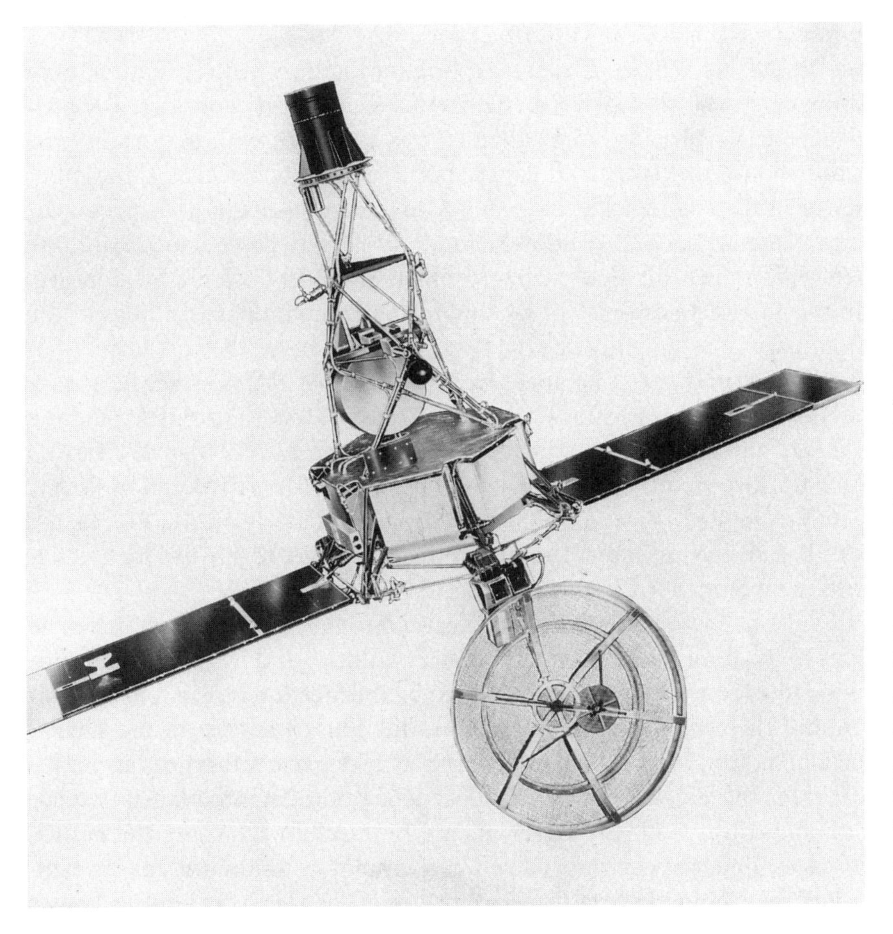

Following its launch on August 27, 1962, NASA's *Mariner 2* became the first robot spacecraft to successfully fly past another planet (Venus). Its technical twin, *Mariner 1,* was lost on July 22, 1962, when range safety destroyed an errant launch vehicle. This picture shows the spacecraft's solar panels and high-gain antenna extended, as displayed during the interplanetary cruise phase of the planetary flyby mission. *(NASA)*

(which appeared in both science and science-fiction literature) that the cloud-shrouded planet was a prehistoric world, mirroring a younger Earth. Except for a few physical similarities like size and surface gravity level, robot spacecraft visits in the 1960s and 1970s continued to show that Earth and Venus were very different worlds. For example, the surface temperature on Venus reaches almost 932°F (500°C), its atmospheric pressure is more than 90 times that of Earth, it has no surface water, and its dense atmosphere, with sulfuric acid clouds and an overabundance of carbon dioxide (about 96 percent), represents a runaway greenhouse of disastrous proportions.

The next Mariner project undertaken by NASA and JPL targeted the planet Mars. Two spacecraft were prepared, *Mariner 3* and its backup *Mariner 4* (an identical twin). *Mariner 3* was launched from Cape Canaveral on November 5, 1964, but the shroud encasing the spacecraft atop its rocket failed to open properly and *Mariner 3* did not get to Mars. Three weeks later *Mariner 4* was launched successfully and sent on an eight-month voyage to the Red Planet. Why was such a quick recovery and new launch possible in so short a time?

In the early days of space exploration, launch vehicle failures were quite common, so aerospace engineers and managers considered it prudent to build two (or more) identical spacecraft for each important mission. Should one spacecraft experience a fatal launch accident, the other spacecraft could quickly be readied to take advantage of a particular interplanetary launch window. If both spacecraft proved successful, the scientific return for that particular mission usually more than doubled. In this fortunate case, scientists could use the preliminary findings of the first space robot to guide the data collection efforts of the second robot as it approached the target planet several weeks later. NASA's three most successful "robot twin missions" of the 1970s were *Pioneer 10* and *11* (flybys), *Viking 1* and *2* (landers and orbiters), and *Voyager 1* and *2* (flybys). Starting in 2004, fortune smiled again when NASA's twin Mars Exploration Rovers (MERs), named *Spirit* and *Opportunity,* arrived safely on the Red Planet and began moving across the surface to inaugurate highly productive scientific investigations.

A launch window is the time interval during which a spacecraft can be sent to its destination. An interplanetary launch window is generally confined to a few weeks each year (or less) by the location of Earth in its orbit around the Sun. Proper timing allows the launch vehicle to use Earth's orbital motion in its overall trajectory. Earth-departure timing is also critical, if the spacecraft is to arrive at a particular point in interplanetary space simultaneously with the target planet. By carefully choosing the launch window, interplanetary spacecraft can employ a minimum energy path called the Hohmann transfer trajectory, after the German engineer Walter

Hohmann (1880–1945), who described this orbital transfer technique in 1925. Orbital mechanics, payload mass, and rocket-vehicle thrust all influence interplanetary travel.

The most energy-efficient launch windows from Earth to Mars occur about every two years. Determining launch windows for missions to the giant outer planets is a bit more complicated. For example, only once every 176 years do the four giant planets (Jupiter, Saturn, Uranus, and Neptune) align themselves in such a pattern that a spacecraft launched from Earth to Jupiter at just the right time might be able to visit the three other giant planets on the same mission, using a technique called *gravity assist.* (*Gravity assist* is discussed in chapter 2) This unique opportunity occurred in 1977, and NASA scientists took advantage of a special celestial alignment by launching two sophisticated robot spacecraft, called *Voyager 1* and *2,* on multiple giant planet encounter missions. As described shortly, *Voyager 1* visited Jupiter and Saturn, while *Voyager 2* took the so-called "grand tour" and visited all four giant planets on the same mission.

In the cold war environment of the early 1960s, a great deal of political emphasis and global attention was given to achievements in space exploration. The superpower that accomplished this or that space exploration "first" earned a central position on the world political stage. So, NASA managers soon recognized that building identical-twin spacecraft (just in case one did not complete the mission) proved to be a relatively inexpensive approach to pursuing major scientific objectives while earning political capital. Superpower competition during the cold war fueled an explosion in space exploration and produced an age of discovery, unprecedented in history. Primarily because of robot spacecraft, more scientific information about the solar system and the universe was collected in between 1960 and 2000 than in all previous human history. This exciting wave of discovery continues in the post–cold war era, as more sophisticated space robots, such as *Cassini/Huygens,* explore the unknown.

Before discussing the spectacular results of the Viking mission or the great journeys of Voyager spacecraft, this chapter returns to the very important *Mariner 4* mission to Mars. *Mariner 4* was successfully launched from Cape Canaveral on November 28, 1964, traveled for almost eight months through interplanetary space, and then zipped past Mars on July 14, 1965. At its closest approach, *Mariner 4* was just 6,120 miles (9,845 km) from the surface of Mars during the flyby. As this space robot encountered Mars, it collected the first close-up images of another planet. These images, played back from a small video recorder over a long period, showed lunar-type impact craters, some of them touched with frost in the chill of the Martian evening. *Mariner 4*'s 21 complete pictures, in addition to 21 lines of a 22nd picture, might be regarded as quite crude when compared to the high-resolution imagery of Mars provided by contemporary robot spacecraft.

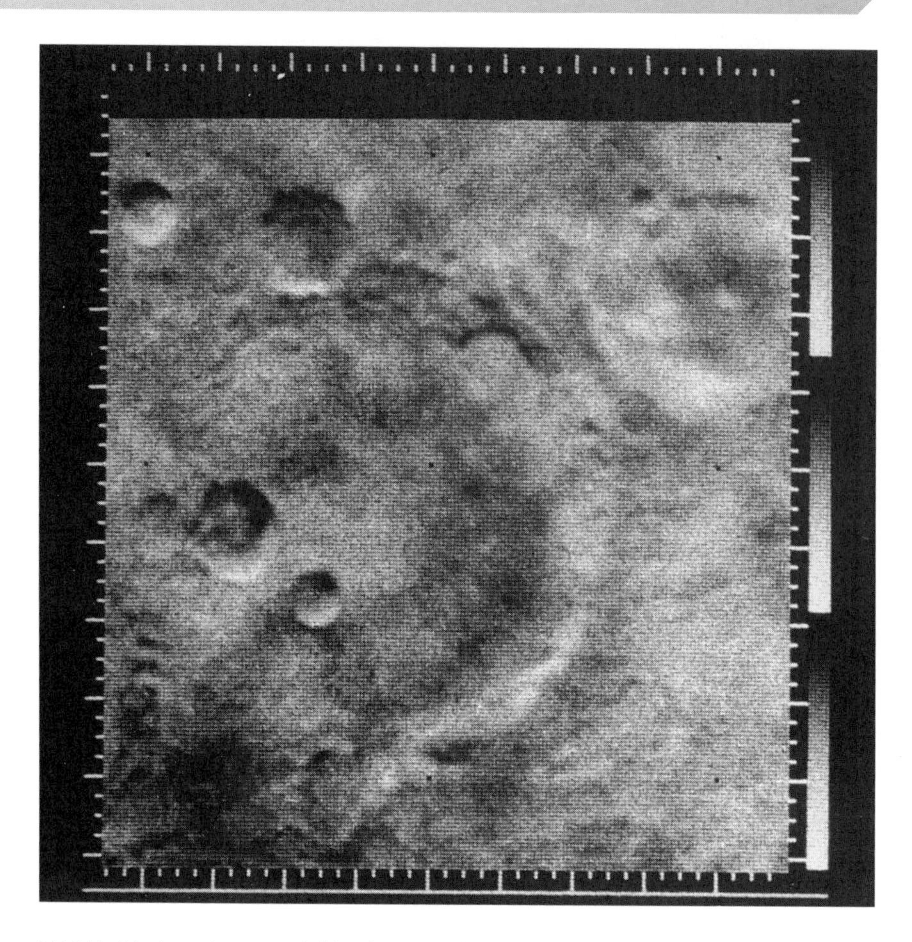

NASA's *Mariner 4* snapped this photograph of Mars at a slant range of 7,800 miles (12,550 km), as the robot spacecraft flew past the Red Planet on July 14, 1965. *(NASA)*

These first images of another world started a revolution that overturned many long-cherished views about the Red Planet, however.

Throughout human history the Red Planet, Mars, has been at the center of astronomical thought. The ancient Babylonians followed the motions of this wandering red light across the night sky and named it after Nergal, their god of war. Later, the Romans, also honoring their own god of war, gave the planet its present name. The presence of an atmosphere, polar caps, and changing patterns of light and dark on the surface caused many pre–space age astronomers and scientists to consider Mars an "Earthlike planet"—the possible abode of extraterrestrial life. The American astronomer Percival Lowell (1855–1916) was one of the most outspoken proponents of the canal theory. In several popular publications, he insisted that Mars was a dying planet whose intelligent inhabitants constructed huge canals to distribute a scarce supply of water around the alien world. Invasions from Mars was one of the popular themes in science-fiction literature and in the entertainment industry. For example, when actor Orson Welles broadcast a radio drama in 1938 based on H. G. Wells's science-fiction classic *The War of the Worlds,* enough people believed the report of invading Martians to create a near panic in some areas of the northeastern United States.

With *Mariner 4* leading the scientific parade, however, a wave of sophisticated robot spacecraft—flybys, orbiters, landers, and rovers—have shattered the canal theory—the persistent romantic myth of a race of ancient Martians struggling to bring water from the polar caps to the more productive regions of a dying world. Spacecraft-derived data have shown that the Red Planet is actually a "halfway" world. Part of the Martian surface is ancient, like the surfaces of the Moon and Mercury, while part is more evolved and Earthlike. Mars remains at the center of intense investigation by a new wave of sophisticated robot spacecraft. The continued search for microbial life (existent or extinct) and the resolution of the intriguing mystery about the fate of liquid water, which appears to have flowed on ancient Mars in large quantities, top the current exploration agenda.

Other successful Mariner missions included *Mariner 5,* launched in 1967 to Venus; *Mariner 6,* launched in 1969 to Mars; *Mariner 7,* launched in 1969 to Mars; and *Mariner 9,* launched in 1971 to Mars. In November 1971, *Mariner 9* became the first artificial satellite of Mars and the first spacecraft of any country to orbit another planet. The robot spacecraft waited patiently for a giant planet-wide dust storm to abate and then compiled a collection of high-quality images of the surface of Mars that provided scientists with their first global mosaic of the Red Planet. *Mariner 9* also took the first close-up images of the two small (natural) Martian satellites, Phobos and Deimos.

Mariner 10 became the first spacecraft to use a gravity-assist boost from one planet to send it to another planet—a key innovation in spaceflight, which enabled exploration of the outer planets by robot spacecraft. *Mariner 10*'s launch from Cape Canaveral in November 1973 delivered the spacecraft to Venus in February 1974, where a gravity-assist flyby allowed it to encounter the planet Mercury in March and September of that year. *Mariner 10* was the first and, thus far, the only spacecraft of any country to explore the innermost planet in the solar system. On August 3, 2004, NASA launched *MESSENGER* from Cape Canaveral and sent the orbiter spacecraft on a long-interplanetary journey to Mercury. In March 2011, *MESSENGER* is set to become the first robot spacecraft to achieve orbit around Mercury.

MESSENGER Mission

MESSENGER is a NASA acronym that stands for the *ME*rcury *S*urface, *S*pace *EN*vironment, *GE*ochemistry and *R*anging mission. The space robot will orbit Mercury following three flybys of that planet. The orbital phase will use data collected during the flybys as an initial guide to conducting its focused scientific investigation of this mysterious world, which remains the least explored of the terrestrial (or inner) planets of the solar system.

On August 3, 2004, the 1,070-pound (485-kg) *MESSENGER* was successfully launched from Cape Canaveral Air Force Station by a Boeing Delta II rocket. During a planned 4.9-billion-mile (7.9-billion-km) interplanetary journey that includes 15 trips around the Sun, *MESSENGER* has flown past Earth once (in August 2005), will fly past Venus twice (in October 2006 and June 2007), and then past Mercury three times (in January 2008, October 2008, and September 2009) before easing into orbit around Mercury.

The Earth and Venus flybys use the gravity-assist maneuver to guide *MESSENGER* toward Mercury's orbit. The three Mercury flybys will help *MESSENGER* match the planet's speed and location for an orbit insertion maneuver in March 2011. The flybys of Mercury also allow *MESSENGER* to gather important data, which scientists will then use to plan the yearlong orbital phase of the mission.

The MESSENGER spacecraft, designed and built for NASA by the Johns Hopkins University Applied Physics Laboratory (JHUAPL) is only the second robot spacecraft sent to Mercury. *Mariner 10* flew past Mercury three times in 1974–75, but, because of orbital mechanics limitations, could only gather detailed data on less than half of the planet's surface.

The MESSENGER mission has an ambitious science plan. The space robot's complement of seven science instruments will determine Mercury's composition; produce color images of the planet's surface on a global basis; map Mercury's magnetic field and measure the properties of the planet's core; examine Mercury's intriguing poles to determine the extent of any water ice or other frozen volatile material deposits in permanently shadowed regions; and characterize Mercury's tenuous atmosphere and Earthlike magnetosphere.

The first intense search for life on Mars was begun in 1975, when NASA launched the agency's Viking missions, consisting of two orbiter and two lander spacecraft. Development of the elaborate robotic mission was divided between several NASA centers and private U.S. aerospace firms. JPL built the Viking orbiter spacecraft, conducted mission communications, and eventually assumed management of the mission. The Viking mission and the search for life on Mars are discussed in subsequent chapters.

Credit for the single space-robot mission that has visited the greatest number of giant planets goes to JPL's Voyager project. Launched in 1977, the twin *Voyager 1* and *Voyager 2* flew by the planets Jupiter (1979) and Saturn (1980–81). *Voyager 2* then went on to have an encounter with

Artist's concept of NASA's Viking mission spacecraft (orbiter and lander combined) approaching Mars in 1976. *(NASA)*

Artist's concept of NASA's far-traveling *Voyager 2* robot spacecraft, as it looks back upon Neptune and its moon Triton, seven hours after its closest approach to the distant planet on August 25, 1989. Artist Don Davis created this painting based on a computer-assembled simulation of the spacecraft's trajectory through the Neptune system. *(NASA/JPL)*

Uranus (1986) and with Neptune (1989). Both *Voyager 1* and *Voyager 2* are now traveling on different trajectories into interstellar space. In February 1998, *Voyager 1* passed the *Pioneer 10* spacecraft to become the most distant human-made object in space. The Voyager Interstellar Mission (VIM) (described in Chapter 12) should continue well into the next decade.

Millions of years from now—most likely when human civilization has completely disappeared from the surface of Earth—four robot spacecraft (*Pioneer 10* and *11*, *Voyager 1* and *2*) will continue to drift through the interstellar void. Each spacecraft will serve as a legacy of human ingenuity and inquisitiveness. By carrying a special message from Earth, each far-traveling robot spacecraft also bears permanent testimony that at least one moment in the history of the human species a few people raised their foreheads to the sky and reached for the stars. Though primarily designed for scientific inquiry within the solar system, these four relatively simple robotic exploring machines are now a more enduring artifact of human civilization than any cave painting, great monument, giant palace, or high-rise city created here on Earth.

A new generation of more sophisticated spacecraft appeared in the late 1980s and early 1990s. These vehicles allowed NASA to conduct much more detailed scientific investigation of the planets and of the Sun. The robot spacecraft used in the Galileo mission to Jupiter and the Cassini mission to Saturn are representative of significant advances in sensor technology, computer technology, and aerospace engineering.

The Galileo mission began on October 18, 1989, when the sophisticated spacecraft was carried into low Earth orbit by the space shuttle *Atlantis* and then launched on its interplanetary journey by means of an inertial upper stage (IUS) rocket. Relying on gravity-assist flybys to reach Jupiter, the *Galileo* spacecraft flew past Venus once and Earth twice. As it traveled through interplanetary space beyond Mars on its way to Jupiter, *Galileo* encountered the asteroids Gaspra (October 1991) and Ida (August 1993). *Galileo*'s flyby of Gaspra on October 29, 1991, provided scientists their first-ever close-up look at a minor planet. On its final approach to Jupiter, *Galileo* observed the giant planet's bombardment by fragments of Comet Shoemaker-Levy-9, which had broken apart. On July 12, 1995, the *Galileo* mother spacecraft separated from its hitchhiking companion (an atmospheric probe) and the two robot spacecraft flew in formation to their final destination.

On December 7, 1995, *Galileo* fired its main engine to enter orbit around Jupiter and gathered data transmitted from the atmospheric probe during that small robot's parachute-assisted descent into the Jovian atmosphere. During its two-year prime mission, the *Galileo* spacecraft performed 10 targeted flybys of Jupiter's major moons. In December 1997, the sophisticated robot spacecraft began an extended scientific mission

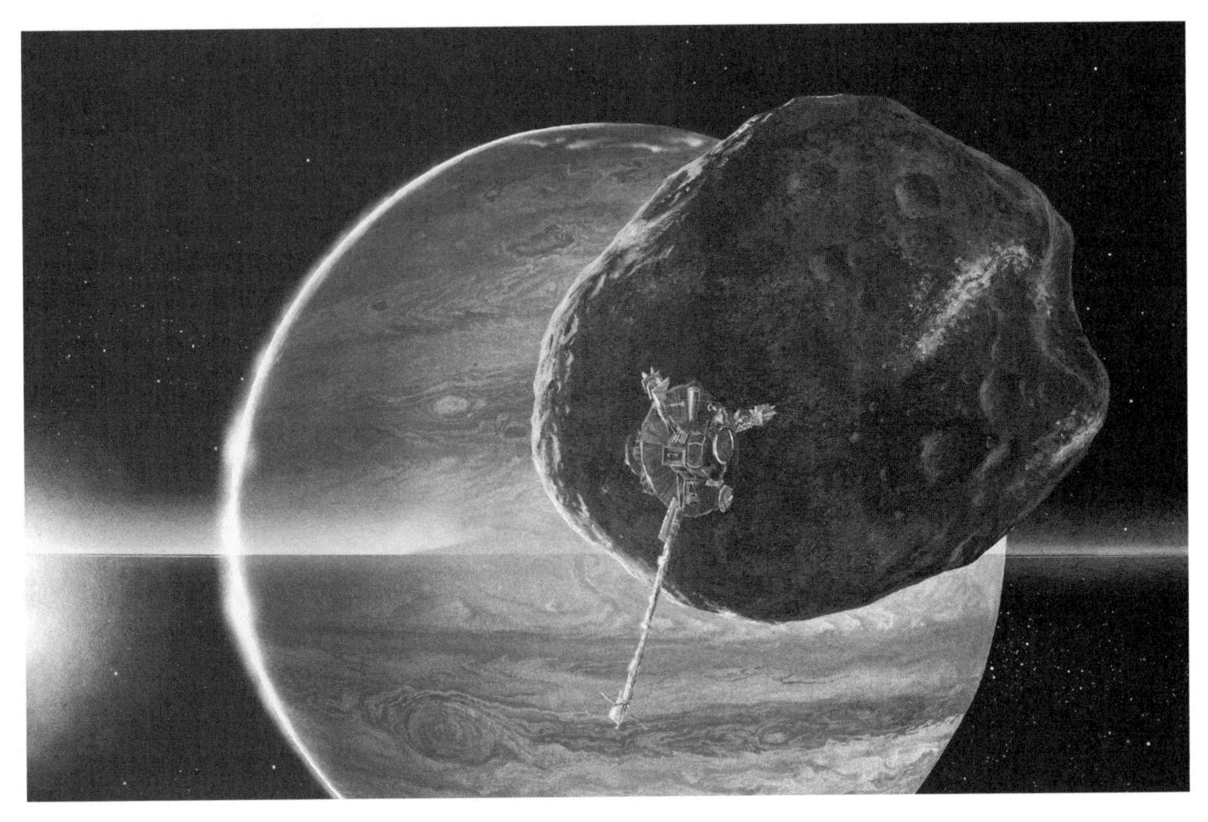

This artist's concept shows NASA's *Galileo* spacecraft as it performed a very close flyby of Jupiter's tiny inner moon Amalthea in November 2002. *(NASA)*

that featured eight flybys of Jupiter's smooth, ice-covered moon Europa and two flybys of the pizza-colored, volcanic Jovian moon, Io.

Galileo started a second extended scientific mission in early 2000. This second extended mission included flybys of the Galilean moons Io, Ganymede, and Callisto, plus coordinated observations of Jupiter with the *Cassini* spacecraft. In December 2000, *Cassini* flew past the giant planet to receive a much-needed gravity assist that enabled the large spacecraft to eventually reach Saturn. *Galileo* conducted its final flyby of a Jovian moon in November 2002, when it zipped past the tiny inner moon, Amalthea.

The encounter with Amalthea left *Galileo* on a course that would lead to an intentional impact with Jupiter in September 2003. NASA mission controllers deliberately crashed the *Galileo* mother spacecraft into Jupiter at the end of the space robot's very productive scientific mission, to avoid any possibility of contaminating Europa with terrestrial microorganisms. As an uncontrolled derelict, the *Galileo* might have eventually crashed into Europa sometime within the next few decades. Many exobiologists suspect that Europa has a life-bearing, liquid-water ocean underneath its icy surface. Since the *Galileo* spacecraft was probably harboring a variety of hitchhiking terrestrial microorganisms, scientists thought it prudent to completely avoid any possibility of contamination of Europa. The easiest way to resolve the potential problem was to simply dispose of the retired *Galileo* in the frigid, swirling clouds of Jupiter. So, NASA and the JPL mission controllers accomplished this task while still maintaining sufficient control over *Galileo*'s behavior and trajectory.

Today, JPL remains heavily engaged in activities associated with deep-space automated scientific missions. Efforts at the Laboratory in Pasadena include subsystem engineering, instrument development, and more automated levels of data reduction and analysis to support deep space missions. The sophisticated *Cassini,* which is now exploring the Saturn system, and the robust *Spirit* and *Opportunity* Mars Exploration Rovers, which are now rolling across the surface of the Red Planet, are examples of successful contemporary JPL missions involving sophisticated robot spacecraft.

On the horizon are such exciting space robot missions as *Dawn*—the first spacecraft ever planned to orbit two different celestial bodies after leaving Earth. *Dawn* will launch in 2007, orbit the large main belt asteroid, Vesta, starting in 2011, and then begin orbiting the largest main belt asteroid, Ceres, in 2015.

✧ Robot Spacecraft in Service to Astronomy

Each portion of the electromagnetic spectrum (that is, radio waves, infrared radiation, visible light, ultraviolet radiation, X-rays, and gamma rays) brings astronomers and astrophysicists unique information about the universe and the objects within it. For example, certain radio frequency (RF) signals help scientists characterize cold molecular clouds. The cosmic microwave background (CMB) represents the fossil radiation from the big bang, the enormous ancient explosion considered by most scientists to have started the present universe about 15 billion years ago. The infrared (IR) portion of the spectrum provides signals that let astronomers observe non-visible objects such as near-stars (brown dwarfs) and relatively cool stars. Infrared radiation also helps scientists peek inside dust-shrouded stellar nurseries (where new stars are forming) and unveil optically opaque regions at the core of the Milky Way Galaxy. Ultraviolet (UV) radiation provides astrophysicists special information about very hot stars and quasars, while visible light helps observational astronomers characterize planets, main sequence stars, nebulae, and galaxies. Finally, the collection of X-rays and gamma rays by space-based observatories brings scientists unique information about high-energy phenomena, such as supernovae, neutron stars, and black holes. The presence of black holes is inferred by intensely energetic radiation emitted from extremely hot material as it swirls in an accretion disk, before crossing the particular black hole's event horizon.

Scientists recognized that they could greatly improve their understanding of the universe if they could observe all portions of the electromagnetic spectrum. As the technology for space-based astronomy matured toward the end of the 20th century, NASA created the Great Observatories Program. This important program involved a series of four highly sophisticated space-based astronomical observatories—each carefully designed with state-of-the-art equipment to gather "light" from a particular portion (or portions) of the electromagnetic spectrum. An observatory spacecraft is a robot spacecraft that does not have to travel to a celestial destination to explore it. Instead, the observatory spacecraft occupies a special orbit around Earth or an orbit around the Sun, from which vantage point it can observe distant targets without the obscuring and blurring effects of Earth's atmosphere. Infrared observatories should also operate in orbits that minimize interference from large background thermal radiation sources, such as Earth and the Sun.

NASA initially assigned each Great Observatory a development name and then renamed the orbiting astronomical facility to honor a famous scientist. The first Great Observatory was the *Space Telescope (ST),* which became the *Hubble Space Telescope (HST).* It was launched by the space shuttle in 1990 and then refurbished while in orbit through a series of subsequent shuttle missions. With constantly upgraded instruments and improved optics, this long-term space-based observatory is designed to gather light in the visible, ultraviolet, and near-infrared portions of the spectrum. This spacecraft honors the American astronomer Edwin Powell Hubble (1889–1953). NASA is now examining plans for another (possibly robotic) refurbishment mission, which would keep the *HST* operating for several more years until its replacement by the *James Webb Space Telescope (JWST)* around 2011.

The second Great Observatory was the *Gamma Ray Observatory (GRO),* which NASA renamed the *Compton Gamma Ray Observatory (CGRO),* following its launch by the space shuttle in 1991. Designed to observe high-energy gamma rays, this observatory started collecting valuable scientific information from 1991 to 1999 about some of the most violent processes in the universe. NASA renamed the observatory to honor the American physicist and Nobel laureate, Arthur Holly Compton (1892–1962). The *CGRO*'s scientific mission officially ended in 1999. The

NASA's *Hubble Space Telescope (HST)* is being unberthed and carefully lifted out of the payload bay of the space shuttle *Discovery* and then placed into sunlight by the shuttle's robot arm, in February 1997. This event took place during the STS-82 mission, which NASA also calls the second *HST* serving mission (HST SM-02). *(NASA/JSC)*

following year, NASA mission managers commanded the massive spacecraft to perform a controlled de-orbit burn. This operation resulted in a safe reentry in June 2000 and the harmless impact of surviving pieces in a remote portion of the Pacific Ocean.

NASA originally called the third observatory in this series the *Advanced X-ray Astrophysics Facility (AXAF)*. NASA renamed this observatory the *Chandra X-ray Observatory (CXO)* to honor the Indian-American astrophysicist and Nobel laureate Subrahmanyan Chandrasekhar (aka Chandra) (1910–95). The observatory spacecraft was placed into a highly elliptical orbit around Earth in 1999. The *CXO* examines X-ray emissions from a variety of energetic cosmic phenomena, including supernovas and the accretion disks around suspected black holes, and should operate until at least 2009.

The fourth and final member of NASA's Great Observatory Program is the *Space Infrared Telescope Facility (SIRTF)*. NASA launched this observatory in 2003 and renamed it the *Spitzer Space Telescope (SST)* to honor the American astrophysicist Lyman Spitzer, Jr. (1914–97). The sophisticated infrared observatory provides scientists a fresh vantage point from which to study processes that have until now remained mostly in the dark, such as the formation of galaxies, stars, and planets. The *SST* also serves as an important technical bridge to NASA's Origins Program—an ongoing attempt to scientifically address such fundamental questions as "Where did we come from?" and "Are we alone?"

How Robot Spacecraft Work

A robot spacecraft is an uncrewed platform that engineers have designed to be placed into an orbit about Earth, or on an interplanetary trajectory to another celestial body or into deep space. The space robot is essentially a combination of hardware that forms a mission-oriented spacecraft. The collection of hardware that makes up a robot spacecraft includes structure, thermal control, wiring, and subsystem functions, such as attitude control, command, data handling, and power.

NASA engineers often refer to a robot spacecraft as a flight system to distinguish it from equipment that remains on Earth as part of the ground system for a particular project or mission. The robot spacecraft itself might contain ten or more subsystems, including an attitude control subsystem (discussed later in this chapter), which in turn contains numerous assemblies, such as reaction wheel assemblies or inertial reference assemblies. In certain instances, like those involving the telecommunications system, there are transmitter and receiver subsystems on both the spacecraft (as part of the flight system) and back on Earth (as part of the ground system). So, the use of system and subsystem nomenclature in the aerospace field can be a bit confusing. There are even times when systems are contained within subsystems, as in, for example, the case of an *imaging subsystem* that contains a *lens system.* Just remember that the hierarchy of aerospace hardware is: system, subsystem, assembly, and component (or part) in that descending order. Because of the complexity of a robot spacecraft and the interdependent nature of many of its systems and subsystems, however, engineers and scientists will often appear very arbitrary in their application of this hardware classification scheme. Fortunately, a little apparent confusion in nomenclature in no way detracts from the quality of the hardware that makes the robot spacecraft function and perform marvelous feats of automated exploration and scientific data collection. And that, after all, is the main reason why these fascinating machines are built in the first place.

Individual robot spacecraft can be very different from one another in design and level of complexity, including the type and number of subsystems and component parts and assemblies found in each individual subsystem. Not all of the different types of robot spacecraft discussed in this chapter need the same subsystems. For example, a robot probe, which descends into a planetary atmosphere on a one-way scientific mission, will generally not have a propulsion subsystem or an attitude-control subsystem. But the probe will carry scientific instruments, need electric power, have a structure, require an effective thermal-control system, use an onboard computer, and transmit the data it collects. This chapter focuses on the basic subsystems that satisfy mission requirements of modern, complex flyby- or orbiter-class robot spacecraft. The treatment is sufficiently broad, however, to embrace the often less complex (from a spacecraft engineering perspective) types of space robots, such as landers and rovers.

Different space robots possess different levels of machine intelligence. A robot's level of machine intelligence determines the degree of autonomous operation possible and the amount of human supervision required. For deep space missions, direct human supervision is usually impractical or impossible; so, a space robot engaged in this type of mission must have an appreciable level of machine intelligence. Specifically, at a great distance from Earth, a robot spacecraft must have the autonomy and machine intelligence to monitor and control itself. When a space robot is light-minutes away from Earth, human members of the mission cannot respond to anomalies in time. All of a robot spacecraft's subsystems should contain and run fault-protection algorithms, which can quickly detect and respond to a problem without direct human assistance. When a fault-protection algorithm detects a problem, it can respond by safing the subsystem in difficulty. Safing is the process by which a spacecraft automatically shuts down or reconfigures components to prevent damage either from within or from changes in the external environment. Many terrestrial machines and home appliances have safing features engineered into the device. A thermal limit switch on the electric motor of an office paper shredder is an example. When the motor works too hard and starts getting a bit too hot, the thermal limit switch shuts down the device before any permanent damage can occur. When the motor cools to a safe level, the thermal limit switch resets and the shredder can be used again. Robot spacecraft have many such safing features engineered into their complex subsystems.

✧ Space Robots in Service to Science

Robot spacecraft come in all shapes and sizes. Each space robot is usually custom-designed and carefully engineered to meet the specific needs and environmental challenges of a particular space exploration mission. For example, lander spacecraft are designed and constructed to acquire scientific data and to function in a hostile planetary surface environment. Since the complexity of space robots varies greatly, engineers and space scientists find it convenient to categorize robot spacecraft according to the missions they are intended to fly. This chapter introduces the broad general classes of robot spacecraft. Chapter 3 provides a complementary historic snapshot of how the size and complexity of robot spacecraft have changed over the last four decades. Subsequent chapters describe the features of important space robots from each of the major broad classes, such as flybys, orbiters, landers, and rovers.

Most interplanetary missions are flown to collect scientific data. However, some space robot missions, like NASA's *Deep Space 1 (DS1),* have as their primary objective the demonstration of new space technologies (see chapter 9). On technology-demonstration missions, the collection of scientific data remains an important, though secondary, objective. When the collection of scientific data is the primary mission of a robot spacecraft, then all the subsystems and components that are onboard the spacecraft are there in support of that single purpose. The space robot is designed and constructed so as to gather scientific data at the target interplanetary location or celestial object.

The robot spacecraft exists to deliver its scientific instruments to a particular interplanetary destination; to allow these instruments to make their measurements, perform their observations, and/or conduct their experiments under the most favorable achievable conditions; and then to return data from the instruments back to scientists on Earth. In the interesting case of a sample return mission (see chapter 7), the robot spacecraft must collect and then return material samples from an alien world. Once the space robot delivers its extraterrestrial cargo to Earth, scientists perform detailed investigations upon the alien materials in a special, biologically isolated (quarantine) facility.

There are many different types of scientific instruments that a robot spacecraft can carry. For convenience, scientists and engineers usually divide these instruments into two general classes: direct-sensing instruments and remote-sensing instruments. A direct-sensing instrument interacts with the phenomenon (of interest) in the immediate vicinity of the instrument. Examples include a radiation-detection instrument and a magnetometer. In contrast, a remote-sensing instrument examines an object or phenomenon at a distance without being in direct contact with that object. The passage of electromagnetic radiation from object to instrument supports information transfer and data collection. Remote-sensing instruments usually form some type of image of the object being studied or else collect characteristic data from the object, such its temperature, luminous intensity, or energy level at a particular wavelength.

Scientists also find it convenient to classify scientific instruments as either passive or active. A passive instrument detects radiation, particles, or other information naturally emitted by the object or phenomenon under study. A magnetometer is a passive, direct-sensing scientific instrument carried by many robot spacecraft to detect and measure the interplanetary magnetic fields in the vicinity of the spacecraft. Imaging instruments are examples of passive remote-sensing instruments, which collect the electromagnetic radiation emitted by, or reflected from, a planetary body. Sunlight serves as the natural source of illumination for the observed reflected radiation from a planetary body. (Passive imaging instruments are discussed shortly). An active instrument supplies its own source of electromagnetic radiation or particle radiation to stimulate a characteristic response from the target being illuminated or irradiated. A synthetic aperture radar, as carried by NASA's *Magellan* orbiter spacecraft, and the alpha proton X-ray spectrometer (APXS) used by NASA's *Mars Pathfinder* rover are examples of active scientific instruments.

✧ General Classes of Scientific Spacecraft

Scientific space robots include: flyby spacecraft, orbiter spacecraft, atmospheric probe spacecraft, atmospheric balloon packages, lander spacecraft, surface penetrator spacecraft, surface rover spacecraft, and observatory spacecraft.

There are three basic possibilities for a robot spacecraft's trajectory when it encounters a planet. The first possible trajectory involves a direct hit or hard landing. This is an impact trajectory (trajectory *a* in the figure). A hard landing involves a relatively high-velocity impact landing of the robot spacecraft on the surface of a planet or moon. This usually destroys

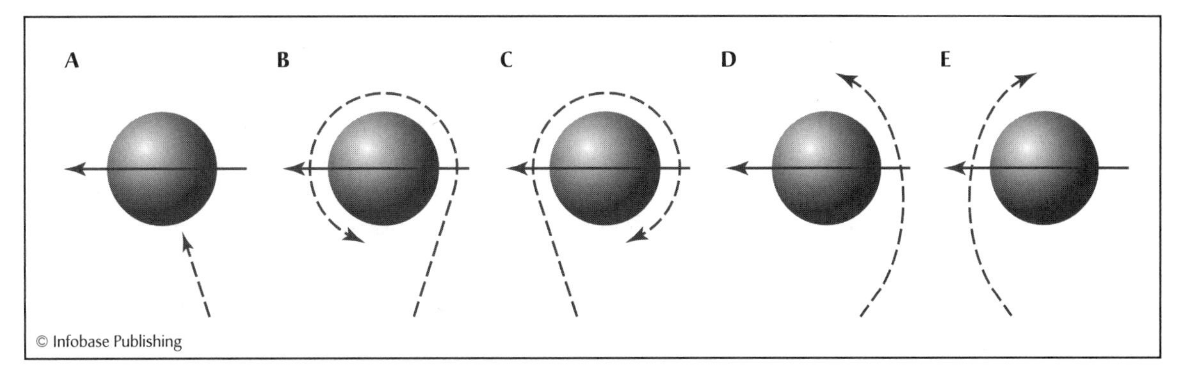

Possible trajectories of a robot spacecraft encountering a planet

all equipment, except perhaps for a very rugged instrument package or payload container. The hard landing can be intentional, as has occurred during NASA's Ranger missions, which were designed to crash into the lunar surface; or unintentional, as when a retrorocket system fails to fire or a parachute system fails to deploy, and the robot lander strikes the planetary surface at an unexpected and unplanned high speed.

Aerospace engineers design lander spacecraft to follow an impact trajectory to a planet's surface. They also want the robot to survive by touching down on the surface at a very low speed. Sometimes, a lander spacecraft is sent on a direct impact trajectory; at other times the robot is carried through interplanetary space by a mother spacecraft and then released on an impact trajectory after the mother spacecraft has achieved orbit around the target planet. Following separation from the orbiting mother spacecraft, the lander travels on a carefully designed impact trajectory to the target planet's surface. NASA's Surveyor spacecraft to the Moon are an example of the former soft-landing mission approach, while the *Viking 1* and *2* lander missions to Mars are an example of the latter design approach.

The *Viking 1* and *2* lander spacecraft placed on the Martian surface in 1976 represent one of the great early triumphs of robotic space exploration. After separation from the Viking orbiter spacecraft, the lander (protected by an aeroshell) descended into the thin Martian atmosphere at speeds of approximately 9,940 miles per hour (16,000 km/hr). As it descended, the lander was slowed down by aerodynamic drag until its aeroshell was discarded. Each robot lander spacecraft was then slowed down further by the release of a parachute. Finally, the robot achieved a gentle landing by automatically firing retrorockets. Of special significance is the fact that both Viking landers successfully accomplished the entire soft-landing sequence automatically, without any direct human intervention or guidance.

In another lander/probe mission scenario, the mother spacecraft releases the lander or robot probe, while the co-joined spacecraft pair is still some distance from the target planetary object. Following release and separation, the robot probe follows a ballistic impact trajectory into the atmosphere and onto the surface of the target body. This scenario occurred when the *Cassini* mother spacecraft released the hitchhiking *Huygens* probe on December 25, 2004, as *Cassini* orbited around Saturn. Following separation, the *Huygens* traveled for about 20 days along a carefully planned ballistic trajectory to Saturn's moon Titan. When it arrived at Titan on January 14, 2005, the *Huygens* entered the moon's upper atmosphere, performed a superb data-collecting descent, and successfully landed on the moon's surface.

The second type of trajectory is an orbital-capture trajectory. The spacecraft is simply captured by the gravitational field of the planet and

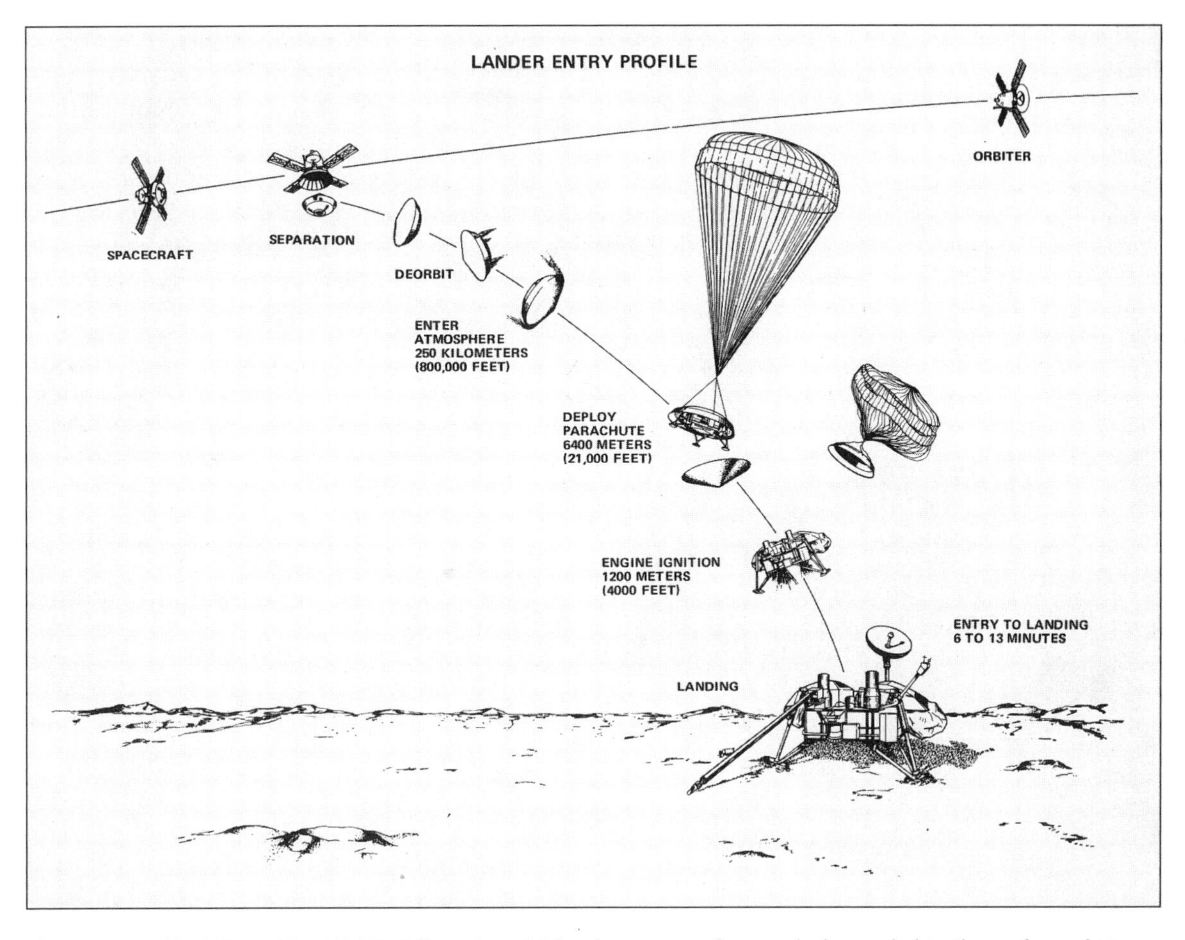

The entry profile followed by NASA's *Viking 1* and *2* lander spacecraft as each descended to the surface of Mars and successfully soft-landed in 1976 *(NASA)*

enters orbit around it. Depending upon its precise speed and altitude (and other parameters), the robot spacecraft can enter this captured orbit from either the trailing edge (trajectory *b* in the figure on page 37) or the leading edge (trajectory *c* in the figure on page 37) of the planet. In the third type of trajectory, called a flyby trajectory, the spacecraft remains far enough away from the planet to avoid capture, but passes close enough to be strongly affected by its gravity. In this case, the speed of the spacecraft will be increased if it approaches from the trailing side of the planet (trajectory *d* in the figure on page 37) and diminished if it approaches from the leading side (trajectory *e* in the figure on page 37). In addition to changing speed, the spacecraft's motion also changes direction.

The increase in speed of the flyby spacecraft actually comes from a decrease in speed of the planet itself. In effect, the spacecraft is being "pulled along" by the planet. Of course, this is a greatly simplified discussion of complex encounter phenomena. A full account of spacecraft trajectories must consider the speed and actual trajectory of the spacecraft and planet, how close the spacecraft will come to the planet, and the size (mass) and orbital speed of the planet, in order to make even a simple calculation. Aerospace engineers make good use of this natural planetary

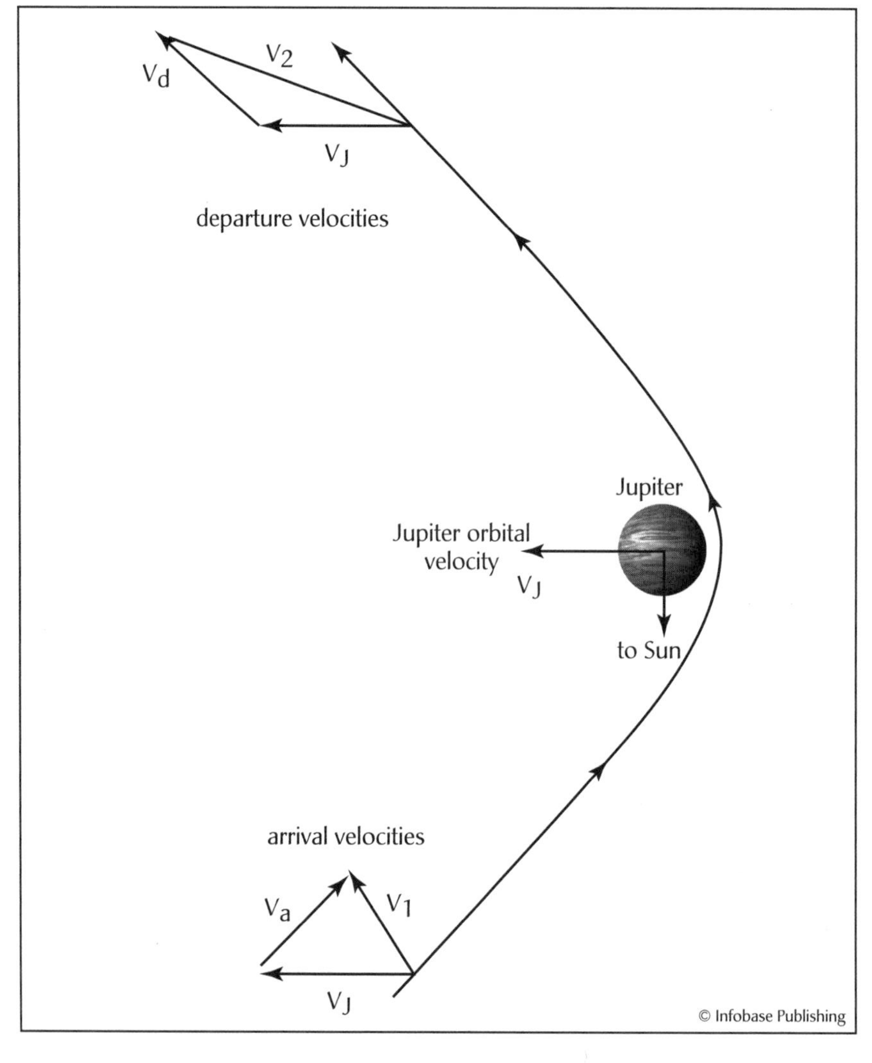

Velocity change experienced by a robot spacecraft during a Jupiter flyby *(NASA)*

tug on a flyby spacecraft, and they call this important orbital-mechanics technique a gravity-assist maneuver.

A better understanding of the gravity-assist is obtained through the use vectors in a slightly more detailed, mathematical explanation. The way in which speed is added to the flyby spacecraft during close encounters with the planet Jupiter is shown in the figure on page 40. During the time that either *Voyager 1* or *2* was near Jupiter, the heliocentric (Sun-centered) path each spacecraft followed in its motion with respect to Jupiter was closely approximated by a hyperbola.

The heliocentric velocity of the spacecraft is the vector sum of the orbital velocity of Jupiter (V_J) and the velocity of the spacecraft with respect to Jupiter (that is, tangent to its trajectory—the hyperbola). The spacecraft moves toward Jupiter along an asymptote, approaching from the approximate direction of the Sun and with asymptotic velocity (V_a). The heliocentric arrival velocity (V_1) is then computed by vector addition: $V_1 = V_J + V_a$. The spacecraft then departs Jupiter in a new direction, determined by the amount of bending that is caused by the effects of the gravitational attraction of Jupiter's mass upon the mass of the spacecraft. The asymptotic departure speed (V_d) on the hyperbola is equal to the arrival speed. Thus, the length of V_a equals the length of V_d. For the heliocentric departure, the velocity is: $V_2 = V_J + V_d$. This vector sum appears in the upper portion of the figure.

During the relatively short period of time that the spacecraft is near Jupiter, the orbital velocity of Jupiter (V_J) changes very little, and so scientists assume that V_J is equal to a constant.

The vector sums in the figure illustrate that the deflection, or bending, of the spacecraft's trajectory caused by Jupiter's gravity results in an increase in the speed of the spacecraft *along its hyperbolic path,* as measured relative to the Sun. For *Voyager 1* and *Voyager 2,* this increase in velocity reduced the total flight time necessary to reach Saturn and points beyond. This indirect type of deep-space mission to the outer planets saves two or three years of flight time, compared to direct-trajectory missions, which do not take advantage of gravity assist.

What happens to Jupiter (or any other planet) as a result of a spacecraft's gravity-assist maneuver? The principle of conservation of momentum is at work here. (In Newtonian mechanics, linear momentum is defined as the product of mass times velocity). While the spacecraft gains momentum (and thus speed) during its encounter with Jupiter, the giant planet loses some of its momentum (and consequently orbital speed) during the encounter—since there is no change in mass for either object. Because of the extreme difference in their masses, however, the change in Jupiter's velocity is negligible.

Flyby spacecraft follow a continuous trajectory and are not captured into a planetary orbit. These spacecraft have the capability to use their

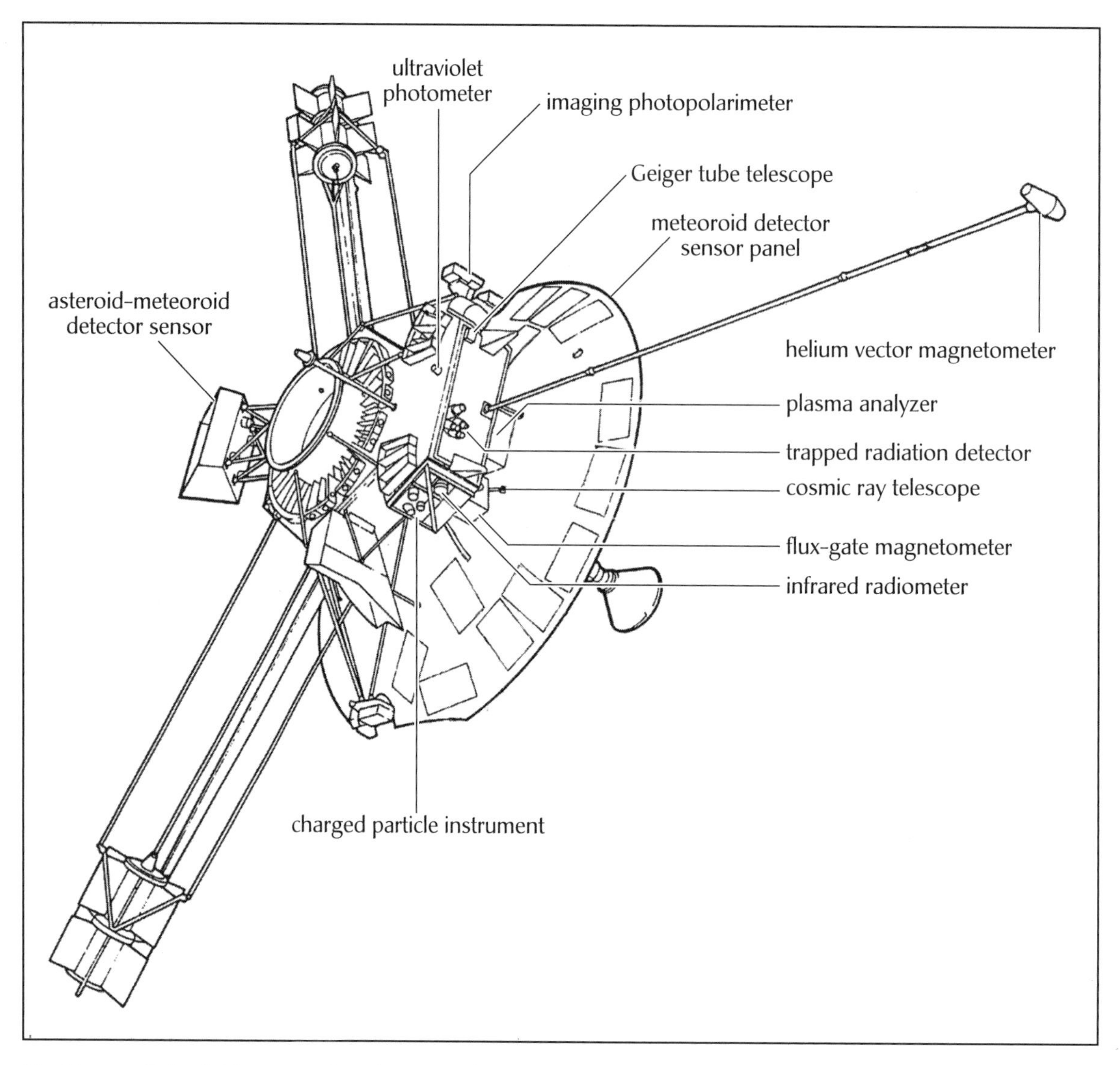

The *Pioneer 10* (and *11*) spacecraft with its complement of scientific instruments. Electric power was provided to the far-traveling robot spacecraft by a long-lived radioisotope thermoelectric generator (RTG). *(NASA)*

onboard instruments to observe passing celestial targets (for example, a planet, a moon, an asteroid), and can even compensate for their target's apparent motion in an optical instrument's field of view. They must be able to transmit data at high rates back to Earth and also capable of storing data on board for those periods when their antennae are not pointing toward Earth. Flyby spacecraft must be capable of surviving in a powered-down, cruise mode for many years of travel through interplanetary space, and then of bringing all their sensing systems to focus rapidly on a target object during an encounter period that may last only for a few crucial hours or minutes. NASA's *Pioneer 10* and *11* and the *Voyager 1* and *2* are examples of highly successful flyby scientific spacecraft. NASA uses the flyby spacecraft during the initial, or reconnaissance, phase of solar-system exploration.

An orbiter spacecraft is designed to travel to a distant planet and then orbit around that planet. This type of scientific spacecraft must possess a substantial propulsive capability to decelerate at just the right moment in

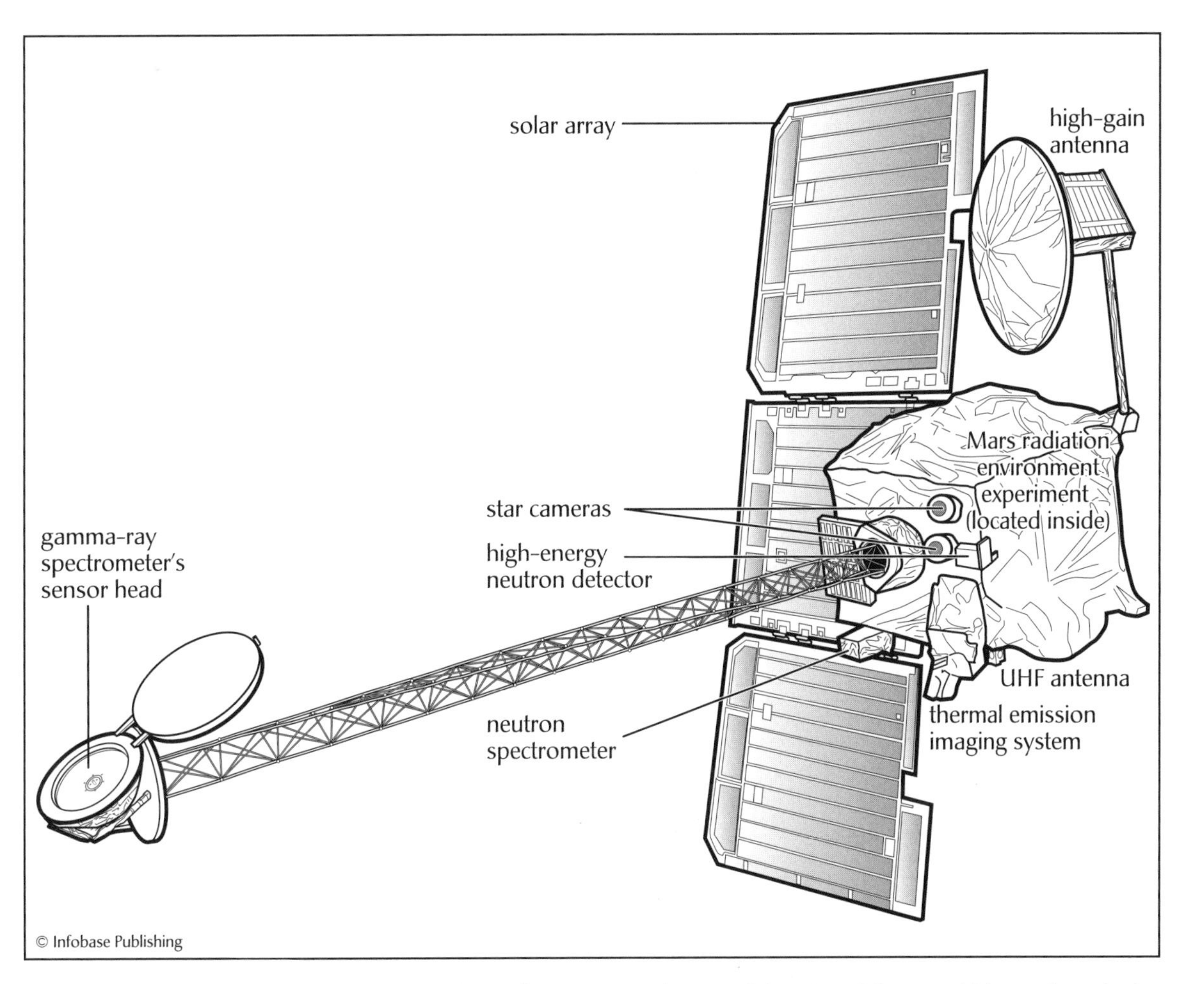

The operational configuration of NASA's robot orbiter spacecraft, named the *Mars Odyssey,* which was launched on April 7, 2001, and arrived at the Red Planet in late October that year. The spacecraft used its complement of scientific instruments to conduct a multiyear examination of Mars, composition, detecting water and shallow buried ice and studying the Martian radiation environment. *(NASA)*

order to achieve a proper orbit insertion. Aerospace engineers designing an orbiter spacecraft recognize the fact that solar occultations will occur frequently as it orbits the target planet. During these periods of occultation, the spacecraft is shadowed by the planet, cutting off solar-array production of electric power and introducing extreme variations of the spacecraft's thermal environment. Generally, a rechargeable battery system augments solar electric power. Active thermal control techniques (e.g., the use of tiny electric-powered heaters) are used to complement traditional passive thermal-control design features. The periodic solar occultations also interrupt uplink and downlink communications with Earth, making

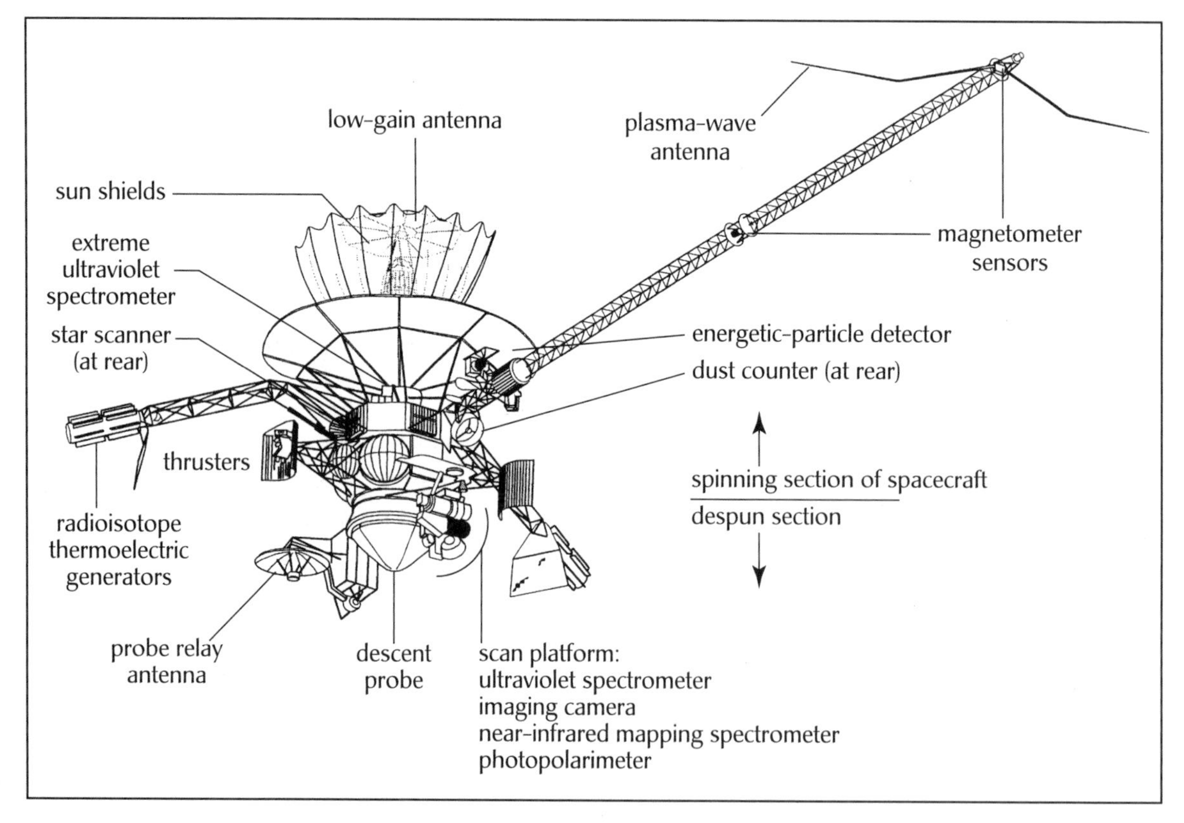

NASA's *Galileo* spacecraft in interplanetary flight configuration with descent probe still attached. On July 12, 1995, the *Galileo* mother spacecraft separated from its atmospheric descent probe and the two robot spacecraft flew in formation to their final destination. On December 7, 1995, *Galileo* fired its main engine to enter Jupiter's orbit and collected data radioed from the probe during its parachute-assisted descent into the planet's atmosphere. *Galileo*'s final flyby, of the small moon Amalthea, left the orbiter spacecraft on course for an intentional, mission-ending plunge into Jupiter's atmosphere in September 2002. *(NASA)*

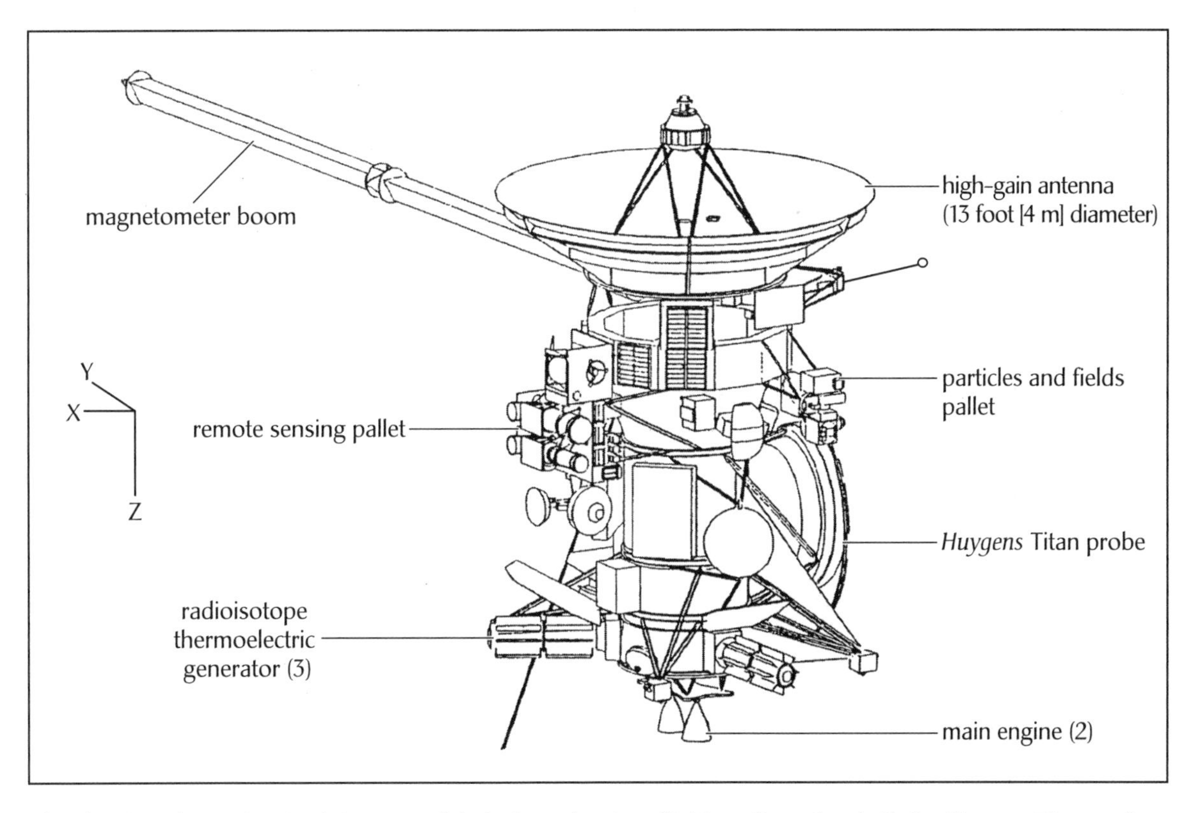

This drawing shows the *Cassini* spacecraft in its interplanetary flight configuration (with the *Huygens* Titan probe still attached) prior to its arrival at Saturn in July 2004. *(NASA)*

onboard data storage a necessity. NASA uses orbiter spacecraft as part of the second, in-depth study phase of solar system exploration. The *Lunar Orbiter, Magellan, Galileo,* and *Cassini* are examples of successful scientific orbiters.

Some scientific-exploration missions involve the use of one or more smaller, instrumented spacecraft, called atmospheric-probe spacecraft. These probes separate from the main spacecraft prior to its closest approach to a planet, in order to study the planet's gaseous atmosphere as they descend through it. Usually, an atmospheric-probe spacecraft is deployed from its mother spacecraft (that is, the main spacecraft) by the release of springs or other devices that simply separate it from the mother spacecraft without significantly modifying the probe's trajectory. Following probe release, the mother spacecraft usually executes a trajectory-correction maneuver to prevent its own atmospheric entry and to help the main spacecraft continue with its flyby or orbiter mission activities. NASA's *Pioneer Venus* (four probes), *Galileo* (one probe), and *Cassini* (*Huygens* probe) missions involved the deployment of a probe or probes into the target planetary body's atmosphere (i.e., Venus, Jupiter, and Saturn's moon Titan, respectively). An aeroshell protects the atmospheric probe spacecraft from the intense heat caused by atmospheric friction during entry. At some point in the descent trajectory, the aeroshell is jettisoned and a parachute then is used to slow the probe's descent sufficiently for it to perform its scientific observations. Data usually are telemetered from the atmospheric probe to the mother spacecraft, which then either relays the data back to Earth in real time or records the data for later transmission to Earth.

Jet Propulsion Laboratory (JPL) technicians clean and prepare the upper equipment module for mating with the propulsion module of the *Cassini* orbiter spacecraft at the Kennedy Space Center in 1997. The large *Cassini/Huygens* spacecraft configuration was successfully launched on October 15, 1997, by a Titan IV-Centaur rocket vehicle. The robot spacecraft arrived at Saturn in July 2004, after a long journey through interplanetary space, which included gravity-assist flybys of Venus (April 1998 and June 1999), Earth (August 1999), and Jupiter (December 2000). *(NASA/JPL)*

An atmospheric balloon package is designed for suspension from a buoyant gas-filled bag that can float and travel under the influence of the winds in a planetary atmosphere. Tracking of the balloon package's progress across the face of the target planet will yield data about the general circulation patterns of the planet's atmosphere. A balloon package needs a power supply and a telecommunications system (to relay data and support tracking). It also can be equipped with a variety of scientific instruments to measure the planetary atmosphere's composition, temperature, pressure, and density.

During their flyby of Venus in June 1985, the Russian *Vega 1* and *2* deployed constant-pressure instrumented balloon aerostats. Each 11-foot- (3.4-m-) diameter balloon has an 11-pound (5-kg) science payload suspended beneath it by a 39-foot- (12-m-) long cable. The aerostats floated at an altitude of approximately 31 miles (50 km), in the most active layer of Venus's three-tiered cloud system. Data (such as temperature, pressure, and wind velocity) from each balloon's science instruments were transmitted directly to Earth for the 47-hour lifetime of the aerostat mission. After two days of operation, floating almost 5,600 miles (9,000 km) through the Venusian atmosphere, each

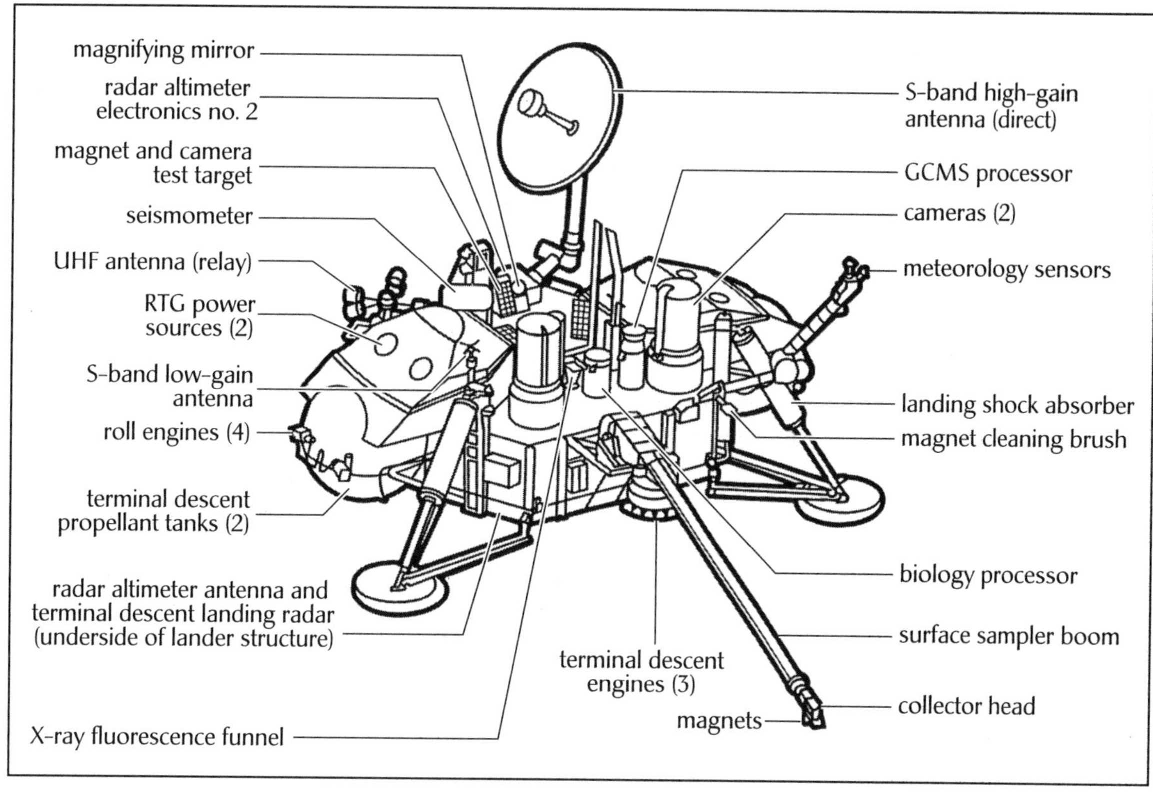

This drawing shows the *Viking* lander spacecraft and its complement of instruments. *(NASA)*

balloon entered the sunlit dayside of the planet, over-expanded due to solar heating, and burst.

Lander spacecraft are designed to reach the surface of a planet and survive at least long enough to transmit back to Earth useful scientific data, such as imagery of the landing site, measurement of the local environmental conditions, and an initial examination of soil composition. For example, the Russian *Venera* lander spacecraft have made brief scientific investigations of the inferno-like Venusian surface. In contrast, NASA's Surveyor lander craft extensively explored the lunar surface at several landing sites in preparation for the human Apollo Project landing missions, while NASA's *Viking 1* and *2* lander craft investigated the surface conditions of Mars at two separate sites for many months.

A surface penetrator spacecraft is designed to enter the solid body of a planet, an asteroid, or a comet. It must survive a high-velocity impact and then transmit subsurface information back to an orbiting mother spacecraft.

NASA launched the *Mars Polar Lander* (*MPL*) in early January 1999. *MPL* was an ambitious mission to land a robot spacecraft on the frigid surface of Mars near the edge of the planet's southern polar cap. Two small penetrator probes (called *Deep Space 2*) piggybacked on the lander spacecraft on the trip to Mars. After an uneventful interplanetary journey, all contact with the *MPL* and the *Deep Space 2* penetrator experiments was lost as the spacecraft arrived at the planet on December 3, 1999. The missing lander was equipped with cameras, a robotic arm, and instruments to measure the composition of the Martian soil. The two tiny penetrators were to be released as the lander spacecraft approached Mars and then follow independent ballistic trajectories, making impact on the surface and then plunging below it in search of water ice.

The exact fate of the lander and its two tiny microprobes remains a mystery. Some NASA engineers believe that the *MPL* might have tumbled down into a steep canyon, while others speculate the *MPL* may have experienced too rough a landing and become disassembled. A third hypothesis suggests the *MPL* may have suffered a fatal failure during its descent through the Martian atmosphere. No firm conclusions could be drawn, because the NASA mission controllers were completely unable to communicate with the missing lander or either of its hitchhiking planetary penetrators.

Finally, a surface rover spacecraft is carried to the surface of a planet, soft-landed, and then deployed. The rover can either be semiautonomous or fully controlled (through teleoperation) by scientists on Earth. Once deployed on the surface, the electrically powered rover can wander a certain distance away from the landing site and take images and perform soil analyses. Data then are telemetered back to Earth by one of several techniques; via the lander spacecraft, via an orbiting mother spacecraft, or (depending on size of rover) directly from the rover vehicle. The Soviet Union deployed two highly successful robot surface rovers (called *Lunokhod 1* and *2*) on the Moon in the 1970s. In December 1996, NASA launched the Mars Pathfinder mission to the Red Planet. From its innovative airbag-protected bounce and role landing on July 4, 1997, until the final data transmission on September 27, the robot lander/rover team returned numerous close-up images of Mars and chemical analyses of various rocks and soil found in the vicinity of the landing site. The *Spirit* and *Opportunity* (2003) Mars Exploration Rovers are the first of many robot rovers that will scamper across the Red Planet this century. As described in chapter 8, NASA plans to continue exploring the surface of Mars with a variety of more sophisticated lander and mobile robots over the next two decades.

An observatory spacecraft is a space robot that does not travel to a destination to explore. Instead, this type of robot spacecraft travels in an orbit around Earth or around the Sun, from which vantage points the

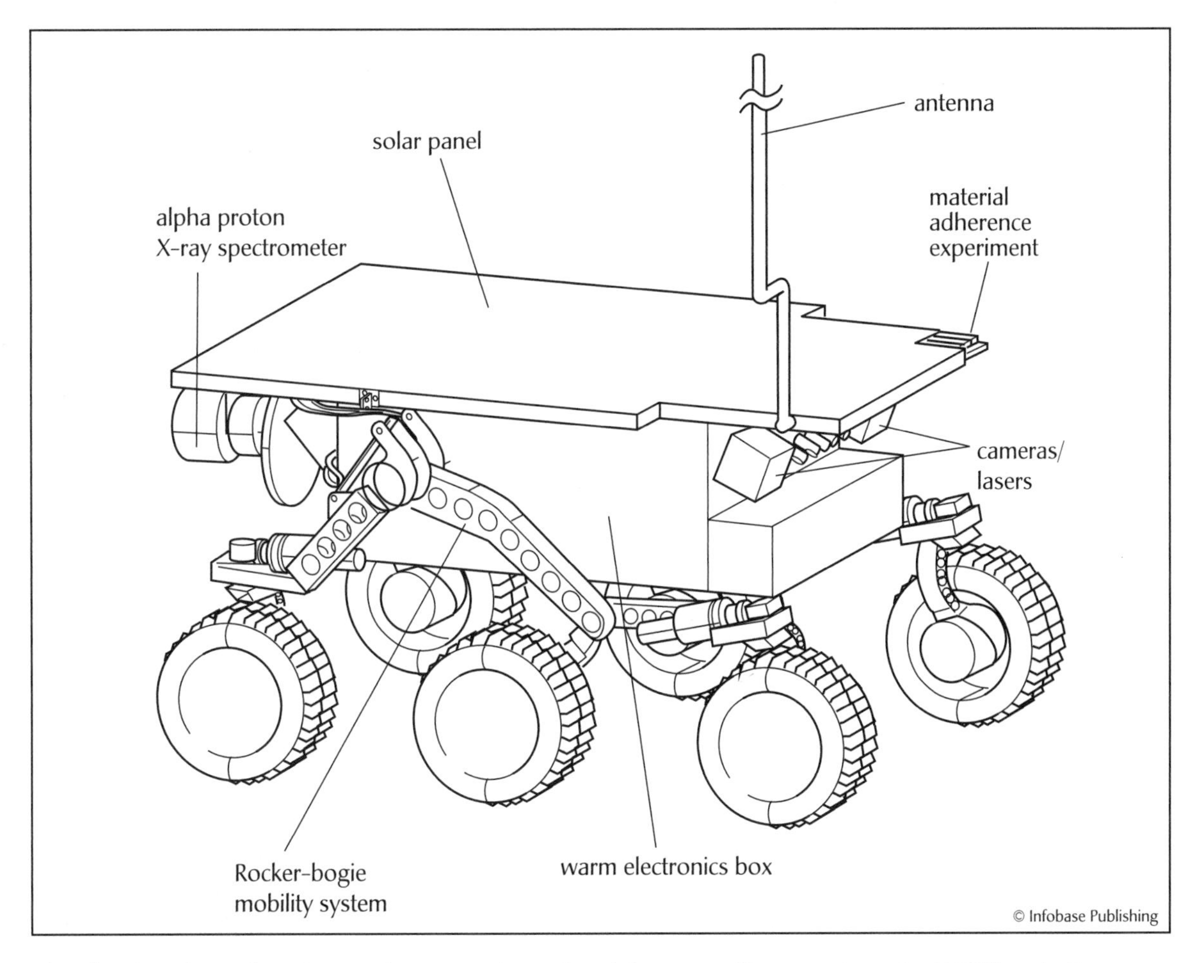

This drawing shows the primary science experiment (an alpha proton X-ray spectrometer [APXS]) and other equipment of the *Mars Pathfinder* rover. In 1997, this tiny robot explored the surface of Mars (in an ancient floodplain known as Ares Vallis), under the watchful teleoperation and supervision of human controllers on Earth. *(NASA/JPL)*

observatory can view distant celestial targets unhindered by the blurring and obscuring effects of Earth's atmosphere. NASA's *Spitzer Space Telescope (SST)* is an example. The *Spitzer Space Telescope* is the final mission in NASA's Great Observatories Program—a family of four orbiting observatories each studying the universe in a different portion of the electromagnetic spectrum. The *Spitzer Space Telescope (SST)*—previously called the *Space Infrared Telescope Facility (SIRTF)*—consists of a 2.8-foot- (0.85-m-) diameter telescope and three cryogenically cooled science instruments. NASA renamed this space-based infrared telescope in honor of the American astronomer Lyman Spitzer, Jr. (1914–97). The *SST* represents the most powerful and sensitive infrared telescope ever launched. The orbiting facility obtains images and spectra of celestial objects at infrared radiation wavelengths between three and 180 micrometers (μm)—an important spectral region of observation mostly unavailable to ground-based telescopes because of the blocking influence of Earth's atmosphere. Following a successful launch (August 25, 2003) from Cape Canaveral, *SST* traveled to an Earth-trailing heliocentric orbit that allowed the telescope to cool rapidly with a minimum expenditure of onboard cryogen (cryogenic coolant). With a projected mission lifetime of at least 2.5 years, *SST* has taken its place alongside NASA's other great orbiting astronomical observatories, and is now collecting high-resolution infrared data that help scientists better understand how galaxies, stars, and planets form and develop. Other missions in this program include the *Hubble Space Telescope (HST)*, the *Compton Gamma Ray Observatory (CGRO)*, and the *Chandra X-ray Observatory (CXO)*.

✧ Functional Subsystems

A robot spacecraft's functional subsystems support the mission-oriented science payload and allow the spacecraft to operate in space, collect data, and communicate with Earth. Aerospace engineers attach all of the other spacecraft components onto the structural subsystem. Aluminum is by far the most common spacecraft structural material. The engineer can select from a wide variety of aluminum alloys, which provide the spacecraft designer with a broad range of physical characteristics, such as strength and machinability. A space robot's structure may also contain magnesium, titanium, beryllium, steel, fiberglass, or low-mass and high-strength carbon composite materials.

How much power does a robot spacecraft need? Engineers have learned from experience that a complex robot spacecraft needs between 300 and 3,000 watts (electric) to properly conduct its mission. Small short-lived robot spacecraft, such as an atmospheric probe and a mini-rover, might need only 25 to 100 watts (electric), which can often be supplied by long-lived batteries. The less power available, however, means the less performance and flexibility the engineers can give the space robot.

The power subsystem must satisfy all of the electric power needs of the robot spacecraft. Engineers commonly use a solar-photovoltaic (solar-cell) system, in combination with rechargeable batteries, to provide a continuous supply of electricity. The spacecraft must also have a well-designed, built-in electric utility grid, which conditions and distributes power to all onboard consumers.

Solar arrays work very well on Earth-orbiting spacecraft and on spacecraft that operate in the inner solar system (within the orbit of Mars and

SOLAR PHOTOVOLTAIC CONVERSION

Solar photovoltaic conversion is the direct conversion of sunlight (solar energy) into electrical energy by means of the photovoltaic effect. A single photovoltaic (PV) converter cell is called a solar cell, while a combination of cells, designed to increase the electric power output, is called a solar array or a solar panel.

Since 1958, solar cells have been used to provide electric power for a wide variety of spacecraft. The typical spacecraft solar cell is made of a combination of *n*-type (negative) and *p*-type (positive) semiconductor materials (generally silicon). When this combination of materials is exposed to sunlight, some of the incidental electromagnetic radiation removes bound electrons from the semiconductor material atoms, thereby producing free electrons. A hole (positive charge) is left at each location from which a bound electron has been removed. Consequently, an equal number of free electrons and holes are formed. An electrical barrier at the *p-n* junction causes the newly created free electrons near the barrier to migrate deeper into the *n*-type material and the matching holes to migrate further into the *p*-type material.

If electrical contacts are made with the *n*- and *p*-type materials, and these contacts connected through an external load (conductor), the free electrons will flow from the *n*-type material to the *p*-type material. Upon reaching the *p*-type material, the free electrons will enter existing holes and once again become bound electrons. The flow of free electrons through the external conductor represents an electric current that will continue as long as more free electrons and holes are being created by exposure of the solar cell to sunlight. This is the general principle of solar photovoltaic conversion.

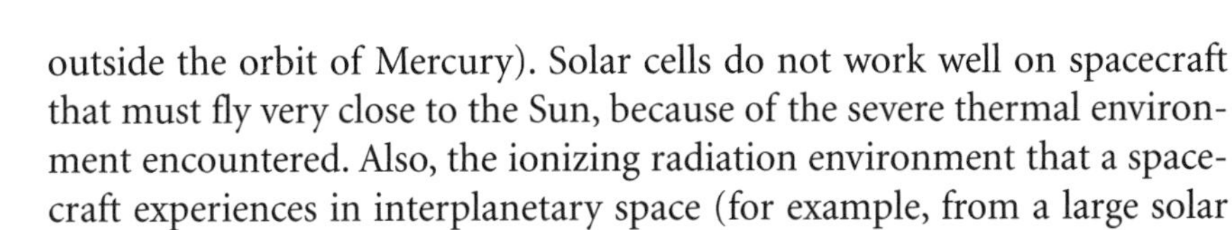

outside the orbit of Mercury). Solar cells do not work well on spacecraft that must fly very close to the Sun, because of the severe thermal environment encountered. Also, the ionizing radiation environment that a spacecraft experiences in interplanetary space (for example, from a large solar flare) or while orbiting in a planet's trapped radiation belt, can damage the solar cells and significantly reduce their useful lifetime.

Some robot spacecraft must operate for years in deep space or in very hostile planetary environments, where a solar-photovoltaic power subsystem becomes impractical if not altogether infeasible. Under these mission circumstances, the engineer selects a long-lived nuclear power supply called a radioisotope thermoelectric generator (RTG). The RTG converts the decay heat from a radioisotope directly into electricity by means of the thermoelectric effect. The United States uses the radioisotope plutonium-238 as the nuclear fuel in its RTGs. (Chapter 10 provides additional discussion about the use of RTGs on space missions.)

A spacecraft's attitude-control subsystem includes the onboard system of computers, low-thrust rockets (thrusters), and mechanical devices

Electro-Optical Imaging Instruments

Two families of detectors perform electro-optical imaging from scientific spacecraft: vidicons and the newer charge-coupled devices (CCDs). Although the detector technology differs, in each case an image of the target celestial object is focused by a telescope onto the detector, where it is converted to digital data. Color imaging requires three exposures of the same target through three different color filters, selected from a filter wheel. Ground processing combines data from the three black-and-white images, reconstructing the original color by using three values for each picture element (pixel).

A vidicon is a vacuum tube resembling a small cathode ray tube (CRT). An electron beam is swept across a phosphor coating on the glass where the image is focused, and its electrical potential varies slightly in proportion to the levels of light it encounters. This varying potential becomes the basis of the video signal produced. Viking, Voyager, and many earlier NASA spacecraft used vidicon-based electro-optical imaging systems to send back spectacular images of Mars (*Viking 1* and *2* orbiter spacecraft) and the outer planets: Jupiter, Saturn, Uranus, and Neptune (*Voyager 1* and *2* flyby spacecraft).

The newer CCD imaging system is typically a large-scale integrated circuit that has a two-dimensional array of hundreds of thousands of charge-isolated wells, each representing a pixel. Light falling on a well is absorbed by a photoconductive substrate (e.g., silicon) and releases a quantity of electrons proportional to the intensity of the incident light. The CCD then detects and stores accumulated electrical charges, which represent the light level on each well. These charges subsequently are read out for conversion to digital data. CCDs are much more sensitive to light over a wider portion of the electro-magnetic spectrum than vidicon tubes; they are also less massive and require less energy to operate. In addition, they interface more easily with digital circuitry, simplifying (to some extent) onboard data processing and transmission back to Earth. The *Galileo*'s solid state imaging (SSI) instrument contained a CCD with an 800 × 800 pixel array.

Not all CCD imagers have two-dimensional arrays. The imaging instrument on NASA's *Mars*

(such as a momentum wheel) used to keep a spacecraft stabilized during flight and to precisely point its instruments in some desired direction. Stabilization is achieved by spinning the spacecraft or by using a three-axis active approach that maintains the spacecraft in a fixed, reference attitude by having it fire a selected combination of thrusters when necessary.

Stabilization can be achieved by spinning the spacecraft, as was done on the *Pioneer 10* and *11* during their missions to the outer solar system. In this approach, the gyroscopic action of the rotating spacecraft mass is the stabilizing mechanism. Propulsion system thrusters are fired to make any desired changes in the spacecraft's spin-stabilized attitude.

Spacecraft also can be designed for active three-axis stabilization, as was done on the *Voyager 1* and *2*, which explored the outer solar system and beyond. In this method of stabilization, small-propulsion system

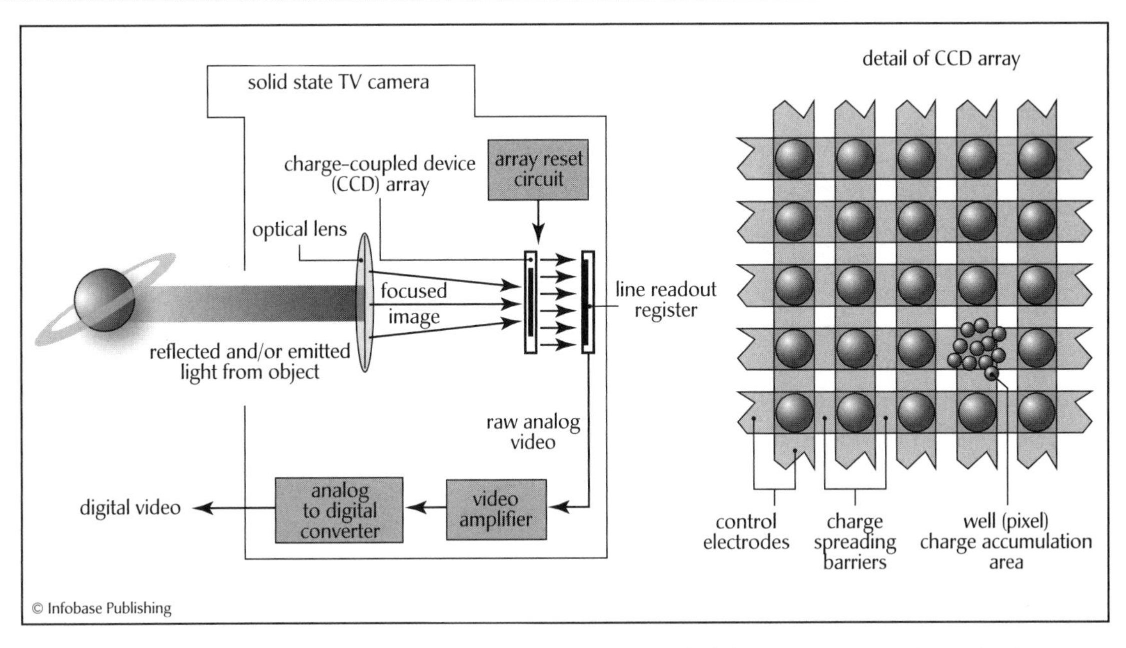

Diagram for a typical solid-state imaging system with charge-coupled device (CCD) array *(NASA/JPL)*

Global Surveyor has a detector, called the Mars Orbiter Camera (MOC), consisting of a single line of CCD sensors. As the spacecraft moves in orbit around Mars, the single line of CCD sensors creates a two-dimensional image of the Martian surface in a push-broom effect caused by the spacecraft's motion.

thrusters gently nudge the spacecraft back and forth within a deadband of allowed attitude error. Another method of achieving active three-axis stabilization is to use electrically powered reaction wheels, which are also called momentum wheels. These massive wheels are mounted in three orthogonal axes onboard the spacecraft. To rotate the spacecraft in one direction, the proper wheel is spun in the opposite direction. To rotate the vehicle back, the wheel is slowed down. Excessive momentum, which builds up in the system due to internal friction and external forces, occasionally must be removed from the system; this usually is accomplished with propulsive maneuvers.

Either general approach to spacecraft stabilization has basic advantages and disadvantages. Spin-stabilized vehicles provide a continuous "sweeping motion" that is generally desirable for fields and particle instruments.

Such spacecraft, however, may then require complicated systems to despin antennae or optical instruments that must be pointed at targets in space. Three-axis controlled spacecraft can point antennae and optical instruments precisely (without the necessity for despinning), but these craft then may have to perform rotation maneuvers to use their field and particle science instruments properly.

Some robot spacecraft have an articulation-control subsystem, which is closely associated with the attitude-control subsystem. The articulation-control subsystem controls the movement of jointed or folded components and assemblies. Examples include a packaged solar array that unfolds following launch; a robot arm that extends from a lander and scoops up soil; and an electro-optical imaging system on a steerable platform, which can track a planetary target during a flyby encounter.

The attitude-control subsystem works closely with a robot spacecraft's propulsion subsystem and makes sure that the space robot points in the right direction before a major rocket-engine burn or a sequence of tiny thruster firings occurs. Minor attitude adjustments usually take place automatically, as a smart space robot essentially drives itself through interplanetary space. Some major rocket-engine burns take place under the supervision of mission controllers on Earth, who uplink precise firing instructions to the spacecraft's computer/clock through the telecommunications subsystem. Other major propulsion system firings, like an orbit injection burn, involve a totally automated sequence of events.

The process of planetary orbit insertion places the robot spacecraft at precisely the correct location at the correct time to enter into an orbit about the target planet. Orbit insertion requires not only the precise position and timing of a flyby mission, but also a controlled deceleration. As the spacecraft's trajectory is bent by the planet's gravity, the command sequence within the onboard computer/clock subsystem fires the spacecraft's retroengine(s) at the proper moment and for the proper duration. Once this retroburn (or retrofiring) has been completed successfully, the spacecraft is captured into orbit by its target planet. If the retroburn fails (or is improperly sequenced), the spacecraft will continue to fly past the planet. It is quite common for this retroburn to occur on the farside of a planet as viewed from Earth—requiring this portion of the orbit insertion sequence to occur essentially automatically (based entirely on onboard commands and machine intelligence) and without any interaction with the flight controllers on Earth.

The thermal-control subsystem regulates the temperature of a robot spacecraft and keeps it from getting too hot or too cool. Thermal control is a complex problem because of the severe temperature extremes a space robot experience during a typical scientific mission. In the vacuum environment of outer space, radiation-heat transfer is the only natural mechanism for exchanging thermal energy (heat) into or out of a spacecraft. Under some special circumstances, a gaseous or liquid working fluid may be dumped from the spacecraft in order to provide a temporary solution to a transient heat load—but this is an extreme exception rather than the generally accepted design approach to thermal control. The overall thermal energy balance for a spacecraft near a planetary body is determined by several factors: thermal energy sources within the spacecraft; direct solar radiation (the Sun has a characteristic blackbody temperature of about 9,925°F (5,770 K); direct thermal (infrared) radiation from the planet (e.g., Earth has an average surface temperature of about 58°F (288 K); indirect (reflected) solar radiation from the planetary body; and thermal radiation emitted from the surface of spacecraft to the low-temperature sink of outer space. (Deep space has a temperature of about -455°F [3 K]).

Under these conditions, thermally isolated portions of a spacecraft in orbit around Earth could encounter temperature variations from about -100°F (200 K), during Earth-shadowed or darkness periods, to 170°F (350 K), while operating in direct sunlight. Spacecraft materials and components can experience thermal fatigue due to repeated temperature cycling during such extremes. Consequently, engineers take great care to provide the proper thermal control for a spacecraft. As previously mentioned, radiation-heat transport is the principal mechanism for heat flow into and out of the spacecraft, while conduction-heat transfer generally controls the flow of heat within the spacecraft.

There are two major approaches to spacecraft thermal control: passive and active. Passive thermal control techniques include: the use of special paints and coatings, insulation blankets, radiating fins, sun shields, heat pipes, as well as careful selection of the spacecraft's overall geometry (that is, both the external and internal placement of temperature-sensitive components). Active thermal control techniques include the use of heaters (including small radioisotope sources) and coolers, louvers and shutters, or the closed-loop pumping of cryogenic materials.

An open-loop flow (or overboard dump) of a rapidly heated working fluid might be used to satisfy a onetime or occasional special mission requirement to remove a large amount of thermal energy in a short period of time. Similarly, a sacrificial ablative surface could be used to handle a singular, large transitory external heat load. But these transitory (essentially one-shot) thermal control approaches are the exception rather than the engineering norm.

For interplanetary spacecraft, engineers often use passive thermal-control techniques such as surface coatings, paint, and insulation blankets to provide an acceptable thermal environment throughout the mission. Components painted black will radiate more efficiently. Surfaces covered with white paint or white thermal blankets will reflect sunlight effectively and protect the spacecraft from excessive solar heating. Engineers also use gold (that is, gold-foil surfaces) and quartz mirror tiles on the surfaces of special components.

Active heating can be used to keep components within tolerable temperature limits. Resistive electric heaters, controlled either autonomously or on command from Earth, can be applied to special components to keep them above a certain minimum allowable temperature during the mission. Similarly, radioisotope heat sources (generally containing a small quantity of plutonium-238) can be installed where necessary to provide at-risk components with a small, essentially permanent supply of thermal energy. The small radioisotope heat sources are especially useful for specific components on lander and rover robots that must stay within certain temperature limits in order to survive the frigid nighttime conditions experienced on the surface of the Moon or Mars.

✧ Spacecraft Clock and Data Handling Subsystem

A modern clock is generally an electronic circuit, often involving a fairly sophisticated integrated circuit, which produces high-frequency timing signals. One common application for high precision electronic clocks is synchronization of the operations performed by a computer- or microprocessor-based system. Typical clock rates in microprocessor circuits are in the megahertz range, with 1 megahertz (1 MHz) corresponding to 1 million cycles per second (10^6 cps).

Aerospace engineers usually make a spacecraft's clock an integral part of the command and data handling subsystem. The spacecraft clock is very important because it meters the passing time during the life of the space robot mission and regulates nearly all activity within the spacecraft. The clock may be very simple (for example, incrementing every second and bumping its value up by one), or it may be much more complex (with several main and subordinate fields of increasing temporal resolution down to milliseconds, microseconds, or less). In aerospace operations, many types of commands that are uplinked to the spacecraft are set to begin execution at specific spacecraft clock counts. In downlinked telemetry, spacecraft clock counts (which indicate the time a telemetry frame was created) are included with engineering and science data to facilitate processing, distribution, and analysis.

The data handling subsystem is the onboard computer responsible for the overall management of a robot spacecraft's activity. Aerospace engineers often refer to this type of multifunctional spacecraft computer as the command and data handling subsystem. This important subsystem is usually the same computer that maintains timing, interprets commands from Earth, collects, processes, and formats the telemetry data that are to be returned to Earth, and manages high-level fault protection and fail-safe routines. Under fail-safe design philosophy, engineers try to design aerospace-system hardware that avoids compounding failures. Should a component fail, the subsystem moves into a predetermined "safe" position, before the failure can cause further damage. Fail-safe design allows a robot spacecraft to sustain a failure and still retain the capability to accomplish most, if not all, of its planned mission.

Fault tolerance is the capability of a robot spacecraft (or one of its major subsystems) to function despite experiencing one or more component failures or software glitches. Engineers use redundant circuits or functions, as well as components that can readily be reconfigured, to construct spacecraft that are very fault-tolerant. For the robot flight system to enjoy effective level of fault tolerance, the spacecraft's main computer must be robust and contain a great deal of internal redundancy. The computer must also possess a high level of machine intelligence, so that it can monitor the health and status of all spacecraft subsystems, quickly detect imminent failures, and then promptly take effective action to curtail the problem—without direct human guidance or supervision. For example, the spacecraft's computer could issue commands to the affected subsystem, activating standby hardware or making software changes—either or both of which steps constitute a viable workaround repair or safe isolation of the fault. Prompt isolation of the troublesome equipment or

Single-Event Upset

A single-event upset (SEU) is a bit flip (that is, a zero (0) is changed to a one (1), or vice versa) in a digital microelectronic circuit. The SEU is caused by the passage of high-energy ionizing radiation through the silicon material of which the semiconductors are made. Space radiation-induced bit flips (or SEUs) can damage data stored in memory, corrupt operating software, or cause the central processing unit (CPU) to write over critical data tables or to halt. An SEU might even trigger an unplanned event involving a computer-controlled subsystem–as, for example, the firing of a thruster–an unplanned event that then adversely affects the mission. Aerospace engineers deal with the SEU problem in a variety of ways. Their remedial techniques include the use of additional shielding around the sensitive electronics components of a robot spacecraft, selecting more radiation-resistant electronic parts, the use of multiple-redundancy memory units and polling electronics, and the regular resetting of the spacecraft's onboard computers.

misbehaving software prevents the original fault from rippling through the spacecraft.

✧ Navigation of a Robot Spacecraft

The navigation of a robot spacecraft has two main aspects. The first is orbit determination—the task involving knowledge and prediction of the spacecraft's position and velocity. The second aspect of spacecraft navigation is flight-path control—the task involving the firing of a spacecraft's onboard propulsion systems (such as a retrorocket motor or tiny attitude-control rockets) to alter the spacecraft's velocity.

Navigating a robot spacecraft in deep space is a challenging operation. For example, no single measurement directly provides mission controllers with information about the lateral motion of a spacecraft as it travels on a mission deep within the solar system. Aerospace engineers define lateral motion as any motion except motion directly toward or away from Earth (this motion is called radial motion). Spacecraft flight controllers use measurements of the Doppler shift of telemetry (particularly a coherent downlink carrier) to obtain the radial component of a spacecraft's velocity relative to Earth. Spacecraft controllers add a uniquely coded ranging pulse to an uplink communication with a spacecraft, and record the transmission time. When the spacecraft receives this special ranging pulse, it returns a similarly coded pulse on its downlink transmission. Engineers know how long it takes the spacecraft's onboard electronics to "turn" the ranging pulse around. For example, the *Cassini* takes 420 nanoseconds (ns) ± 9 ns to turn the ranging pulse around. There are other known and measured (calibrated) delays in the overall transmission process, so when the return pulse is received back on Earth—at NASA's Deep Space Network (DSN), for example—spacecraft controllers can then calculate how far (radial distance) the spacecraft is away from Earth. Mission controllers also use angular quantities to express a spacecraft's position in the sky.

Robot spacecraft that carry electro-optical imaging instruments can use these instruments to perform optical navigation. For example, they can observe the destination (target) planet or moon against a known background star field. Mission controllers will often carefully plan and uplink appropriate optical navigation images as part of a planetary encounter command sequence uplink. When the spacecraft collects optical navigation images, it immediately downlinks (transmits) these images to the human navigation team at mission control. The mission controllers rapidly process the optical imagery and use these data to obtain precise information about the spacecraft's trajectory as it approaches its celestial target.

Once a spacecraft's solar or planetary orbital parameters are known, these data are compared to the planned mission data. If there are discrepancies, mission controllers plan for, and then have the spacecraft execute, an appropriate trajectory correction maneuver (TCM). Similarly, small changes in a spacecraft's orbit around a planet may become necessary to support the scientific mission. In that case, the mission controllers plan for and instruct the spacecraft to execute an appropriate orbit trim maneuver (OTM). This generally involves having the spacecraft fire some of its low-thrust, attitude-control rockets. Trajectory correction and orbit trim maneuvers use up a spacecraft's onboard propellant supply, which is often a very carefully managed, mission-limiting consumable.

✧ Telecommunications

Aerospace space engineers use the word *telecommunications* to describe the flow of data and information (usually by radio signals) between a spacecraft and an Earth-based communications system. A robot spacecraft generally has only a limited amount of power available to transmit a signal that sometimes must travel across millions or even billions of miles (kilometers) of space before reaching Earth. A deep space exploration spacecraft often has a transmitter that has no more than 20 watts of radiating power.

One part of aerospace engineering solution to this problem is to concentrate all available signal-radiating power into a narrow radio beam and then to send this narrow beam in just one direction, instead of broadcasting the radio signal in all directions. Often this is accomplished by using a parabolic dish antenna on the order of three to 15 feet (1 to 5 m) in diameter. Even when these concentrated radio signals reach Earth, however, they have very small power levels. The other portion of the solution to the telecommunications problem is to use special, large-diameter radio receivers on Earth, such as found in NASA's Deep Space Network, which is discussed in the next section. These sophisticated radio antennae are capable of detecting the very-low-power signals from distant spacecraft.

In telecommunications, the radio signal transmitted to a spacecraft is called the uplink. The transmission from the spacecraft to Earth is called the downlink. Uplink or downlink communications may consist of a pure radio-frequency (RF) tone (called a carrier), or these carriers may be modified to carry information in each direction. Engineers sometimes refer to commands transmitted to a spacecraft as an upload. Communications with a spacecraft involving only a downlink are called one-way communications (OWC). When the spacecraft is receiving an uplink-signal at the same time that a downlink signal is being received on Earth, the telecommunications mode is often referred to as two-way communications (TWC).

Engineers usually modulate spacecraft carrier signals by shifting each waveform's phase slightly at a given rate. One scheme is to modulate the carrier with a frequency, for example, near one megahertz (MHz). This

1-MHz modulation is then called a subcarrier. The subcarrier is modulated to carry individual phase shifts that are designated to represent binary ones (1s) and zeros (0s)—the spacecraft's telemetry data. The amount of phase shift used in modulating data onto the subcarrier is referred to as the modulation index and is measured in degrees. This same type of communications scheme is also on the uplink. Binary digital data modulated onto the uplink are called command data. They are received by the spacecraft and either acted upon immediately or stored for future use or execution. Data modulated onto the downlink are called telemetry and include science data from the spacecraft's instruments and spacecraft state-of-health data from sensors within the various functional subsystems (such as power, propulsion, thermal control, etc.).

Demodulation is the process of detecting the subcarrier and processing it separately from the carrier, detecting the individual binary phase shifts, and registering them as digital data for further processing. The device used for this is called a modem, which is short for *mo*dulator/*dem*odulator. These same processes of modulation and demodulation are often used with Earth-based computer systems and facsimile (fax) machines to transmit data back and forth over a telephone line. Before the era of high-speed cable connections, when people used personal computers to chat over the Internet, their dial-up modem would employ a familiar audio frequency carrier that the telephone system could handle.

The dish-shaped *high-gain antenna* (HGA) is the type of antenna frequently used by robot spacecraft for communications with Earth. The amount of gain achieved by an antenna refers to the amount of incoming radio signal power it can collect and focus into the spacecraft's receiver(s). In the frequency ranges used by spacecraft, the high-gain antenna incorporates a large parabolic reflector. Such an antenna may be fixed to the spacecraft bus or steerable. The larger the collecting area of the high-gain antenna, the higher the gain, and the higher the data rate it will support. The higher the gain, however, the more highly directional the antenna becomes. Therefore, when a spacecraft uses a high-gain antenna, the antenna must be pointed within a fraction of a degree of Earth for communications to occur. Once this accurate antenna-pointing is achieved, communications can take place at a rapid rate, using a highly focused radio signal.

The low-gain antenna (LGA) provides wide-angle coverage at the expense of gain. Coverage is nearly omni-directional, except for areas that may be shadowed by the spacecraft structure. The low-gain antenna is designed for relatively low data rates. It is useful as long as the spacecraft is relatively close to Earth (for example, within a few astronomical units). Sometimes a spacecraft is given two low-gain antennae to provide full omnidirectional coverage, since the second LGA will avoid the spacecraft-structure blind spots experienced by the first LGA. Engineers often mount the low-gain antenna on top of the high-gain antenna's subreflector.

The medium-gain antenna (MGA) represents a design compromise in spacecraft engineering. Specifically, the MGA provides more gain than the low-gain antenna and has wider-angle antenna-pointing accuracy requirements (typically 20 to 30 degrees) than the high-gain antenna.

✧ Deep Space Network

The majority of NASA's scientific investigations of the solar system are accomplished through the use of robot spacecraft. The Deep Space Network (DSN) provides the two-way communications link that guides and controls these spacecraft and brings back to Earth the spectacular planetary images and other important scientific data they collect.

The DSN consists of telecommunications complexes strategically placed on three continents—providing almost continuous contact with scientific spacecraft traveling in deep space as Earth rotates on its axis. The Deep Space Network is the largest and most sensitive scientific telecommunications system in the world. It also performs radio and radar astronomy observations in support of NASA's mission to explore the solar system and the universe. The Jet Propulsion Laboratory (JPL) in Pasadena, California, manages and operates the Deep Space Network for NASA.

The Jet Propulsion Laboratory established the predecessor to the DSN. Under a contract with the U.S. Army in January 1958, the Laboratory deployed portable radio tracking stations in Nigeria (Africa), Singapore (Southeast Asia), and California to receive signals from, and plot the orbit of, *Explorer 1*—the first American satellite to successfully orbit Earth. Later that year (on December 3, 1958), as part of the emerging new federal civilian space agency, JPL was transferred from U.S. Army jurisdiction to that of NASA. At the very onset of the nation's civilian space program, NASA assigned JPL responsibility for the design and execution of robotic lunar and planetary exploration programs. Shortly afterward, NASA embraced the concept of the DSN as a separately managed and operated telecommunications facility that would accommodate all deep-space missions. This management decision avoided the need for each spaceflight project to acquire and operate its own specialized telecommunications network.

Today, the DSN features three deep-space communications complexes placed approximately 120 degrees apart around the world: at Goldstone in California's Mojave Desert; near Madrid, Spain; and near Canberra, Australia. This global configuration ensures that, as Earth rotates, an antenna is always within sight of a given spacecraft, day and night. Each complex contains up to 10 deep-space communication stations equipped with large parabolic reflector antennae.

Every deep-space communications complex within the DSN has a 230-foot- (70-m-) diameter antenna. These antennae, the largest and most sensitive in the DSN network, are capable of tracking spacecraft that are more than 10 billion miles (16 billion km) away from Earth. The 41,450-square-foot (3,850-m^2) surface of the 230-foot- (70-m-) diameter reflector must remain accurate within a fraction of the signal wavelength, meaning that the dimensional precision across the surface is maintained to within 0.4-inch (1 cm). The dish and its mount have a mass of nearly 15.8 million pounds (7.2 million kg).

A view of the 230-foot- (70-m-) diameter antenna of the Canberra Deep Space Communications Complex, located outside Canberra, Australia. NASA's Deep Space Network (DSN) is made up of three complexes, one of which is this facility. The other complexes are located in Goldstone, California, and Madrid, Spain. The national flags representing the three DSN sites appear in the foreground of this image. *(NASA)*

There is also a 112-foot- (34-m-) diameter high-efficiency antenna at each complex, which incorporates advances in radio frequency antenna design and mechanics. The reflector surface of the 112-foot- (34-m-) diameter antenna is precision-shaped for maximum signal-gathering capability.

The most recent additions to the DSN are several 112-foot (34-m) beam waveguide antennae. On earlier DSN antennae, sensitive electronics were centrally mounted on the hard-to-reach reflector structure, making upgrades and repairs difficult. On beam waveguide antennae, the sensitive electronics are now located in a below-ground pedestal room. Telecommunications engineers bring an incident radio signal from the reflector to this room through a series of precision-machined radio frequency reflective mirrors. Not only does this architecture provide the advantage of easier access for maintenance and electronic equipment enhancements, but the new configuration also accommodates better thermal control of critical electronic components. Furthermore, engineers can place more electronics in the antenna to support operation at multiple frequencies. Three of these new 112-foot (34-m) beam waveguide antennae have been constructed at the Goldstone, California, complex, along with one each at the Canberra and Madrid complexes.

There is also one 85-foot- (26-m-) diameter antenna at each complex for tracking Earth-orbiting satellites, which travel primarily in orbits 100 miles (160 km) to 620 miles (1,000 km) above Earth. The two-axis astronomical mount allows these antennae to point low on the horizon to acquire (pick up) fast-moving satellites as soon as they come into view. The agile 85-foot- (26-m-) diameter antennae can track (slew) at up to three degrees per second. Finally, each complex also has one 36-foot- (11-m-) diameter antenna to support a series of international Earth-orbiting missions under the Space Very Long Baseline Interferometry (SVLBI) project.

All of the antennae in the DSN network communicate directly with the Deep Space Operations Center (DSOC) at JPL in Pasadena, California. The DSOC staff directs and monitors operations, transmits commands, and oversees the quality of spacecraft telemetry and navigation data delivered to network users. In addition to the DSN complexes and the operations center, a ground communications facility provides communications that link the three complexes to the operations center at JPL, to spaceflight control centers in the United States and overseas, and to scientists around the world. Voice and data communications traffic between various locations is sent via landlines, submarine cable, microwave links, and communications satellites.

The Deep Space Network's radio link to scientific robot spacecraft is basically the same as other point-to-point microwave communications systems, except for the very long distances involved and the very low radio frequency signal strength received from the robot spacecraft. The total signal power arriving at a network antenna from a typical robot spacecraft encounter among the outer planets can be 20 billion times weaker than the power level in a modern digital wristwatch battery.

The extreme weakness of this radio-frequency signal results from restrictions placed on the size, mass, and power supply of a particular spacecraft by the payload volume and mass-lifting limitations of its launch vehicle. Consequently, the design of the radio link is the result of engineering trade-offs between spacecraft transmitter power and antenna diameter, and the signal sensitivity that engineers can build into the ground-receiving system.

Typically, a spacecraft signal is limited to 20 watts, or about the same amount of power required to light the bulb in a refrigerator. When the spacecraft's transmitted radio signal arrives at Earth—from, for example, the neighborhood of Saturn—it has spread over an area with a diameter equal to about 1,000 Earth diameters. (Earth has an equatorial diameter of 7,928 miles [12,756 km]). As a result, the ground antenna is able to receive only a very small part of the signal power, which is also degraded by background radio noise, or static.

Radio noise is radiated naturally from nearly all objects in the universe, including Earth and the Sun. Noise is also inherently generated in all electronic systems including the DSN's own detectors. Since noise will always be amplified along with the signal, the ability of the ground-receiving system to separate noise from the signal is critical. The DSN uses state-of-the art, low-noise receivers and telemetry-coding techniques to create unequaled sensitivity and efficiency.

Telemetry is basically the process of making measurements at one point and transmitting the data to a distant location for evaluation and use. A robot spacecraft sends telemetry to Earth by modulating data onto its communications downlink. Telemetry includes state-of-health data about the spacecraft's subsystems and science data from its instruments. A typical scientific spacecraft transmits its data in binary code, using only the symbols 1 and 0. The spacecraft's data handling subsystem (telemetry system) organizes and encodes these data for efficient transmission to ground stations back on Earth. The ground stations have radio antennae and specialized electronic equipment to detect the individual bits, decode the data stream, and format the information for subsequent transmission to the data user (usually a team of scientists).

Data transmission from a robot spacecraft can be disturbed by noise from various sources that interferes with the decoding process. If there is a high signal-to-noise ratio, the number of decoding errors will be low. But if the signal-to-noise ratio is low, then an excessive number of bit errors can occur. When a particular transmission encounters a large number of bit errors, mission controllers will often command the spacecraft's telemetry system to reduce the data transmission rate (measured in bits per second) in order to give the decoder (at the ground station) more time to determine the value of each bit.

To help solve the noise problem, a spacecraft's telemetry system might feed additional or redundant data into the data stream. These additional data are then used to detect and correct bit errors after transmission. The information theory equations used by telemetry analysts in data evaluation are sufficiently detailed to allow the detection and correction of individual and multiple bit errors. After correction, the redundant digits are eliminated from the data, leaving a valuable sequence of information for delivery to the data user.

Error-detecting and encoding techniques can create a data rate many times greater than that of transmissions that are not coded for error detection. DSN coding techniques have the capability to reduce transmission errors in spacecraft science information to less than one in a million.

Telemetry is a two-way process, with a downlink as well as an uplink. Robot spacecraft use the downlink to send scientific data back to Earth, while mission controllers on Earth use the uplink to send commands, computer software, and other crucial data to the spacecraft. The uplink portion of the telecommunications process allows human beings to guide spacecraft on their planned missions, as well as to enhance mission objectives through such important activities as upgrading a spacecraft's onboard software while the robot explorer is traveling through interplanetary space. When large distances are involved, human supervision and guidance is limited to non–real time interactions with the robot spacecraft. That is why deep-space robots must possess high levels of machine intelligence and autonomy.

Data collected by the DSN are also very important in precisely determining a spacecraft's location and trajectory. Teams of human beings (called the mission navigators) use these tracking data to plan all the maneuvers necessary to ensure that a particular scientific spacecraft is properly configured and at the right place (in space) to collect its important scientific data. Tracking data produced by the DSN let mission controllers know the location of a spacecraft that is billions of miles (kilometers) away from Earth to an accuracy of just a few feet (meters).

NASA's Deep Space Network is also a multi-faceted science instrument that scientists can use to improve their knowledge of the solar system and the universe. For example, scientists use the large antennae and sensitive electronic instruments of the DSN to perform experiments in radio astronomy, radar astronomy, and radio science. The DSN antennae collect information from radio signals emitted or reflected by natural celestial sources. Such DSN-acquired radio frequency data are compiled and analyzed by scientists in variety of disciplines, including astrophysics, radio astronomy, planetary astronomy, radar astronomy, Earth science, gravitational physics, and relativity physics.

In its role as a science instrument, the DSN provides the information needed to select landing sites for space missions; determine the composition of the atmospheres and/or the surfaces of the planets and their moons; search for biogenic elements in interstellar space; study the process of star formation; image asteroids; investigate comets, especially their nuclei and comas; search the permanently shadowed regions of the Moon and Mercury for the presence of water ice; and confirm Albert Einstein's theory of general relativity.

The DSN radio science system performs experiments that allow scientists to characterize the atmospheres and ionospheres of planets, determine the compositions of planetary surfaces and rings, look through the solar corona, and determine the mass of planets, moons, and asteroids. It accomplishes this by precisely measuring the small changes that take place in a spacecraft's telemetry signal as radio waves are scattered, refracted, or absorbed. These effects are caused by particles and gases near celestial objects within the solar system. The DSN makes its facilities available to qualified scientists as long as their research activities do not interfere with spacecraft mission support.

Robot Spacecraft Come in All Shapes and Sizes

The last four decades of the 20th century witnessed a marvelous transformation in the technology of robot spacecraft. These interesting exploring machines evolved from rather primitive, cumbersome, and often quite unreliable systems into long-lived, complex, science-oriented platforms—bristling with sophisticated instruments and governed by onboard computers that demonstrated the early levels of machine intelligence needed for autonomous operation at great distances from Earth.

Aerospace engineers have traditionally used several figures to assess the merits of and to compare robot spacecraft. These basic performance indices include mass (especially the amount allocated to science instruments), physical size (dimensions and overall volume), the amount of electric power available on board, and the anticipated lifetime of the flight system. The spacecraft's physical dimensions and mass are very important parameters, because these quantities determine which launch vehicle mission planners can use to send the space robot on its interplanetary journey. More recently, other factors, such as the robot system's level of machine intelligence, autonomy, fault tolerance, redundancy, and capacity for self-repair or fault isolation, have also become significant for spacecraft engineering. It makes no sense to send a space robot on a planned 10-year journey into deep space if the failure of a tiny, inexpensive component takes place five years into the mission and compromises the entire effort. That is why modern aerospace engineers make special efforts to build redundancy, fault tolerance, and resilience into each new robot spacecraft, while still respecting rigidly enforced mass and dimensional (volume) constraints.

Scientists view robot spacecraft as a mobile science platform and generally focus on improving the instruments the space robot can accommodate, the performance of these instruments during the scientifically useful portion a spacecraft's flight, and the quality and quantity of the science data collected and delivered back to Earth. Generally, the more sophisticated a particular instrument is (within rigidly constrained mass, volume and power consumption guidelines), the more science data it can collect per mission. On a more sophisticated robot spacecraft, however, the scientific instruments sometimes compete for available power, attitude or pointing priorities, and data-processing support. So, mission managers must carefully coordinate the collection of scientific instruments carried by a particular robot spacecraft. The goal is to have the instruments complement each other rather than compete for spacecraft resources. When the tasks of spacecraft design, science instrument selection, and mission planning are properly balanced, the space robot becomes a magnificent exploring machine. The *Galileo* is an outstanding example.

This chapter contains a parade of American robot spacecraft, starting with early lunar probes from the cold war era and ending with some of the most sophisticated spacecraft that explored the solar system in the 20th century. All the photographs in the chapter have been carefully selected to include one or several human beings. The presence of aerospace engineers and technicians near each robot spacecraft provides an easily interpretable, visual reference as to the space robot's approximate physical size. Traditional engineering figures of merit, scientific contributions, and significant mission milestones also accompany the discussions about each of the spotlighted space robots.

✧ *Pioneer 3* Spacecraft

Following the disappointing, unsuccessful attempts by the U.S. Air Force and NASA in the *Pioneer 0, 1,* and *2* lunar probe missions in 1958, a collaborative team of aerospace engineers and rocket scientists from the U.S. Army Ballistic Missile Agency (ABMA), NASA, and the Jet Propulsion Laboratory (JPL) tried to reach the Moon with another family of robot probes that year. The team designed and constructed two small robot probes for this effort: the 12.9-pound (5.9-kg) *Pioneer 3* spacecraft and its nearly identical technical twin, the *Pioneer 4* spacecraft.

Pioneer 3 was a spin-stabilized, cone-shaped probe 22.8 inches (58 cm) high and 9.9 inches (25 cm) in diameter at its base. The cone consisted of a thin fiberglass shell coated with gold wash to make the surface electrically conducting. The external surface of the cone was also painted with white stripes for thermal control. Engineers wanted to keep the interior portion of *Pioneer 3* between 50°F (10°C) and 122°F (50°C) throughout the mission.

There was a small probe at the tip of the cone, which combined with the cone itself to act as an antenna. *Pioneer 3* had a ring of mercury batteries at the base of the cone to supply electric power. A photoelectric sensor protruded from the center of the ring. Engineers designed this sensor with two photocells. In concept, when the probe was within 18,650 miles (30,000 km) of the Moon, the sensor would be triggered by the light of the Moon (actually by sunlight reflected from the lunar surface). This sensor was just a test device and was intended to serve as a camera trigger mechanism for a planned camera system—although the space probe carried no camera.

Pioneer 3's single science experiment involved the detection of cosmic radiation. At the center of the cone there was a voltage supply tube and two Geiger-Mueller radiation detection tubes. The mission plan had *Pioneer 3* flying to the Moon and returning science data about the radiation environment between Earth and the Moon.

Pioneer 3 had a 1.1-pound (0.5-kg) transmitter that delivered a 0.1-watt (W) phase-modulated signal at a frequency of 960.05 megahertz (MHz). The modulated carrier wave power was 0.08 W, and the total effective radiated power was 0.18 W.

Wearing clean-room attire, aerospace technicians perform a prelaunch inspection of the *Pioneer 3* probe. The 12.9-pound (5.9-kg) robot spacecraft was launched from Cape Canaveral by a Juno II rocket on December 6, 1958. Due to a launch vehicle failure, *Pioneer 3* failed to achieve escape velocity and never reached the Moon. Following an unplanned ballistic trajectory, the probe reentered Earth's atmosphere on December 7. *(NASA)*

On December 6, 1958, a Juno II rocket lifted off from Cape Canaveral, Florida carrying the *Pioneer 3* spacecraft. The flight plan called for the robot probe to pass close to the Moon after about 34 hours and then to continue on, entering an orbit around the Sun. However, *Pioneer 3* never achieved escape velocity—the speed necessary to break free of Earth's gravitational embrace. Apparently there was a depletion of rocket propellant, which caused the first-stage rocket engine to shut down 3.7 seconds early. In addition, the spacecraft's injection angle was about 71 degrees instead of the planned 68 degrees. As a result, the *Pioneer 3* spacecraft traveled along a high ballistic trajectory, reaching an altitude of 63,615 miles (102,360 km) before falling back to Earth. The robot probe reentered Earth's atmosphere and burned up over Africa on December 7.

Recognizing the spacecraft's failure to achieve escape velocity, flight planners quickly revised *Pioneer 3*'s mission objectives to emphasize measurement of radiation levels high above Earth. The probe returned telemetry for approximately 25 hours of its 38-hour journey. The other 13 hours corresponded to telemetry blackout periods owing to the location of the two tracking stations. Telemetry data about the probe's internal temperature showed that the spacecraft's thermal-control measures were effective. The interior temperature of the spacecraft remained at about 109°F (43°C) over most of the flight. Despite the probe's failure to reach the vicinity of the Moon, *Pioneer 3*'s cosmic radiation data proved scientifically significant, indicating that the Van Allen belt contained two distinct regions of trapped radiation.

✧ Ranger Project

NASA's Ranger Project in the early to mid-1960s was the first focused American effort to explore another planetary body (the Moon) with robot spacecraft. The Ranger spacecraft were designed to relay pictures and other data as they approached the Moon and finally crash-landed into the surface.

Starting in 1959, engineers designed the Ranger spacecraft in three distinct phases, called blocks. Each block had different mission objectives and promoted progressively more advanced system design. NASA and Jet Propulsion Laboratory (JPL) mission managers and spacecraft engineers planned multiple launches in each block. They believed that this development approach would maximize the engineering experience and the scientific value of the overall project. Another goal inherent in this stepwise development strategy was to assure at least one successful flight to the Moon.

Block 1 consisted of the *Ranger 1* and *2* spacecraft, which were launched into orbit around Earth in 1961. These launches were intended to test the Atlas/Agena launch vehicle and spacecraft equipment. There was no attempt to send either of these two early Ranger spacecraft to the Moon.

Why spend time and money putting a spacecraft into orbit around Earth, if the overall objective of this project was to send a robot probe on an impact trajectory to the Moon? In the early 1960s, many of the elements of spacecraft technology used today were virtually unknown or untested before the Ranger Project. Perhaps the most important of these spacecraft technologies was three-axis attitude stabilization. This technique means that the spacecraft maintains an attitude that is fixed in relation to space instead of being stabilized by spinning. A robot spacecraft that enjoyed three-axis attitude stabilization could point its solar panels at the Sun; point a large antenna at Earth; and point cameras and other directional-science instruments at their appropriate targets. Ranger spacecraft also tested the use of onboard thrusters (low-thrust rockets) to accurately target a robot spacecraft at the Moon. Finally, mission controllers needed to learn how to combine sequences performed by the spacecraft's on-board computer with commands sent from the ground. Operational techniques still untried were two-way communications (TWC) with, and closed-loop tracking of, a robot spacecraft during spaceflight.

Ranger 1 was a 674-pound (306-kg) robot spacecraft, which had the primary mission of testing the performance of those spacecraft components and operational functions necessary for carrying out subsequent lunar and planetary missions, using essentially the same spacecraft design. The robot spacecraft also had a secondary, science-related, objective to study the nature of particles and (magnetic) fields in interplanetary space.

Ranger 1 was a Block 1 design spacecraft and consisted of a hexagonal base 4.92 feet (1.5 m) across, upon which engineers mounted a 13.1-foot- (4-m-) high cone-shaped tower of aluminum struts and braces. Two solar panel wings, measuring 17.1 feet (5.2 m) from tip to tip, extended from the base. Engineers attached a high-gain directional dish antenna to the bottom of the base. Spacecraft equipment and experiments were mounted on the base and the tower. Science instruments included a magnetometer, medium-energy range particle detectors, a cosmic-ray ionization chamber, cosmic dust collectors, and solar X-ray scintillation counters.

Ranger 1's telecommunications subsystem comprised a high-gain antenna, an omni-directional medium-gain antenna, and two transmitters—one operating at 960.1 megahertz (MHz) with a power output of 0.25 watt and the other operating at 960.05 MHz with a power output of 3 watts. Electric power was supplied to the spacecraft by 8,680 solar cells mounted on the two solar panels, a 125-pound (57-kg) silver-zinc battery, and several smaller batteries associated with some of the experiments. Attitude control for the spacecraft was provided by a combination

NASA's *Ranger 1* spacecraft undergoing a prelaunch inspection at Cape Canaveral in July 1961. The primary mission of this early robot spacecraft was to perform on-orbit tests that prepared the way for subsequent lunar photography missions. An Atlas-Agena B rocket vehicle successfully placed *Ranger 1* into the planned Earth parking orbit on August 23, 1961. The Agena rocket then failed to restart in space, so when *Ranger 1* separated from the Agena, the spacecraft went into low Earth orbit and began tumbling. *Ranger 1* reentered Earth's atmosphere on August 30, 1961. *(NASA)*

of a solid-state timing controller, Sun and Earth sensors, and pitch and roll jets (small thrusters). Engineers used a combination of passive thermal-control techniques, including gold plating, white paint, and polished aluminum surfaces.

Mission planners had designed the *Ranger 1* to go into a parking orbit around Earth and then to perform a restart of the attached Agena B upper stage rocket, so as to place the spacecraft in a 37,290-mile- (60,000-km-)

by-690,000-mile (1,110,000-km) orbit around Earth. As *Ranger 1* traveled in this highly elliptical orbit around Earth, the spacecraft would be able to demonstrate some of the critical hardware systems and operational techniques necessary for future lunar missions.

Ranger 1 was successfully launched from Cape Canaveral, Florida, on August 23, 1961, by an Atlas-Agena B rocket vehicle combination. The spacecraft achieved the desired parking orbit, but the Agena B rocket failed to restart and place *Ranger 1* into the planned higher-altitude, high-elliptic orbit around Earth. Instead, when *Ranger 1* separated from the Agena B rocket stage, the spacecraft went into a low Earth orbit and began tumbling. The distressed spacecraft reentered Earth's atmosphere on August 30, 1961. Given lemons and deciding to make lemonade, mission controllers called the *Ranger 1* flight a partial success, because several operational procedures and spacecraft equipment were tested during spaceflight. For the scientists, however, the mission was a major disappointment, because little scientific data were returned.

Ranger 2, launched on November 18, 1961, by an Atlas-Agena B rocket from Cape Canaveral, also encountered difficulties. The spacecraft was placed in a parking orbit around Earth, but the Agena B stage failed to restart. This left *Ranger 2* stranded in low Earth orbit upon separation from the Agena B stage. The orbit of *Ranger 2* continued to decay until it reentered Earth's atmosphere on November 20.

As a result of engine restart problems with the Agena B stage, both Block 1 Ranger spacecraft found themselves in short-lived low-Earth orbits in which the spacecraft could not stabilize themselves, gather solar power, or survive for very long.

Block 2 of NASA's Ranger Project involved the launch of three robot spacecraft to the Moon in 1962. These spacecraft carried a television camera, a radiation detector, and a seismometer in a separate capsule, which was slowed by a rocket motor and packaged to survive a low-velocity impact on the lunar surface. Misfortune continued to plague the Block 2 missions. When viewed collectively, the three Block 2 missions demonstrated acceptable performance of the Atlas-Agena B launch vehicle and the adequacy of the Ranger design, but unfortunately these successes did not all occur at once during the same mission.

On January 26, 1962, *Ranger 3* was launched into deep space, but an inaccuracy put the space robot off course, and it missed the Moon entirely. *Ranger 4* had a perfect launch on April 23, 1962, but an onboard computer failure completely disabled the spacecraft, and the derelict craft impacted on the farside of the Moon on April 26 after 64 hours of flight without returning any scientific data. NASA launched *Ranger 5* on October 18, 1962. Because of an unknown malfunction, *Ranger 5* ran out of electric power and ceased operation. The disabled spacecraft flew past the Moon within 450 miles (725 km) of the lunar surface. Scientists could not harvest any significant data from the three Block 2 missions.

Block 3 of the Ranger Project involved four spacecraft missions to the Moon in the period 1964–65. This new series of spacecraft carried a television instrument designed to observe the lunar surface during approach and to transmit high-resolution images throughout the final minutes of flight, right up to the moment of impact. The Block 3 Ranger spacecraft carried no other experiments.

On January 30, 1964, NASA launched *Ranger 6* to the Moon. The spacecraft carried a collection of six television vidicon cameras. The space probe's flight to the Moon was flawless. The camera system experienced an in-flight anomaly, however, causing it to fail. So, 65.5 hours after launch, *Ranger 6* dutifully slammed into the lunar surface near the eastern edge of Mare Tranquillitatis (Sea of Tranquility) without returning a single image of the Moon.

Fortune finally smiled on this early NASA space robot project. The next three Ranger missions, carrying a redesigned camera system, proved completely successful and helped pave the way for the Apollo Project human landings. NASA launched *Ranger 7* on July 28, 1964. The robot spacecraft had a flawless flight to the Moon and then successfully photographed its way down to lunar impact in a plain south of the crater Copernicus. *Ranger 7* transmitted more than 4,300 pictures of the lunar surface. The robot spacecraft's images revealed that craters caused by impact were the dominant features of the lunar surface, even in apparently smooth and empty plains. Scientists eagerly examined the *Ranger 7* photographs, which revealed tiny craters and impact marks as small as 20 inches (50 cm) across.

NASA sent *Ranger 8* to the Moon on February 17, 1965. This robot spacecraft swept an oblique course over the south of Oceanus Procellarum and Mare Nubium, before crashing in Mare Tranquillitatis, where the *Apollo 11* astronauts would land four and one-half years later. *Ranger 8* returned more than 7,000 images of the Moon. About a month later (on March 21, 1965), NASA sent *Ranger 9* to the Moon. *Ranger 9,* the last spacecraft in this robot family, impacted inside the crater Alphonsus. The robot spacecraft performed superbly and transmitted 5,814 images of the lunar surface during the last 19 minutes of its flight. *Ranger 9* slammed into the Moon at an impact velocity of 1.66 miles per second (2.67 km/s). NASA treated the general public to the experience of "real-time space exploration" by providing a live broadcast of the spacecraft's transmitted television images.

The trouble-plagued beginning of the Ranger Project taught spacecraft engineers a great deal, but provided scientists with virtually nothing. The last three Ranger missions, however, were flawless examples of using robot spacecraft to explore planetary bodies. *Ranger 7, 8,* and *9* provided lunar scientists with important new knowledge about the Moon's surface and paved the way for a more sophisticated family of lander robots (called Surveyor), as well as the human explorers who would arrive by the end of the decade.

✧ *Lunar Prospector* Spacecraft

A little more than three decades after the last Ranger spacecraft smashed onto the Moon's surface, NASA's *Lunar Prospector* spacecraft gathered some important data that promises to shape the future of lunar exploration and settlement in the 21st century. A NASA Discovery Program spacecraft, *Lunar Prospector* was a modern space robot designed for a low-altitude, polar orbit investigation of the Moon. The spacecraft was launched successfully from Cape Canaveral Air Force Station, Florida, by a Lockheed Athena II vehicle (formerly called the Lockheed Martin Launch Vehicle) on January 6, 1998. After swinging into orbit around the Moon on January 11, the *Lunar Prospector* used its complement of instruments to perform a detailed study of surface composition. The 277-pound (126-kg) spacecraft (dry mass without propellant) also searched for resources—especially suspected deposits of water ice in the permanently shadowed regions of the lunar poles. The orbiter spacecraft carried a gamma ray spectrometer, a neutron spectrometer, a magnetometer, an electron reflectometer, an alpha particle spectrometer, and a Doppler gravity experiment.

The *Lunar Prospector* was shaped like a drum, 4.25 feet (1.3 m) high, with a diameter of 4.5 feet (1.4 m). The spacecraft's three instrument-carrying masts were each eight feet (2.4 m) long. Body-mounted solar cells complemented by nickel-cadmium (NiCd) rechargeable batteries supplied the spacecraft with approximately 200 watts of electric power.

NASA's *Lunar Prospector* spacecraft in a clean room at Cape Canaveral in late December 1997. This robot spacecraft was successfully launched on January 6, 1998, traveled to the Moon, and then performed an important orbital mission that mapped the Moon's elemental composition. One of the most significant results of the *Lunar Prospector* mission was confirmation that there could be large quantities of frozen water near the Moon's polar regions. *(NASA/Ames Research Center)*

Lunar Prospector was a simple, reliable, spin-stabilized spacecraft. Unlike most sophisticated space robots, *Lunar Prospector* did not have an onboard computer for control and mission. Rather, engineers installed a simple electronics box, called the Command and Data Handling Unit, which accepted a maximum of 60 commands from mission controllers at the NASA Ames Research Center in Mountain View, California (San Francisco Bay Area).

The data from this mission complemented the detailed imagery data from the Department of Defense's Clementine mission of 1994. In particular, tantalizing data from the *Lunar Prospector*'s neutron spectrometer suggested the presence of significant amounts of water ice in the Moon's polar regions. While still subject to confirmation and detailed analysis, the presence of large quantities of water ice at the lunar poles would make the Moon a valuable supply depot for any future human settlement of the Moon and regions beyond.

Lunar Prospector's Neutron Spectrometer

Lunar Prospector is the very first interplanetary spacecraft to use neutron spectroscopy to detect water. The robot spacecraft's neutron spectrometer (NS) is the instrument designed to detect minute amounts of water ice that may exist on the Moon. The Moon has a number of permanently shadowed craters near its poles with continuous temperatures of -310°F (-190°C) or less. These craters may act as cold-traps, permanently storing any water ice scattered into these frigid areas by comets or asteroids that made impact eons ago.

The neutron spectrometer does not detect hydrogen directly, since during its primary science mission the spacecraft operated in a 62-mile- (100-km-) altitude polar orbit above the Moon's surface. Instead, this instrument searched for what nuclear scientists call "cool" neutrons—energetic neutrons that have bounced off hydrogen atoms somewhere in the lunar crust and are now much slower and less energetic. When cosmic rays collide with atoms in the Moon's crust, the cosmic rays cause violent nuclear reactions that release neutrons, gamma rays, and a host of other subatomic particles. Some of these cosmic ray–generated neutrons escape directly to space as hot or "fast" neutrons. Other neutrons shoot off and collide with atoms in the crust, bouncing around like a pinball and losing some kinetic energy with each collision. If the colliding neutrons run into heavy atoms, such as iron (Fe), they do not lose very much energy in each collision. Consequently, the neutrons in this particular group are still close to their original (rapid) speed when each experiences the final collision that bounces it off into space. Such neutrons will still be warm (physicists say "epithermal"), when they reach the orbiting *Lunar Prospector* and are detected by the neutron spectrometer.

Scientists know that the most effective way of slowing down a speeding neutron is to have the nuclear particle collide with something its own size. There is only one atom the same size as a neutron: hydrogen, the lightest of all the elements. If the Moon's crust contains a lot of hydrogen at

As reported in 1998, mission scientists were able to establish the existence of significant concentrations of hydrogen at the lunar poles, based on telltale dips in the epithermal neutron energy spectra sent back to Earth by the spacecraft's neutron spectrometer (NS) instrument. If, as some scientists suspect, this excess hydrogen exists as part of frozen water molecules buried in permanently shadowed craters at the lunar poles, there could then be as much as 75 billion gallons of water (260 million metric tons of water ice) on the Moon.

After a highly successful 19-month scientific mapping mission, flight controllers decided to turn the spacecraft's originally planned end-of-life crash into the lunar surface into an impact experiment that might possibly confirm the presence of water ice on the Moon. Therefore, as its supply of attitude control fuel neared exhaustion, the spacecraft was directed to crash into a crater near the Moon's south pole on July 31, 1999. Observers from Earth attempted to detect signs of water in the impact plume, but

a certain location—for example, a permanently-shadowed crater with water ice—any neutron that bounces around in this part of the crust will lose energy rapidly (that is, cool off) before bouncing out into space as a very slow, or thermal, neutron. So when the *Lunar Prospector*'s neutron spectrometer flew off polar region craters (suspected of containing water ice), scientists anticipated and saw a surge in the number of thermal neutrons detected, and a corresponding dip in the number of warm, or epithermal, neutrons.

The neutron spectrometer was an 8.6-pound (3.9-kg) instrument, consisting of two canisters—each of which contains helium-3 gas and an energy counter. Any neutrons colliding with the spectrometer's helium-3 atoms will give off a characteristic energy signature, which is then easily detected and counted. In order to take advantage of neutron physics, one of the canisters was wrapped in cadmium and the other in tin. The cadmium screens out low-energy or slow-moving (thermal) neutrons, while the tin does not. So if there was an extensive concentration of hydrogen (such as water ice) at the lunar poles, its presence would be revealed by telltale dips in the epithermal neutron-energy spectra sent back to Earth by this instrument. Differences in the counts between the two canisters of helium-3 gas would indicate the thermal neutron flux detected by the *Lunar Prospector*'s neutron spectrometer.

In summary, *Lunar Prospector* measured cosmic ray–generated neutrons leaving the lunar surface with three different spectrometers. Thermal and epithermal neutrons were detected with a pair of 2.2-inch- (5.7-cm-) diameter by 7.9-inch- (20-cm-) long helium-3 gas-filled proportional counters. One of these detectors was covered with a sheet of cadmium and was sensitive only to epithermal neutrons. The second detector was covered by a sheet of tin (Sn) and measured both epithermal and thermal neutrons. Since the two proportional counters were otherwise identical, scientists attributed any difference in counts between the two to thermal neutrons. The anticoincidence shield (ACS) of the *Lunar Prospector*'s gamma-ray spectrometer (GRS) measured fast neutrons.

no such signal was found. This impromptu impact experiment should be regarded only as a long-shot opportunity, however, and not as a carefully designed scientific procedure. In contrast, the *Lunar Prospector*'s science data have allowed scientists to construct a detailed map of the Moon's surface composition. These data have also greatly improved knowledge of the origin, evolution, and current inventory of lunar resources.

✧ *Magellan* Spacecraft

The Magellan mission was a NASA solar system exploration mission to the planet Venus. On May 4, 1989, the 7,810-pound (3,550-kg) *Magellan* spacecraft was delivered to Earth orbit by the space shuttle *Atlantis* during the STS 30 mission. The large robot explorer was then sent on an interplanetary trajectory to the cloud-shrouded planet by a solid-fueled inertial upper-stage (IUS) rocket system. *Magellan* was the first interplanetary spacecraft to be launched by the space shuttle. On August 10, 1990, *Magellan* was inserted into orbit around Venus and began initial operation of its very successful radar-mapping mission. The spacecraft was named after the famous 16th-century Portuguese explorer Ferdinand Magellan (1480–1521), who was the first person to circumnavigate the globe.

Built partially with spare parts from other missions, the Magellan spacecraft was 15.1 feet (4.6 m) long and topped with a 12-foot- (3.7-m-) diameter high-gain antenna. The high-gain antenna was used for both communications back to Earth and for the radar-mapping mission of Venus. This large antenna was a spare from NASA's Voyager mission to the outer planets, as was *Magellan*'s 10-sided main structure and a set of hydrazine thrusters. The spacecraft's command-data computer system, attitude-control unit, and power-distribution unit were spare parts from the Galileo mission to Jupiter. Finally, *Magellan*'s medium-gain antenna was a spare part from the *Mariner 9* spacecraft.

Magellan was a three-axis stabilized spacecraft that used three reaction wheels. Electric power was provided by two square solar panels, each measuring 8.2 feet (2.5 m) on a side. At the beginning of the mission, the solar panels supplied 1,200 watts of electric power. Over the course of the mission, the solar panels gradually degraded, as spacecraft engineers had anticipated. By the end of the mission in the fall of 1994, mission controllers found it necessary to manage power usage carefully to keep the spacecraft operating. The solar panels were used in conjunction with two rechargeable nickel-cadmium (NiCd) batteries.

Magellan's prime science payload was a single radar instrument, which operated simultaneously as a synthetic-aperture radar (SAR), altimeter, and radiometer. The radar frequency was 2.385 gigahertz (GHz), a peak power of 325 watts, and a swath width (variable) of 15.5 miles (25 km).

Because a dense, opaque atmosphere shrouds Venus, scientists cannot use conventional optical cameras to make images of the planet's surface. Instead, *Magellan*'s imaging radar used bursts of microwave energy in a manner somewhat like a camera flash to illuminate the planet's surface.

Magellan's high-gain antenna sent out millions of pulses each second toward the planet. The antenna then collected the echoes returned to the spacecraft when the radar pulses bounced off Venus's surface. The radar pulses were not sent directly downward but rather at a slight angle to the side of the spacecraft. Because of this feature, engineers sometimes refer to *Magellan*'s radar system as a side-looking radar. In addition, special processing techniques were used on the radar data to create higher resolution images—as if the radar had a much larger antenna or aperture. Engineers call this clever technique synthetic-aperture radar (SAR).

From 1990 to 1994, *Magellan* used its sophisticated imaging radar system to make the most detailed map of Venus ever captured. After concluding its radar-mapping mission, *Magellan* made global maps of the Venusian gravity field. During this phase of the mission, the spacecraft did not use its radar mapper, but instead transmitted a constant radio signal back to Earth. When it passed over an area on Venus with higher than normal gravity, the spacecraft would speed up slightly in its orbit. This movement then would cause the frequency of Magellan's radio signal to change very slightly, owing to the Doppler effect. Because of the ability of the radio receivers in the NASA Deep Space Network to measure radio frequencies extremely accurately, scientists were able to construct a very detailed gravity map of Venus. In fact, during this phase of its mission, the spacecraft provided high-resolution gravity data for about 95 percent of the planet's

NASA's *Magellan* spacecraft with its attached inertial upper-stage (IUS) rocket in the payload bay of the space shuttle *Atlantis* prior to launch in April 1989. On May 4, the space shuttle *Atlantis* delivered and deployed *Magellan* into low Earth orbit. The inertial upper-stage rocket then sent the robot orbiter spacecraft on an interplanetary trajectory to Venus. From 1990 to 1994, *Magellan* used its sophisticated imaging radar system to make the most detailed map of the cloud-shrouded planet ever captured. *(NASA/Kennedy Space Center)*

surface. Flight controllers also tested a new maneuvering technique called aerobraking—a technique that uses a planet's atmosphere to slow or steer a spacecraft.

The craters revealed by *Magellan*'s detailed radar images suggested to planetary scientists that the surface of Venus is relatively young—perhaps recently resurfaced or modified about 500 million years ago by widespread volcanic eruptions. The planet's current harsh environment has persisted at least since then. No surface features were detected that suggest the presence of oceans or lakes at any time in the past on Venus. Furthermore, scientists found no evidence of plate tectonics, that is, the movements of huge crustal masses.

Magellan's mission ended with a dramatic plunge through the dense atmosphere to the planet's surface. This was the first time an operating planetary spacecraft has ever been crashed intentionally. Contact was lost with the spacecraft on October 12, 1994, at 10:02 universal time (3:02 A.M. Pacific Daylight Time). The purpose of this last maneuver was to gather data on the Venusian atmosphere before the spacecraft ceased functioning during its fiery descent. Although much of the *Magellan* is believed to have been vaporized by atmospheric heating during this final plunge, some sections may have survived and hit the planet's surface intact.

✧ *Galileo* Spacecraft

NASA's *Galileo* spacecraft was designed to study the large, gaseous planet Jupiter, its moon, and its surrounding magnetosphere. The spacecraft was named in honor of the Italian scientist Galileo Galilei (1564–1642). In 1610, Galileo inaugurated the use of telescopic astronomy when he fashioned and used a primitive optical telescope to observe the heavens. Viewing Jupiter with this device, for example, Galileo discovered the giant planet's four major moons and named them Io, Europa, Ganymede, and Callisto. Today, astronomers often refer to these interesting bodies as the Galilean satellites.

Galileo's primary mission at Jupiter began when the large robot spacecraft entered orbit in December 1995 and its descent probe, which had been released five months earlier, dove into the giant planet's atmosphere. The orbiter spacecraft's primary mission included a 23-month, 11-orbit tour of the Jovian system, including 10 close encounters with Jupiter's major moons. Although the primary mission was completed in December 1997, NASA decided to extend the mission three times. As a result of these extended missions, the *Galileo* spacecraft experienced 35 encounters with Jupiter's major moons—11 with Europa, eight with Callisto, eight with Ganymede, seven with Io, and one with Amalthea.

The December 2002 flyby of Amalthea also brought the spacecraft closer to Jupiter than at any time since it began orbiting the giant planet on December 7, 1995. On February 28, 2003, the NASA flight team terminated its operation of the *Galileo* spacecraft. The human directors sent a final set of commands to the far-traveling robot spacecraft, putting it on a course that resulted in its mission-ending plunge into Jupiter's atmosphere on September 21, 2003. This action was taken to prevent an abandoned *Galileo* spacecraft from wandering in orbit through the Jovian system and possibly crashing into Europa—contaminating this possibly life-bearing moon with hitchhiking terrestrial microorganisms.

The *Galileo* spacecraft and its two-stage inertial upper stage (IUS) rocket were carried into Earth orbit on October 18, 1989, by the space shuttle *Atlantis* as part of NASA's STS-34 mission. After deployment from the shuttle's cargo bay, the IUS fired and then accelerated the spacecraft out of Earth's orbit toward the planet Venus and the first of three planetary gravity-assist maneuvers designed to boost *Galileo* toward Jupiter. After flying past Venus at an altitude of about 10,000 miles (16,000 km) on February 10, 1990, the robot spacecraft swung past Earth at an altitude of 597 miles (960 km) on December 8, 1990. That flyby increased *Galileo*'s speed enough to send it on a two-year elliptical orbit around the Sun. The spacecraft returned for a second Earth swingby on December 8, 1992, at an altitude of 188 miles (303 km). With this final gravity-assist maneuver, *Galileo* left Earth and headed for Jupiter.

In April 1991, the robot spacecraft was scheduled to deploy its 16-foot- (4.8-m-) diameter high-gain antenna, as *Galileo* moved away from the Sun and the risk of overheating ended. The large antenna, however, failed to deploy fully. A special team of engineers and technicians performed extensive tests here on Earth and determined that a few (possibly three) of the antenna's 18 ribs were held by friction in the closed position. Despite exhaustive efforts to free the ribs, the high-gain antenna would not deploy. From 1993 to 1996, NASA personnel rescued the mission by developing extensive new flight and ground software and by enhancing the ground stations of the Deep Space Network to perform the mission using the spacecraft's low-gain antennae.

On October 29, 1991, *Galileo* became the first spacecraft ever to encounter an asteroid, when it passed by Gaspra, flying within 1,000 miles (1,602 km) of the stony asteroid's center at a relative speed of about 5 miles per second (8 km/s). Images and other data collected by *Galileo* revealed that Gaspra was a cratered, complex, irregular body about 12.4 by 7.4 by 6.8 miles (20 by 12 by 11 km), with a thin covering of dust and rubble.

Galileo performed its second asteroid encounter on August 28, 1993, when it flew past a more distant asteroid called Ida. Ida is about 34 miles

This image shows NASA's *Galileo* spacecraft being prepared for mating with its inertial upper stage (IUS) rocket in the Vertical Processing Facility (VPF) at the Kennedy Space Center in August 1989. The robot spacecraft and upper-stage rocket were then loaded into the payload bay of the space shuttle *Atlantis,* which carried and deployed the combination into low Earth orbit in October 1989. Once the deployed payload combination was at a safe distance from the space shuttle, the IUS rocket fired and sent the *Galileo* spacecraft on its six-year interplanetary journey to Jupiter. *(NASA/Kennedy Space Center)*

(55 km) long and 15 miles (24 km) wide. Imagery data collected by *Galileo* also indicated that Ida has its own tiny moon, named Dactyl, which has a diameter of 0.9 mile (1.5 km).

On July 13, 1995, *Galileo*'s hitchhiking descent probe was released from the mother spacecraft and began a five-month free fall toward Jupiter. Since the 750-pound- (340-kg-) mass robot probe had no thrusters, its ballistic flight path was completely established by carefully pointing the *Galileo* mother spacecraft before the probe was released.

The *Galileo* spacecraft arrived at Jupiter on December 7, 1995. Upon arrival, the *Galileo* orbiter spacecraft flew past two of Jupiter's major moons—Europa and Io. Galileo passed Europa at an altitude of 20,000 miles (32,200 km), while the encounter with Io was at an altitude of just 600 miles (965 km). About four hours after leaving Io, the *Galileo* orbiter spacecraft made its closest approach to Jupiter, experiencing 25 times more ionizing radiation (from its intense trapped radiation belts) than the level considered deadly for human beings.

Eight minutes later, the *Galileo* orbiter started receiving data from the descent probe, which slammed into the top of the Jovian atmosphere at a comet-like speed of 29.4 miles per second (47.3 km/s). In the process, this hardy probe withstood temperatures twice as hot as the Sun's surface. The probe slowed by aerodynamic braking for about two minutes before deploying its parachute and jettisoning a heat shield.

The wok-shaped probe then floated down about 125 miles (200 km) through Jupiter's upper clouds, transmitting data to the *Galileo* orbiter on pressure, temperature, winds, lightning, atmospheric composition, and solar heat flux. Fifty-eight minutes into its descent, high temperatures and crushing pressures silenced the probe's transmitters. The hardy robot probe sent data from a depth with a pressure 23 times that of the average on Earth's surface. The well-designed probe had functioned up to a pressure found in Jupiter's atmosphere, which was more than twice the mission's requirement.

An hour after receiving the last transmission from the probe, at a point about 130,000 miles (209,000 km) above the planet, the *Galileo* spacecraft fired its main engine to brake into orbit around Jupiter. During this first orbit around Jupiter, new software was installed (via telecommunications from mission controllers on Earth), which gave the robot orbiter extensive new onboard data-processing capabilities. The new software permitted data compression, enabling the spacecraft to transmit up to 10 times the number of images and other measurements than would have been possible otherwise. In addition, hardware changes on the ground and adjustments to the spacecraft-to-Earth communications system increased the average telemetry rate tenfold. Although the problem with *Galileo*'s high-gain antenna kept a few of the mission's original scientific objectives from being met, the great majority were accomplished through human ingenuity and a spacecraft design that accommodated workarounds and the use of backup systems (such as low-gain antennae) in productive ways not previously envisioned. The *Galileo* robot spacecraft mission to Jupiter is a prime example of humans and robots working together to explore space.

The *Galileo* orbiter spacecraft had a mass of 4,891 pounds (2,218.5 kg) at launch and measured 17 feet (5.3 m) from the top of the low-gain antenna to the bottom of the Jovian atmosphere-descent probe. (See figures on pages 44 and 82.) The robot orbiter featured an innovative dual-spin design. Most interplanetary spacecraft are stabilized in flight either by spinning around a major axis, or by maintaining a fixed orientation in space, usually referenced by the Sun and another star. As the first dual-spin interplanetary spacecraft, *Galileo* combined both of these stabilization techniques. *Galileo* had a spinning section that rotated at three revolutions per minute (rpm), and a "despun" section, which was counter-rotated to provide a fixed orientation for cameras and other remote sensors. A star scanner on the spinning side determined orientation and spin rate; gyroscopes on the despun side of the spacecraft provided the basis for measuring turns and pointing instruments.

The power supply, propulsion module, and most of the computers and control electronics were mounted on the spinning section. The spinning section also carried direct sensing instruments to study magnetic fields and charged particles. The science instruments included magnetometer sensors mounted on a 36-foot- (11-m-) long boom to minimize interference from the spacecraft's electronics; a plasma detector to measure low-energy charged particles; and a plasma-wave detector to study electromagnetic waves generated by the particles. The *Galileo* orbiter also carried a high-energy particle detector and a detector to measure dust in interplanetary space (cosmic dust) as well as near the giant planet (Jovian dust). Finally, the spacecraft's heavy ion counter was there to assess any potentially hazardous charged-particle (ionizing radiation) environment that the space robot flew through, especially during close flybys of Jupiter and intense trapped radiation belts.

Galileo's despun section carried instruments that needed to be held steady. These instruments included the camera system; the near-infrared mapping spectrometer to make multispectral measurements for atmospheric and surface chemical analysis; the ultraviolet spectrometer to study gases; and the photopolarimeter-radiometer to measure radiant and reflected energy. The camera system obtained images of Jupiter's satellites at resolutions from 20 to 1,000 times better than the best possible images obtained by NASA's *Voyager 1* and *2* spacecraft. The primary reason for this dramatic improvement in imagery-data collection was the fact that *Galileo* used a more sensitive solid-state video-imaging system with an array of charged-coupled devices (CCDs). The despun section on *Galileo* also carried a dish antenna that picked up signals from the descent probe as it plunged into Jupiter's atmosphere.

The spacecraft's propulsion module consisted of 12 2.25-pound-force (10-newton) thrusters and a single, 90-pound-force (400-newton) rocket engine, which used monomethyl-hydrazine fuel and nitrogen-tetroxide oxidizer.

Because radio signals (which travel at the speed of light) take more than one hour to travel from Earth to Jupiter and back, the *Galileo* spacecraft was designed to operate from computer instructions sent to it in advance and stored in the spacecraft's computer memory. With *Galileo,* a single master sequence of commands could cover a period ranging from weeks to months of quiet operations between flybys of Jupiter's moons. During busy encounter operations, however, one sequence of commands covered only about a week. These command sequences operated through flight software installed in the spacecraft's computer subsystem, with built-in automatic fault-protection software designed to put *Galileo* in a safe state in case of a computer glitch or some other unanticipated circumstance.

The *Galileo* spacecraft received its electric power from two radioisotope thermoelectric generator (RTG) units. Heat liberated by the natural radioactive decay of the radioisotope plutonium-238 was directly converted to electricity to operate the orbiter spacecraft's equipment. At launch (in 1989), the RTG units generated a total of electric power level of 570 watts; at the end of the spacecraft's mission (in 2003), these units were still generating 485 watts of electricity. NASA has used such long-lived radioisotope generator units on other successful robot spacecraft missions, including the *Viking 1* and *2* landers on Mars, the *Voyager 1* and *2* and *Pioneer 10* and *11* flyby missions to the other planets, the *Ulysses* spacecraft to study the Sun's poles, and *Cassini* to Saturn. (Chapter 10 contains additional discussion about NASA's use of space nuclear power systems.)

✧ NASA's Soccer-Ball Space Robot

The *Autonomous Extravehicular Activity Robotic Camera (AERCam Sprint)* provided experimental demonstration of the use of a prototype free-flying television camera that (in the future) could be used by astronauts to conduct remote inspections of the exterior of the space shuttle or *International Space Station* from inside the pressurized crew cabin or module. The *AERCam Sprint* free-flyer is a 14-inch- (35.6-cm-) diameter, 35-pound (15.9-kg) robot sphere that contains two television cameras, an avionics system, and 12 small nitrogen gas-powered thrusters. The sphere, which looks like an oversized soccer ball, was released by astronaut Winston Scott during an extravehicular activity as part of the STS-87 mission in December 1987. The perky little round robot flew freely in the forward area of the space shuttle *Columbia*'s large cargo bay, as another astronaut (Steve Lindsey) remotely controlled the robot from the shuttle's aft flight deck. During this in-orbit demonstration experiment, Astronaut Lindsey used a hand controller, two laptop computers, and a window-mounted antenna to control the free-flying robotic camera.

Engineers designed the *AERCam Sprint* to fly very slowly at a rate slower than three inches per second (7.6 cm/s). Remote control of the *AERCam Sprint* is accomplished through two-way ultra-high frequency (UHF) communications, with data regarding the status of the free-flyer's systems transmitted back to the human operator. Television images are transmitted back to the operator by means of a one-way S-band communications link. During the experiment performed on the STS-87 shuttle mission, live television images were relayed from the *Columbia* to NASA's

As if he were playing with a high-tech soccer ball, astronaut Winston Scott reaches out and retrieves the free-flying *Autonomous EVA Robotic Camera (AER Cam)*, during Winston's space walk in the payload bay of the space shuttle *Columbia*. These interesting astronaut-robot spacecraft interactions took place in low Earth orbit, during the STS-87 shuttle mission in December 1997. *(NASA/Johnson Space Center)*

human spaceflight mission control center (MCC) at the Johnson Space Center, Texas.

The *AERCam Sprint* contains two miniature color television cameras. The exterior of the robot sphere is covered with a 0.6-inch- (1.5-cm-) thick layer of felt to cushion any inadvertent contact with a spacecraft surface and to prevent damage during an inspection. The robot sphere is powered by lithium batteries. The batteries and supply of nitrogen gas (for its small thrusters) are intended to last at least seven hours, which correlates to the maximum amount of time NASA astronauts spend during a normal extravehicular activity (EVA). The *AERCam Sprint* sphere has a small floodlight built into it, which is identical to the floodlight used on the helmet of an astronaut's spacesuit. The robot sphere has six, small, flashing yellow-light-emitting diode (LED) lights that make it easier for the human operator or EVA astronaut to see where the free-flying robot is in darkness or shadowed operations. As another visual aid for the human operator or EVA astronaut working with the flying robot, engineers mark the front of the sphere with stripes and arrows, while the back is marked with dots. The *AERCam Sprint* sphere also has a small fabric strap, which serves as a handhold for the EVA astronaut who is deploying or retrieving the robot camera system.

The flight test of the free-flying round robot system lasted about 75 minutes and went very well. Astronaut Winston Scott, who was the NASA astronaut performing an EVA, released the robot in the shuttle's cargo bay and then retrieved it after the test was completed. Astronaut Steve Lindsey controlled the free-flying robot using a combination rotational and translational hand controller from inside the orbiter's crew cabin. He monitored the television images from the robot on an internal video display at the aft flight deck. Lindsey also used the orbiter's aft flight-deck windows to remain in direct visual contact with the robot free-flyer, as it floated and moved around the cavernous cargo bay, which had its doors opened to outer space.

Following this successful experiment, engineers at the NASA Johnson Space Flight Center went on to design, develop and test a nanosatellite-class free-flying robot, which is intended for use during future human spaceflight activities that require remote inspection of a human-crewed spacecraft or station. Use of a small free-flying inspection robot would avoid dangerous extravehicular activities by human crewmembers. NASA engineers named the new spherical-shaped free-flyer the *Mini AERCam.* In comparison to the larger version, the second-generation robot-free-flyer is just 7.5 inches (19 cm) in diameter, has a mass of 11 pounds (5 kg), uses a rechargeable pressurized xenon propulsion system (12 thrusters), and a rechargeable lithium-ion battery. The spherical mini-robot carries two high-quality video cameras and a digital-inspection camera for

Using Robots or Humans in Space Exploration

One debate that has persisted in the American civil space program since 1958 involves the basic question: Should machines or humans be used to explore the solar system? The ideal response, of course, is that both should be used in partnership. This approach was taken most effectively during the Apollo Project when robot spacecraft, such as Ranger and Surveyor types, served as precursors to the human landing missions.

Following the remarkable success of the Apollo Project, however, many space experts began to reconsider the role of humans in space exploration. Healthy debates occurred throughout the 1970s and early 1980s concerning future strategies for NASA's space exploration program. Some long-range planners concluded that most, if not all, future exploration goals could best be served by robot spacecraft, weighing the factors of cost, risk, potential scientific return, and schedule. One major point brought up in favor of robot systems was that human space travelers require extensive and expensive life-support systems. In contrast, robots can survive long journeys into deep space and accomplish exploration goals just as well as humans. Furthermore, the loss of a robot spacecraft does not cause the same national numbness and paralyzing impact as when a human crew is lost during space exploration.

Other aerospace industry experts sharply disagreed during these post–Apollo Project debates. They argued that humans are, and will remain, very important in space exploration. These experts further proclaimed that robots and humans are not interchangeable. The proponents for human spaceflight also pointed out that human beings are far more adaptable than robots and can react better to unexpected events. When things go wrong, human beings can use their creativity and intellect to make innovative repairs.

The tragic loss of the space shuttle *Challenger* and its crew in January 1986, followed 17 years later by the loss of *Columbia* and its crew in February 2003, has revived the robots-versus-humans-for-space-exploration debate. Today, however, there is no real debate about using robot spacecraft to explore remote regions of interplanetary space—the regions beyond the main asteroid belt or the innermost portions of solar system—namely, Venus, Mercury, or areas near the Sun's corona. The present debate centers on the following specific question: Should a human expedition to Mars occur in this century or should the detailed exploration of the Red Planet be assigned to a series of progressively more complex space robots? Fanning the flames of this debate are the obvious health and life risks to the astronaut crew on a three-year interplanetary mission, as well as its enormously large projected price tag. Some studies have estimated the total actual cost of a human expedition to Mars (with a crew of 10) would be about $1 trillion. In comparison, the U.S. government spent a total of about $25 billion to send human explorers to the Moon.

In response to a directive from the White House regarding the exploration of space in the 21st century, NASA planners are now developing strategies for a return to the Moon and then

making still photographs. In orbital-simulation tests, the *Mini AERCam* performed rotational and translation maneuvers, including automatic attitude hold, automatic translation hold, and point-to-point maneuver-

human spaceflight to Mars. The goal is no longer humans or robots. It is humans *and* robots working together. Each brings complementary capabilities that will support the detailed exploration and future settlement of these two worlds. A well-planned and organized robot-human partnership is essential for the successful return of humans to the Moon and for the construction of the first permanent lunar base. A dynamic, well-functioning human-robot partnership is just as crucial if human explorers are to travel to Mars successfully later in this century.

Robotic systems on the space shuttle and the *International Space Station (ISS)* provide a glimpse of how the robot-human partnership in space exploration and operations will grow over the next few decades. The space shuttle's remote manipulator system (RMS) is an excellent example of how this partnership should work. The 49-foot- (15-m-) long robot arm (also called the Canadarm because it was designed and constructed in Canada) is mounted near the forward end of the port side of the orbiter's payload bay. The device has seven degrees of freedom (DOF). Like a human arm, it has a shoulder joint that can move in two directions; an elbow joint; a wrist joint that can roll, pitch, and yaw; and a gripping device. The gripping device is called an end effector. The RMS's end effector is a snare device that closes around special posts, called grapple fixtures. The grapple fixtures are attached to the objects that the RMS is trying to grasp. Astronauts have made extensive use of the RMS during a wide variety of shuttle missions.

The *ISS*, currently under construction in Earth orbit, will have several robotic systems to help astronauts complete their tasks. The assembly and maintenance of the *ISS* rely heavily on the use of extravehicular robotic systems. When fully assembled, the *ISS* robotics complement will include three main manipulators, two small dextrous arms, and a mobile base and transporter system.

The most complex robotic system on the *ISS* is the mobile servicing system (MSS). Jointly developed by Canada and NASA, the MSS is composed of five subsystems: the space station remote manipulator system (SSRMS), the mobile base system (MBS), mobile transporter (MT), the special purpose dextrous manipulator (SPDM), and the robotic workstation (RWS). The SSRMS is a 17-foot- (5.2-m-) long manipulator consisting of two booms, seven joints, and two latching end effectors. Astronauts can control and monitor the SSRMS from one of two modular workstations.

NASA's strategic planners envision an expanded role for robots in the development and operation of a permanent lunar base, and in assisting human explorers on Mars. Some of these future space robots will serve in precursor roles, such as performing focused exploration of candidate sites. Other space robots will be sent ahead to prepare a candidate site on the Moon (and eventually Mars) for the arrival of human beings. Finally, another group of space robots will work in direct partnership with astronauts, as they return to the Moon or explore Mars later in this century. Space robot systems will display an entire spectrum of behavior characteristics from dextrous, teleoperated devices to fully autonomous machines capable of performing their jobs without direct human supervision or guidance.

ing. Expanded human-crew–robot free-flyer tests and demonstrations of small free-flyers are anticipated when NASA's space shuttle resumes scheduled flights and when the construction of the *International Space Station*

(ISS) is completed. Small flying robots, like *Mini AERCam,* could serve many functions around the *ISS,* such as visual inspections in remote areas, close-up observation of human EVAs and robot manipulator activities, and mobile communications relays during complicated space assembly or repair activities.

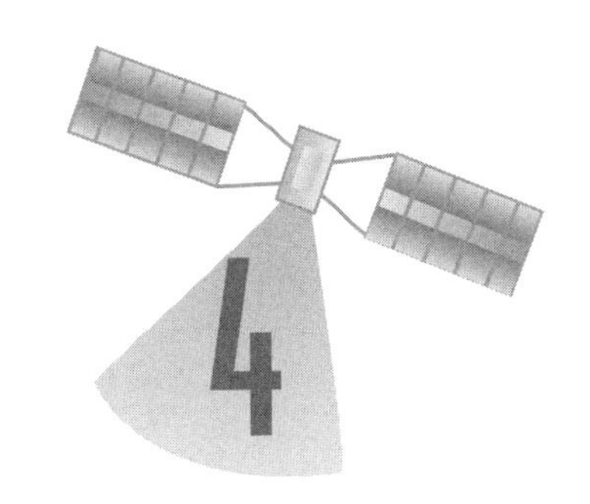

Flyby Spacecraft

A flyby mission is an interplanetary or deep space mission in which the robot spacecraft passes close to its celestial target (typically a distant planet, moon, asteroid, or comet), but does not impact the target or go into orbit around it. Flyby spacecraft follow a continuous trajectory to avoid being captured into a planetary orbit. Once the spacecraft has flown past its target (often after years of travel through deep space), it cannot return to recover lost data. So flyby operations often are planned years in advance of the encounter and refined and practiced in the months prior to the encounter date. Flyby operations are conveniently divided into four phases: observatory phase, far-encounter phase, near-encounter phase, and post-encounter phase.

The observatory phase of a flyby mission is defined as the period when the celestial target can be better resolved in the spacecraft's optical instruments than it can from Earth-based instruments. The phase generally begins a few months prior to the date of the actual planetary flyby. During this phase, the spacecraft is totally involved in making observations of its target, and ground resources become completely operational in support of the forthcoming encounter.

The far-encounter phase includes time when the full disk of the target planet (or other celestial object) can no longer fit within the field of view of the spacecraft's instruments. Observations during this phase are designed to focus on parts of the planetary body (for example, the Caloris Crater on Mercury, Jupiter's Red Spot, or the cantaloupe-like surface features on Triton), rather than the entire planet, and to take advantage of the higher resolution available.

The near-encounter phase includes the period of closest approach to the target. It is characterized by intensely active observations by all of the spacecraft's science experiments. This phase of the flyby mission provides scientists the opportunity to obtain the highest resolution data about the target.

Finally, the post-encounter phase begins when the near-encounter phase is completed and the spacecraft is receding from the target. This phase is characterized by day-after-day observations of a diminishing, thin crescent of the planet just encountered. It provides an opportunity to make extensive observations of the night side of the planet. After the post-encounter phase is over, the spacecraft stops observing the target and returns to the less intense activities of its interplanetary cruise phase—a phase in which scientific instruments are powered down and navigational corrections made to prepare the spacecraft for an encounter with another celestial object of opportunity or for a final journey of no return into deep space. Some scientific experiments, usually concerning the properties of interplanetary space, can be performed in this cruise phase.

This chapter highlights three of the most spectacular scientific flyby missions performed by NASA's robot spacecraft in the 20th century. The first mission discussed is the *Mariner 10* mission to Mercury, by way of a gravity-assist from Venus. The second mission is *Pioneer 11*'s trailblazing flyby encounter of Saturn, following its earlier close encounter with Jupiter, which provided an important gravity-assist. Finally, no discussion of far-traveling robot spacecraft on flyby missions is complete without mention of the amazing journey of the *Voyager 2* spacecraft. *Voyager 2*'s grand-tour mission took the hardy space robot past all of the giant planets. The mission controllers (who diligently monitored its progress during its epic journey through the outer regions of the solar system), the aerospace engineers (who tirelessly worked to reconfigure subsystems and improvise workarounds), and the planetary scientists (who carefully guided data collection during this once in two centuries encounter opportunity), all came to affectionately call the *Voyager 2* the "little robot spacecraft that could."

✧ *Mariner 10*—First Spacecraft to Mercury

Mariner 10 was the seventh successful launch in NASA's Mariner series. (*Mariner 1* and *Mariner 8* experienced launch failures, while *Mariner 3* ceased transmission nine hours after launch and went into orbit around the Sun.) The *Mariner 10* spacecraft was the first to use the gravitational pull of one planet (Venus) to reach another planet (Mercury). It passed Venus on February 5, 1974, at a distance of 2,610 miles (4,200 km). The robot spacecraft then crossed the orbit of Mercury at a distance of 437 miles (704 km) from the surface on March 29, 1974. A second encounter with Mercury occurred on September 21, 1974, at an altitude of about 29,210 miles (47,000 km). A third and final Mercury encounter took place on March 16, 1975, when the spacecraft passed the planet at an altitude of 203 miles (327 km). Many images of the planet's surface were acquired during these flybys, and magnetic field measurements were performed. When the supply of attitude-control gas became depleted on March 24, 1975, this highly successful mission was terminated.

The *Mariner 10* spacecraft structure was an eight-sided magnesium frame with eight electronics compartments. The space robot measured

After a complete prelaunch checkout, aerospace technicians prepare to encapsulate NASA's *Mariner 10* spacecraft in 1973. Successfully launched on November 3 of that year from Cape Canaveral by an Atlas-Centaur rocket vehicle, *Mariner 10* traveled through interplanetary space, encountering Venus on February 5, 1974. With a trail-blazing gravitational assist from Venus, the robot spacecraft then went on to perform three flyby encounters (two in 1974 and one in 1975) of Mercury, the innermost planet in the solar system. *(NASA)*

4.56 feet (1.39 m) diagonally and 1.5 feet (0.46 m) in depth. Two solar panels, each 8.83 feet (2.69 m) long and 3.18 feet (0.97 m) wide, were attached at the top, supporting 54.9 square feet (5.1 square meters) of solar-cell area. Fully deployed and cruising in international space, *Mariner 10* measured 26.2 feet (8 m) across the solar panels and 12.1 feet (3.7 m) from the top of the low-gain antenna to the bottom of the heat shield. Engineers mounted a scan platform with two degrees of freedom on the anti-solar face of the spacecraft structure. A 19-foot- (5.8-m-) long hinged magnetometer boom extended from one of the octagonal sides of the spacecraft structure.

The spacecraft had a total launch mass of 1,106 pounds (503 kg), of which 64 pounds (29 kg) were propellant and attitude-control gas. The total mass of onboard instruments was 175 pounds (79 kg). The spacecraft carried science instruments to measure the atmospheric, surface, and physical characteristics of Venus and Mercury. Experiments included television-photography, infrared radiometers, and ultraviolet spectroscopy.

The *Mariner 10*'s rocket engine was a 50-pound-force (222-newton) liquid monopropellant hydrazine motor located below a spherical propellant tank, which was mounted in the center of the structural framework. The rocket nozzle protruded through a sunshade. Engineers used a total of six (two sets of three orthogonal pairs) pressurized nitrogen gas reaction-jets (thrusters), which they mounted on the tips of the solar panels to achieve three-axis stabilization of the spacecraft. Command and control of these thrusters were the responsibility of an on-board computer.

Finally, *Mariner 10* carried a motor-driven high-gain dish antenna, with a 4.5-foot- (1.37-m-) diameter parabolic reflector made of aluminum honeycomb sandwich material. This high-gain antenna was mounted on a boom on the side of the spacecraft. The spacecraft also had a low-gain, omnidirectional antenna, which was mounted at the end of a 9.35-foot- (2.85-m-) long boom, extending from the anti-solar face of the spacecraft.

Mariner 10 was the first and, thus far, the only spacecraft of any country to explore the innermost planet in the solar system. On August 3, 2004, NASA launched *MESSENGER* from Cape Canaveral and sent the orbiter spacecraft on a long interplanetary journey to Mercury. In March 2011, *MESSENGER* is set to become the first robot spacecraft to achieve orbit around the planet.

✧ *Pioneer 11*—First Space Robot to Saturn

NASA's *Pioneer 11* spacecraft (and its technical twin *Pioneer 10*), as the names imply, are true deep-space explorers—the first spacecraft to navigate the main asteroid belt; the first spacecraft to encounter Jupiter

(*Pioneer 10*) and its fierce radiation belts; the first to encounter Saturn (*Pioneer 11*); and the first human-made object (*Pioneer 10*) to leave the planetary boundary of the solar system. As they flew through interplanetary space, these spacecraft also investigated magnetic fields, cosmic rays, solar wind, and interplanetary dust concentrations.

The *Pioneer 11* spacecraft consisted of several distinct subsystems: a general structure, an attitude control and propulsion system, a communications system, thermal control system, electric power system, navigation system, and a science payload (containing 11 instruments).

NASA's *Pioneer 11* spacecraft awaits installation of its protective shroud at Cape Canaveral in 1973. This robot flyby spacecraft was launched on April 5, 1973, by an Atlas-Centaur rocket vehicle, and swept by Jupiter on December 2, 1974, at an encounter distance of only 26,720 miles (43,000 km). Then, on September 1, 1979, *Pioneer 11* flew past Saturn, demonstrating a safe flight path through the rings for the more sophisticated Voyager spacecraft to follow. *(NASA/Kennedy Space Center)*

The 570-pound (259-kg) *Pioneer 11* spacecraft was carefully designed to fit within the 10-foot- (3-m-) tall shroud of the Atlas-Centaur launch vehicle. The *Pioneer 11* was stowed with its booms retracted and its antenna dish facing upward (that is, upward on the launchpad). Basically, the *Pioneer 11* (and its *Pioneer 10* technical twin) had to be extremely reliable and lightweight. The spacecraft needed a communications system capable of transmitting data over extremely large distances. Since each spacecraft would operate so far from the Sun, engineers chose a nuclear (non-solar) power source for electric power generation.

From its cone-shaped, medium-gain antenna to the adapter ring that fastened the spacecraft to the third stage of its launch vehicle, *Pioneer 11* was 9.5 feet (2.9 m) long. The spacecraft structure centered around a 14-inch- (36-cm-) deep, flat equipment compartment, the top and bottom of which consisted of regular hexagons with sides 28 inches (71 cm) long. Attached to one side of this hexagon was a smaller "squashed" hexagon compartment that carried most of the spacecraft's scientific instruments.

Engineers attached a nine-foot- (2.74-m-) diameter, 18-inch- (46-cm-) deep, parabolic, dish-shaped high-gain antenna made of aluminum honeycomb sandwich material to the front of the equipment compartment. The feed of the high-gain antenna was topped with a medium-gain antenna mounted on three struts, which projected about four feet (1.2 m) forward. Spacecraft engineers also mounted a 2.5-foot- (0.76-m-) diameter low-gain, omnidirectional antenna below the dish of the high-gain antenna. (It may prove helpful to look back at the figure on page 42.)

The *Pioneer 11* (and *Pioneer 10*) spacecraft had three appendages that extended after launch. Two of these appendages were three-rod trusses that each held radioisotope thermoelectric generator (RTG) units about 10 feet (3 m) from the center of the spacecraft. The third appendage was a single rod boom that held a magnetometer sensor about 21.5 feet (6.6 m) from the center of the spacecraft.

The robot spacecraft had three reference sensors to support interplanetary navigation: a star (Canopus) sensor and two Sun sensors. Attitude position could be calculated from the reference direction to Earth and the Sun, with the known direction to the star Canopus used as a backup. *Pioneer 11* had three pairs of rocket thrusters, which could be fired on command either steadily or in pulses. Three pairs of rocket thrusters located near the rim of the antenna dish were used to direct the spin axis of the spacecraft, to keep the spacecraft spinning at the desired rate of 4.8 revolutions per minute (rpm), and to change the velocity of the spacecraft for in-flight maneuvers. The spacecraft's six thrusters could be commanded to fire steadily or in pulses. Each thruster developed its propulsive jet from the decomposition of liquid hydrazine by a catalyst in a small rocket thrust chamber to which the nozzles of the thruster were attached.

The *Pioneer 11* spacecraft carried two identical receivers. The omnidirectional and medium-gain antennae operated together and were connected to one receiver, while the high-gain antenna was connected to the other. The receivers did not operate at the same time, but were interchanged by command or, if there was a period of inactivity, were switched automatically. This clever fail-safe design feature made sure that if a receiver had failed during the mission, the other would automatically take over.

As part of its communications subsystem, the spacecraft had two traveling-wave-tube (TWT) power amplifiers, each of which produced eight watts of transmitted power at S-band. The frequency uplink from Earth to the spacecraft was at 2,110 megahertz (MHz), the downlink to Earth at 2,292 MHz. NASA's Deep Space Network supported telecommunications across great interplanetary distances. From launch, *Pioneer 11* successfully operated on its backup transmitter.

The spacecraft contained two radioisotope thermoelectric generator (RTG) units as its electric power. When *Pioneer 11* reached the vicinity of Jupiter, the RTGs provided 144 watts of electric power for use on the spacecraft, but this level decreased to 100 watts when *Pioneer 11* reached Saturn.

The spacecraft's thermal-control system kept the temperatures inside the science instrument compartment between -10°F (-23°C) and +100°F (+38°C). The rest of the spacecraft was designed to maintain temperatures compatible with the operation of the science instruments. Spacecraft engineers designed the thermal-control system to adapt to the gradual decrease in solar heating as *Pioneer 11* moved away from the Sun. It was also constructed to survive those frigid periods when the spacecraft passed through Earth's shadow (after launch), followed by Jupiter's shadow, and then Saturn's shadow, during the planetary encounters.

As *Pioneer 11* moved through interplanetary space on its way to Jupiter and then Saturn, some of its 11 onboard scientific instruments investigated magnetic fields, cosmic rays, solar wind, and interplanetary dust concentrations—especially those found in the asteroid belt between Mars and Jupiter. At Jupiter and again at Saturn, *Pioneer 11* invested the planetary systems in four main ways: by measuring particles, fields, and ionizing radiation, by spin-scan imaging of the gaseous giant planets and some of their moons, by accurately observing the path of *Pioneer 11* and measuring the gravitational forces of the planets and major satellites acting on the spacecraft, and, finally, by observing changes in the frequency of the S-band radio signal before and after occultation by each planet, to study the structure of its ionosphere and atmosphere.

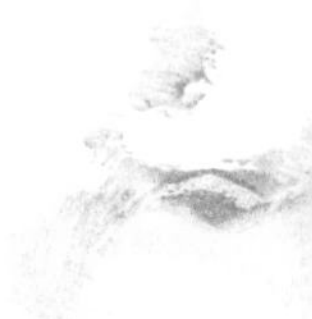

The *Pioneer 10* and *11* spacecraft were identical in construction. In NASA's overall space-exploration strategy, *Pioneer 10* was to blaze the trail to Jupiter. If the asteroid belt or the Jovian magnetosphere proved too hazardous for *Pioneer 10,* then *Pioneer 11* was to serve as the backup spacecraft and complete the primary mission of examining Jupiter during a scientific flyby. NASA's mission planners also reserved the capability, however, of retargeting *Pioneer 11* (using a gravity-assist maneuver at Jupiter) to fly by Saturn. This option was based on the results of *Pioneer 10*'s encounter with Jupiter. So to fully appreciate the flight of *Pioneer 11,* a brief discussion of the performance of its twin (*Pioneer 10*) must first be examined.

The *Pioneer 10* spacecraft was launched from Cape Canaveral Air Force Station, Florida, by an Atlas-Centaur rocket on March 2, 1972. It became the first spacecraft to cross the main asteroid belt and the first to make close-range observations of the Jovian system. Sweeping past Jupiter on December 3, 1973 (its closest approach to the giant planet), it discovered no solid surface under the thick layer of clouds enveloping the giant planet—an indication that Jupiter is a liquid hydrogen planet. *Pioneer 10* also explored the giant Jovian magnetosphere, collected close-up pictures of the intriguing Red Spot, and observed at relatively close range the Galilean satellites Io, Europa, Ganymede, and Callisto. When *Pioneer 10* flew past Jupiter, it acquired sufficient kinetic energy to travel completely out of the solar system.

The *Pioneer 11* spacecraft was launched on April 5, 1973, and swept by Jupiter at an encounter distance of only 26,725 miles (43,000 km) on December 2, 1974. The spacecraft provided additional detailed data and pictures of Jupiter and its moons, including the first views of Jupiter's polar regions. Then, on September 1, 1979, *Pioneer 11* flew by Saturn, demonstrating a safe flight path through the rings for the more sophisticated *Voyager 1* and *2* spacecraft to follow. *Pioneer 11* (by then officially renamed *Pioneer Saturn*) provided the first close-up observations of Saturn, its rings, satellites, magnetic field, radiation belts, and atmosphere. It found no solid surface on Saturn, but discovered at least one additional satellite and ring. After rushing past Saturn, *Pioneer 11* also headed out of the solar system toward the distant stars.

Both Pioneer spacecraft carried a special message (the Pioneer Plaque) for any intelligent alien civilization that might find them wandering through the interstellar void millions of years from now. (The interstellar travels of *Pioneer 10* and *11* are discussed in chapter 12, along with the interesting plaque each carries.)

✧ The Grand Tour of *Voyager 2*

Once every 176 years, the giant outer planets—Jupiter, Saturn, Uranus, and Neptune—align themselves in such an orbital pattern that a spacecraft launched from Earth to Jupiter at just the right time might be able to visit the other three planets on the same mission, by using a gravity-assist.

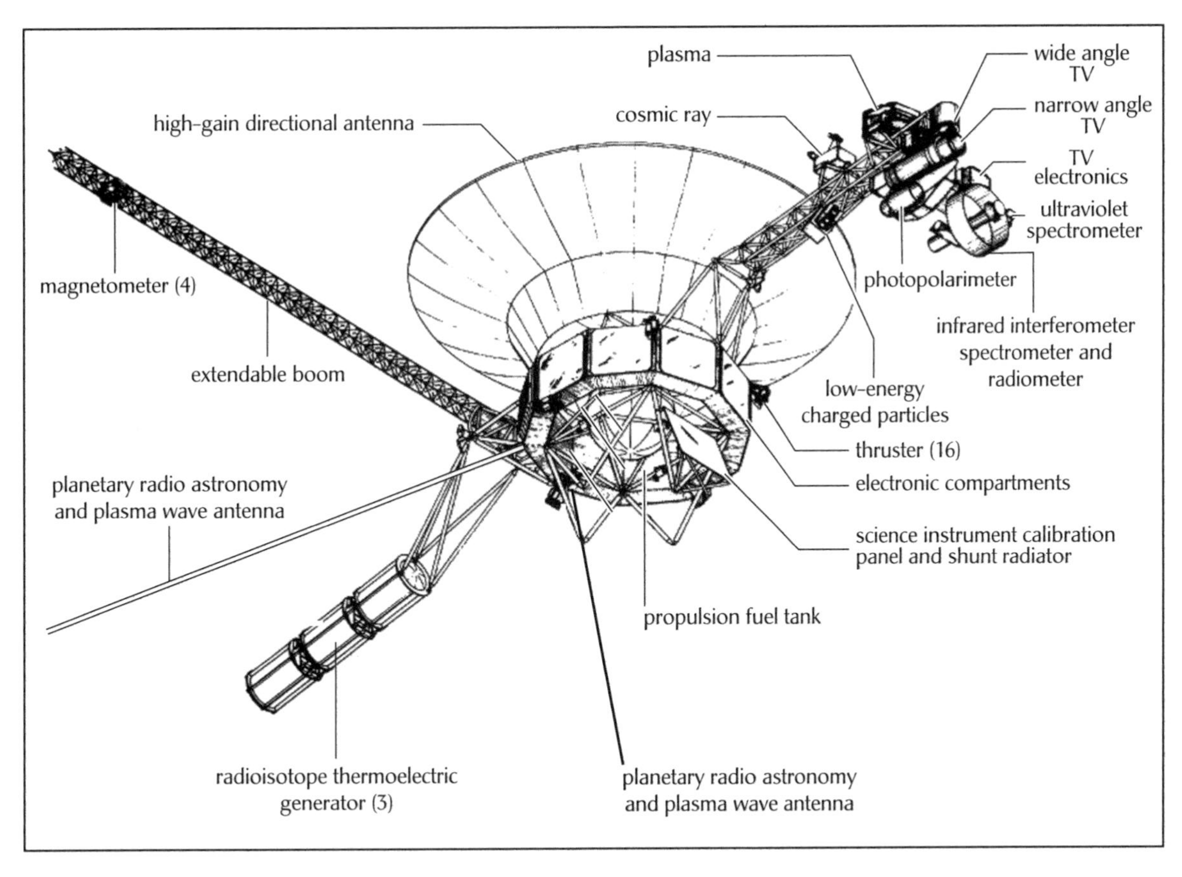

This drawing shows the *Voyager 1* (and *2*) spacecraft with the complement of sophisticated scientific instrument. *(NASA)*

NASA space scientists refer to this multiple-gravity-assist, giant-planet-encounter mission as the Grand Tour. A very special robot spacecraft called *Voyager 2* took advantage of a unique celestial-alignment opportunity in 1977 and made planetary exploration history.

Credit for the single space-robot mission that has visited the most planets goes to JPL's Voyager Project. Launched in 1977, the twin *Voyager 1* and *Voyager 2* spacecraft flew by the planets Jupiter (1979) and Saturn (1980–81). *Voyager 2* then went on to have an encounter with Uranus (1986) and with Neptune (1989). Both *Voyager 1* and *Voyager 2* are now traveling on different trajectories into interstellar space. In February 1998, *Voyager 1* passed the *Pioneer 10* spacecraft to become the most distant human-made object in space. The Voyager Interstellar Mission (VIM) (described in chapter 12) should continue well into the next decade.

Each Voyager spacecraft had a mass of 1,815 pounds (825 kg) and carried a complement of scientific instruments to investigate the outer

NEPTUNE AND TRITON

Neptune is the outermost of the gaseous giant planets and the first planet to be discovered using theoretical predictions. The discovery of Neptune took place on September 23, 1846, and was made by the German astronomer Johann Gottfried Galle (1812–1910), while working at the Berlin Observatory. His discovery was based on independent orbital-perturbation (disturbance) analyses by the French astronomer Urbain-Jean-Joseph Leverrier (1811–77) and the British scientist John Couch Adams (1819–92). Then, just 17 days later on October 10, 1846, the wealthy British amateur astronomer William Lassell (1799–1880) discovered Triton, the planet's largest moon.

Because of Neptune's great distance from Earth, little was known about this majestic blue gaseous planet or its largest moon, Triton, until the *Voyager 2* spacecraft swept through the Neptunian system on August 25, 1989. Neptune's characteristic blue color comes from the selective absorption of red light by the methane (CH_4) found in its atmosphere—an atmosphere consisting primarily of hydrogen (more than 89 percent) and helium (about 11 percent) with minor amounts of methane, ammonia ice, and water ice.

At the time of the *Voyager 2* encounter, Neptune's most prominent surface feature was called the Great Dark Spot (GDS), which was somewhat analogous in relative size and scale to Jupiter's Red Spot. Unlike Jupiter's Red Spot, however, which has been observed for at least 300 years, Neptune's GDS, which was located in the planet's southern hemisphere in 1989, had disappeared by June of 1994, when the *Hubble Space Telescope* looked for it. Then, a few months later, a nearly identical spot appeared in Neptune's northern hemisphere. Neptune is an extraordinarily dynamic planet that continues to surprise space scientists. The *Voyager 2* encounter also revealed the existence of six additional satellites and an interesting ring system.

Voyager 2 approached within 14,920 miles (24,000 km) of Triton's surface, and this flyby provided astronomers with virtually all they currently know about this icy world. Triton has a diameter of 1,680 miles (2,700 km) and travels in a retrograde orbit around Neptune at a distance of 220,600 miles (355,000 km). It is the only large moon in the solar system to travel in this type of reverse, or westward, orbit around its primary.

Triton is one of the coldest objects yet discovered in the solar system. Because of its highly inclined (20 degrees), retrograde orbit, rock-and-ice composition, and frost-covered surface, space scientists like to consider Triton a cousin to the planet Pluto. The exact origin of Triton is not clear, but two hypotheses are popular among astronomers. Either some ancient object had a catastrophic collision with Neptune, leading to the moon's formation, or Neptune captured an icy Kuiper belt object.

If the origin of Triton still remains uncertain, its long-term future is not. Triton's retrograde orbit and the tidal bulge it creates on Neptune is causing the large moon to spiral inward toward Neptune rather than away from its primary (the way the Moon orbits around Earth). Astronomers consider Triton to be a doomed world that within 100 million years (or less) will travel inside Neptune's Roche limit and disintegrate.

As postulated by the French mathematician Edouard-Albert Roche (1820–83) in the 19th century, the Roche limit is the smallest distance from a planet at which gravitational forces can hold together a natural satellite or moon that has about the same average density as the primary body. If the moon's orbit falls within the Roche limit, it will be torn apart by tidal forces. The mean density of Neptune is 102 pounds per cubic foot (1,638 kg/m^3) and that of Triton is 74.9 pounds per cubic foot (1,200 kg/m^3). In the distant future, after Triton is torn apart, Neptune will develop a prominent set of rings to rival the magnificent ring system observed today around Saturn.

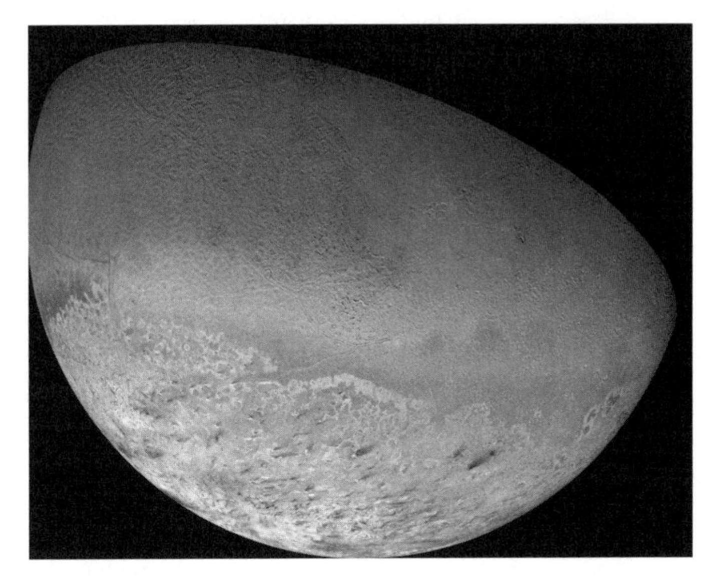

The global mosaic image of Neptune's largest moon, Triton, collected by NASA's *Voyager 2* spacecraft during its flyby of the Neptune system in 1989. Frigid Triton is only one of three objects in the solar system known to have a nitrogen-dominated atmosphere—the two other bodies being Earth and Saturn's large moon, Titan. *Voyager 2's* encounter with Neptune completed this far-traveling exploration robot's epoch journey (sometimes called the Grand Tour mission) that took it past all the gaseous giant planets (Jupiter, Saturn, Uranus, and Neptune) in the outer solar system. *Voyager 2,* and its twin *Voyager 1,* now travel on different trajectories into interstellar space. *(NASA/JPL)*

Voyager 2 images have revealed some of Triton's remarkable geologic history, including geyser-like eruptions that have spewed nitrogen gas and somewhat mysterious dark dust several miles (kilometers) into space. Instead of an ancient, heavily cratered surface, the *Voyager 2* images showed that Triton possessed a relatively smooth and young surface—characterized by winding fissures, frozen lakes, and knobby-surfaced plains that scientists call *cantaloupe terrain* because of the resemblance to the skin of that fruit.

Since Triton is situated about 2.8 billion miles (4.5 billion km) from the Sun and has a fairly reflective surface (which most likely consists of water ice), the surface temperature on this moon is estimated to be a frigid -393°F (-37 K). The moon has a wisp of a nitrogen atmosphere, speculated to be about 100,000 times thinner than Earth's nitrogen-rich atmosphere. Because Triton's surface does not indicate a significant number of craters, planetary scientists believe there has been recent surface activity that has wiped out most evidence of past impacts.

planets and their many moons and intriguing ring systems. These instruments, provided with electric power by a long-lived nuclear power system called a radioisotope thermoelectric generator (RTG), recorded spectacular close-up images of the giant outer planets and their interesting moon systems, explored complex ring systems, and measured properties of the interplanetary medium.

Taking advantage of the 1977 Grand Tour launch window, the *Voyager 2* spacecraft lifted off from Cape Canaveral, Florida, on August 20, 1977, on board a Titan-Centaur rocket. (NASA called the first Voyager spacecraft launched *Voyager 2,* because the second Voyager spacecraft to be launched eventually would overtake it and become *Voyager 1.*) *Voyager 1* was launched on September 5, 1977. This spacecraft followed the same trajectory as its *Voyager 2* twin and overtook its sister ship just after entering the asteroid belt in mid-December 1977.

Voyager 1 made its closest approach to Jupiter on March 5, 1979, and then used Jupiter's gravity to swing itself to Saturn. On November 12, 1980, *Voyager 1* successfully encountered the Saturn system and then was flung up out of the ecliptic plane on an interstellar trajectory.

The *Voyager 2* spacecraft successfully encountered the Jupiter system on July 9, 1979 (closest approach), and then used the gravity-assist technique to follow *Voyager 1* to Saturn. On August 25, 1981, *Voyager 2* encountered Saturn and then went on to successfully encounter both Uranus (January 24, 1986) and Neptune (August 25, 1989). Space scientists consider the end of *Voyager 2*'s encounter with the Neptunian system as the end of a truly extraordinary epoch in planetary exploration. In the first 12 years after they were launched from Cape Canaveral, these incredible robot spacecraft contributed more to scientific knowledge about the giant outer planets of the solar system than had been accomplished in three millennia of Earth-based observations (both naked-eye and telescopic). Following its encounter with the Neptunian system, *Voyager 2* also was placed on an interstellar trajectory and (like its *Voyager 1* twin) now continues to travel outward from the Sun.

Orbiters, Probes, and Penetrators

The same types of very precise navigation- and course-correction procedures used in flyby missions are also applied during the cruise phase of a planetary orbiter mission. The process of planetary-orbit insertion places the spacecraft in precisely the correct location at the correct time to enter into an orbit around the target planet. Orbit insertion requires not only the precise position and timing of a flyby mission but also a controlled deceleration.

As the spacecraft's trajectory is bent by a planet's gravity, the command sequence aboard the spacecraft fires its retroengine(s) at the proper moment and for the proper duration. Once this retroburn (or reverse firing) has been completed successfully, the spacecraft is captured into orbit by the target planet. If the retroburn fails (or is improperly sequenced), the spacecraft will continue to fly past the planet. It is quite common for this retroburn to occur on the farside of a planet as viewed from Earth—requiring this portion of the orbit-insertion sequence to occur automatically (based either on stored onboard commands or some level of artificial intelligence), without any interaction with the flight controllers on Earth.

Once safely in orbit around the target planet, a planetary spacecraft can engage in two general categories of orbital operations: exploration of the planetary system and mapping of the planet. Exploring a planetary system includes making observations of the target planet and its system of satellites and rings. A mapping mission generally is concerned with acquiring large amounts of data about the planet's surface features.

An orbit of low inclination at the target planet usually is well suited to a planetary system exploration mission, since it provides repeated exposure to satellites (moons) orbiting with the equatorial plane as well as adequate coverage of the planet and its magnetosphere. An orbit of high inclination is much better suited for a mapping mission, however, because the target planet will fully rotate below the spacecraft's orbit, thereby providing eventual exposure to every portion of the planet's surface. During either type of mission, the orbiting spacecraft is involved in an extended encounter period with the target planet and requires continuous (or nearly continuous) support from the flight team members at mission control on Earth. The *Cassini* spacecraft is an example of a planetary system exploration mission, while the *Viking 1* and *2* orbiter spacecraft are examples of a planetary mapping mission.

Orbit trim maneuvers (OTMs) are performed in a spacecraft's orbit around a planet for the purpose of adjusting an instrument's field-of-view, improving sensitivity of a gravity field survey, or preventing too much orbital decay. To make a change increasing the altitude of periapsis (the orbiting spacecraft's closest approach to the planet), engineers design the orbit trim maneuver to increase the spacecraft's velocity when the space vehicle is at apoapsis (the orbiting spacecraft's greatest distance from the planet). To decrease the apoapsis altitude, an OTM is performed at periapsis, reducing the spacecraft's velocity. Slight changes in the orbital plane's orientation also can be made with orbit trim maneuvers. The magnitude of such changes is necessarily small, however, owing to the limited amount of maneuver propellant typically carried by a spacecraft.

This chapter also discusses two other types of robot spacecraft: probes and penetrators. A probe is an instrumented robot spacecraft that moves through the atmosphere of a planetary body and/or impacts on its surface. Scientists use probes for the purpose of obtaining atmospheric data during descent and/or surface data on landing. The *Huygens* spacecraft and the *Pioneer Venus Multiprobe* spacecraft are examples of successful planetary probe missions.

Planetary scientists suggest that experiments performed using a network of instrumented penetrators can provide many of the essential facts they need to start understanding the evolution, history, and nature of a planetary body, such as Mars. The scientific measurements performed by an instrumented penetrator might include seismic, meteorologic, and local site characterization studies involving heat flow, soil moisture content, and geochemistry.

A typical penetrator system consists of four major subassemblies: the launch tube, the deployment motor, the decelerator (usually a two-stage device), and the penetrator itself. Scientists can also use a less-sophisticated penetrator to study a small celestial body (such as an asteroid or comet). By impacting the target at great relative velocity, the sacrificial penetrator, or impactor, destroys itself and dislodges a huge plume of surface materials, which can then be analyzed by remote sensing instruments. Chapter 9 describes how NASA's Deep Impact mission used a sacrificial 816-pound (370-kg) penetrator to investigate Comet Tempel 1 in July 2004.

✧ *Mariner 9* Spacecraft

NASA had originally planned the Mariner Mars 71 mission to consist of two spacecraft orbiting Mars on complementary missions. The *Mariner 8* orbiter was to map about 70 percent of the surface of Mars, while the *Mariner 9* orbiter was to study changes in the atmosphere and surface of Mars over an extended period of time. When *Mariner 8* was lost because of a launch failure, however, *Mariner 9* inherited a combined set of mission objectives. For the survey portion of the Mariner Mars 71 mission, *Mariner 9* was now assigned the task of mapping the surface of the Red Planet to the same spatial resolution as originally planned—although the resolution of the images of the polar regions would be decreased due to the increased slant range. NASA mission managers also changed the variable features experiments from studies of six given areas every five days to studies of smaller regions every 17 days.

The compromises and trade-offs worked very well. *Mariner 9* became the first artificial (human-made) satellite of Mars. The robot spacecraft also provided scientists with the first global map of the planet's surface, including the first detailed views of the Martian volcanoes, polar caps, and Valles Marineris. *Mariner 9* also provided the first close-up look at Mars's two tiny natural satellites: Phobos and Deimos.

NASA's *Mariner 9* spacecraft undergoes its final checkout prior to encapsulation and launch at Cape Canaveral, Florida. An Atlas-Centaur rocket successfully launched the Mars-bound spacecraft in May 1971. Achieving orbit around the Red Planet in November 1971, the robot orbiter spacecraft patiently waited for a great dust storm to subside and then compiled a global mosaic of high-quality images of Mars. *Mariner 9* also provided scientists with their first close-up pictures of the two small Martian moons, Phobos and Deimos. *(NASA/Kennedy Space Center)*

Engineers constructed the *Mariner 9* spacecraft on an octagonal magnesium frame, 1.5 feet (0.46 m) deep and 4.6 feet (1.4 m) across a diagonal. Four solar panels, each 7.1 feet (2.2 m) by 3.0 feet (0.9 m), extended out from the top of the frame. Each set of two solar panels spanned 22.6 feet (6.9 m) from tip to tip. Also mounted on top of the frame were two propulsion tanks, the maneuver engine, a 4.72-foot- (1.44-m-) long low-gain antenna mast and a parabolic high-gain antenna. A scan platform was mounted on the bottom of the frame, to which engineers attached the mutually bore-sighted scientific instruments (wide- and narrow-angle television cameras, infrared radiometer, ultraviolet spectrometer, and infrared interferometer spectrometer).

Mariner 9 had an overall height of 7.48 feet (2.28 m). The spacecraft had a launch mass of 2,195 pounds (998 kg), including 966 pounds (439 kg) of expendables. Engineers placed the spacecraft's communications and command and control subsystems within the magnesium frame. The scientific instruments of the spacecraft had a total mass of 139 pounds (63 kg).

Mariner 9's four solar panels generated a total of 500 watts of electric power in orbit around Mars. A rechargeable nickel-cadmium (NiCd) battery provided backup electric power when the spacecraft was in the shadowed portions of its orbit. Propulsion of the spacecraft was provided by a gimbaled rocket engine, which was capable of producing a thrust of 302 pounds of force (1,340 newtons) and of being restarted up to five times. This rocket engine used monomethyl hydrazine and nitrogen tetroxide as its liquid propellants. The spacecraft also had two sets of six pressurized-nitrogen, attitude-control jets, which were mounted on the ends of the solar panels. Reference data for three-axis stability attitude control were provided by a Sun sensor, a Canopus star tracker, an inertial reference unit, an accelerometer, and gyroscopes.

Engineers used louvers on eight sides of the spacecraft's frame and thermal blankets to achieve the necessary level of thermal control. A central computer and sequencer controlled the spacecraft. Telecommunications were achieved by means of two transmitters and a single receiver, through the high-gain parabolic antenna, the medium-gain horn antenna, or the low-gain omnidirectional antenna.

On May 30, 1971, an Atlas-Centaur launch vehicle sent *Mariner 9* on a direct-ascent trajectory to Mars from Cape Canaveral, Florida. The spacecraft's launch mass was nearly doubled by the onboard rocket propellant needed to thrust it into an orbit around the Red Planet, but otherwise closely resembled the earlier Mariner spacecraft. Achieving orbit around Mars in November 1971, the spacecraft arrived just as a great dust storm globally obscured the entire surface of the planet. The spacecraft had simple flight computers with limited memory and used a digital tape-recorder rather than film to store images and other scientific data. With these improvements in space technology, *Mariner 9* was able to wait in orbit until the storm abated, the dust settled, and the surface was clearly visible before compiling its global mosaic of high-quality images of the surface of Mars. The robot spacecraft also provide the first close-up pictures of the two small irregularly shaped Martian moons, Phobos and Deimos. After depleting its supply of attitude-control gas, the spacecraft was turned off by NASA mission managers. *Mariner 9* was left in orbit around Mars, which should not decay for at least 50 years—after which it will enter into the Martian atmosphere.

✧ *Viking 1* and *2* Orbiter Spacecraft

The Viking Project was the culmination of an initial series of American missions to explore Mars in the 1960s and 1970s. This series of interplanetary missions began in 1964 with *Mariner 4,* continued with the *Mariner 6* and *7* flyby missions in 1969, and then the *Mariner 9* orbital mission in 1971 and 1972.

Viking was designed to orbit Mars and to land and operate on the surface of the Red Planet. Two identical spacecraft, each consisting of a lander and an orbiter, were built. The primary mission objectives were to obtain high-resolution images of the Martian surface, characterize the structure and composition of the atmosphere and surface, and search for evidence of life.

Viking 1 was launched on August 20, 1975, from Cape Canaveral, Florida; *Viking 2* was launched on September 9, 1975. As previously mentioned, each spacecraft consisted of an orbiter and a lander. (Technical details about the *Viking 1* and *2* landers appear in chapter 6.) After orbiting Mars and returning images that scientists used to make landing-site selections, each lander detached from its companion orbiter spacecraft and descended through the atmosphere to make an automated soft landing in the area of Mars selected by mission managers.

The orbiters carried the following scientific instruments: (1) a pair of cameras that performed a systematic search for landing sites, then looked at and mapped almost 100 percent of the Martian surface (cameras onboard the *Viking 1* and *Viking 2* orbiters took more than 51,000 photographs of Mars); (2) a water detector that mapped the Martian atmosphere for water vapor and tracked seasonal changes in the amount of water vapor; and (3) an infrared thermal mapper that measured the temperatures of the surface, polar caps, and clouds; it also mapped seasonal changes.

In addition, although the Viking orbiter radios were not considered scientific instruments, they were used as such. By measuring the distortion of radio signals as these signals traveled from each of the Viking orbiter spacecraft to Earth, scientists were able to measure the density of the Martian atmosphere.

The lander spacecraft were sterilized before launch to prevent contamination of Mars by terrestrial microorganisms. These spacecraft spent nearly a year in transit to the Red Planet. The *Viking 1* achieved Mars orbit on June 19, 1976; and *Viking 2* began orbiting Mars on August 7, 1976.

The Viking mission was planned to continue for 90 days after landing. Each orbiter and lander, however, operated far beyond its design lifetime. For example, the *Viking 1* orbiter exceeded four years of active flight operations in orbit around Mars.

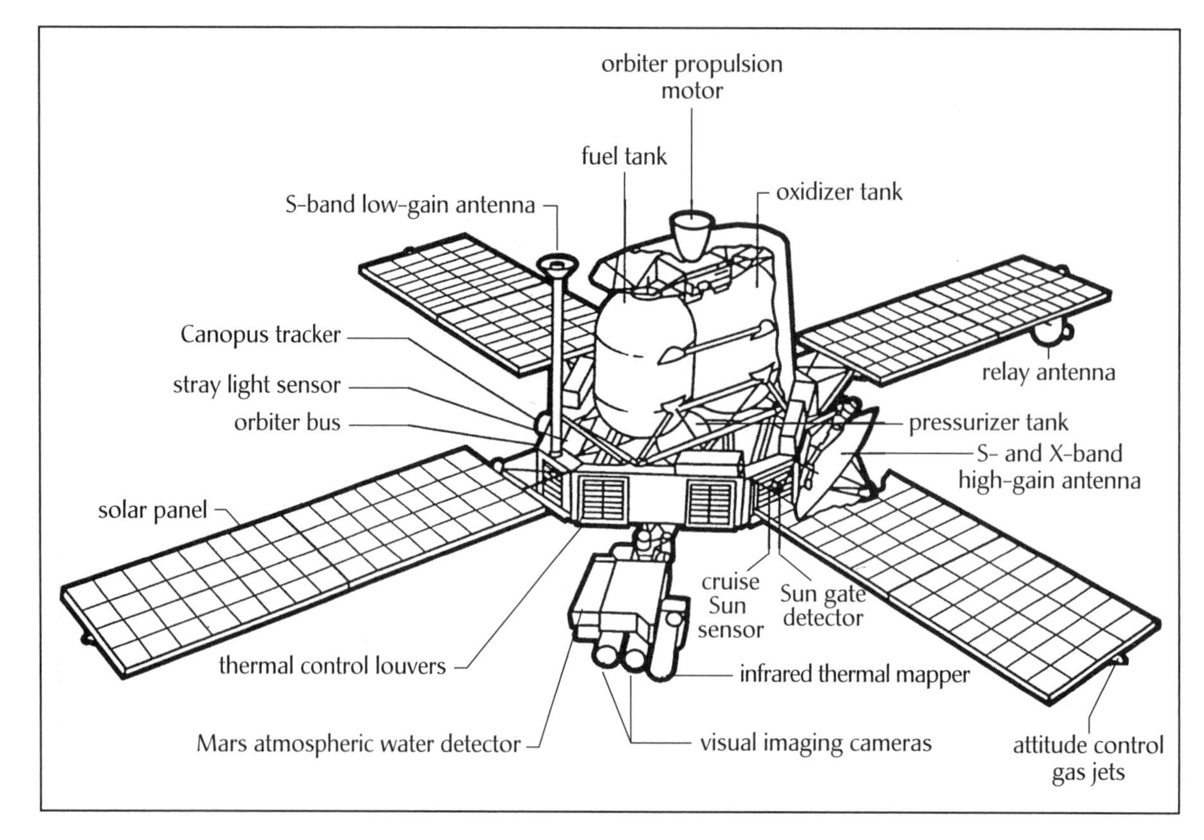

This drawing shows the *Viking 1* (and *2*) orbiter spacecraft and its complement of scientific instruments. *(NASA)*

This two-photographic image mosaic was collected by NASA's *Viking 1* orbiter spacecraft in June 1980. The composite image shows an interesting portion of the Martian surface, etched and grooved by wind erosion in what is an otherwise heavily cratered region. *(NASA/JPL)*

The Viking Project's primary mission ended on November 15, 1976, just 11 days before Mars passed behind the Sun. (This is an astronomical event called a *superior conjunction.*) After conjunction, in mid-December 1976, telemetry and command operations were reestablished and extended mission operations began.

The *Viking 2* orbiter mission ended on July 25, 1978, on account of exhaustion of attitude-control system gas. The *Viking 1* orbiter spacecraft also began to run low on attitude-control system gas, but, through careful planning, it was possible to continue collecting scientific data (at a reduced level) for another two years. Finally, with its control gas supply exhausted, the *Viking 1* orbiter's electrical power was commanded to turn off on August 7, 1980.

✧ *Mars Global Surveyor (MGS)* Spacecraft

NASA launched the *Mars Global Surveyor (MGS)* from Cape Canaveral, Florida, on November 7, 1996, using a Delta II expendable launch vehicle. The safe arrival of this robot spacecraft at Mars on September 12, 1997, represented the first successful mission to the Red Planet in two decades. *MGS* was designed as a rapid, low-cost recovery of the Mars Observer (MO) mission objectives. After a year-and-a-half trimming its orbit from a looping ellipse to a circular track around the planet, the spacecraft began its primary mapping mission in March 1999.

Using a high-resolution camera, the *MGS* spacecraft observed the planet from a low-altitude, nearly polar orbit over the course of one complete Martian year, the equivalent of nearly two earth years. Completing its primary mission on January 31, 2001, the spacecraft entered an extended mission phase.

The *MGS* scientific instruments include a high-resolution camera, a thermal emission spectrometer, a laser altimeter, and a magnetometer/electron reflectometer. With these instruments, the spacecraft successfully studied the entire Martian surface, atmosphere, and interior, returning an enormous amount of valuable scientific data in the process. Among the key scientific findings of this mission so far are high resolution images of gullies and debris flow features, which suggest there may be current sources of liquid water, similar to an aquifer, at or near the surface of the planet.

Magnetometer readings indicate that the Martian magnetic field is not generated globally in the planet's core, but appears to be localized in particular areas of the crust. Data from the spacecraft's laser altimeter have provided the first three-dimensional views of the northern ice cap on Mars. Finally, new temperature data and close-up images of the Martian moon Phobos suggest that its surface consists of a powdery material at

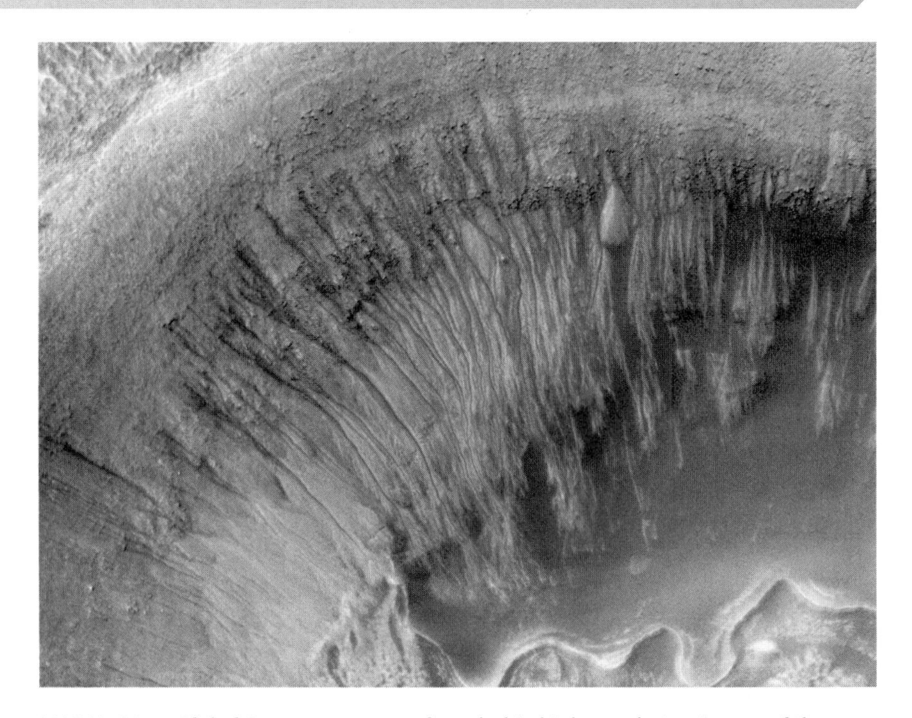

NASA's *Mars Global Surveyor* spacecraft took this high-resolution image of the north wall of a smaller crater located in the southwestern quarter of Newton Crater—a major surface feature on Mars. Scientists hypothesize that Newton Crater, a large basin about 178 miles (287 km) across, was probably formed by an asteroid impact more than three billion years ago. The small crater's north wall has many narrow gullies eroded into it. To some scientists, the presence of these gullies suggests that water and debris once flowed in ancient times on the surface of Mars. *(NASA/JPL/Malin Space Science Systems)*

least three feet (1 m) thick—most likely the result of millions of years of meteoroid impacts.

✧ *Mars Climate Orbiter (MCO)*— Lost in Space Due to Human Error

Originally called part of the Mars Surveyor '98 mission, NASA launched the *Mars Climate Orbiter (MCO)* on December 11, 1998, from Cape Canaveral Air Force Station, Florida, using a Delta II expendable launch vehicle.

The mission of this orbiter spacecraft was to circle Mars and serve as both an interplanetary weather satellite and a communications satellite, relaying data back to Earth from the other part of the Mars Surveyor '98 mission—a lander called the *Mars Polar Lander (MPL),* which also became lost in space. The *MCO* carried two scientific instruments: an atmospheric

MARS OBSERVER (MO) MISSION

NASA's Mars Observer (MO), the first of the Observer series of planetary missions, was designed to study the geoscience of Mars. The primary science objectives for the mission were to: (1) determine the global elemental and mineralogical character of the surface; (2) define globally the topography and gravitational field of the planet; (3) establish the nature of the Martian magnetic field; (4) determine the temporal and spatial distribution, abundance, sources, and sinks of volatiles (substances that readily evaporate) and dust over a seasonal cycle; and (5) explore the structure and circulation of the Martian atmosphere. The 2,240-pound (1,018-kg) robot spacecraft was launched successfully from Cape Canaveral on September 25, 1992.

Unfortunately, for unknown reasons, contact with the *Mars Observer* was lost on August 22, 1993, just three days before scheduled orbit insertion around Mars. Contact with the spacecraft was not reestablished, and it is not known whether this spacecraft was able to follow its automatic programming and go into Mars orbit or if it flew by Mars and is now in a heliocentric orbit. Although none of the primary objectives of the mission were achieved, cruise mode (that is, interplanetary space) data were collected up to the moment of loss of contact. What happened to the *Mars Observer* remains one of the great mysteries of space exploration.

NASA's *Mars Climate Orbiter (MCO)* spacecraft is shown here undergoing acoustic tests prior to delivery to Cape Canaveral and launch. Despite a successful launch on December 11, 1998, and an otherwise uneventful interplanetary journey to the Red Planet, all contact with this space exploration robot was suddenly lost on September 23, 1999, when the spacecraft arrived at Mars. NASA engineers concluded that because of human error in programming the spacecraft's final trajectory, the doomed spacecraft attempted to enter planetary orbit too deep in the Martian atmosphere and consequently burned up. *(NASA/JPL)*

sounder and a color imager. Unfortunately, when the *Mars Climate Orbiter* arrived at the Red Planet, all contact with the spacecraft was lost on September 23, 1999. NASA managers and engineers conducted a post-flight investigation and now believe that the robot spacecraft burned up in the Martian atmosphere due to a fatal error in its arrival trajectory. This human-induced computational error caused the spacecraft to enter too deep into the planet's atmosphere and to encounter destructive aerodynamic heating.

✧ *Mars Odyssey 2001* Spacecraft

Undaunted by the disappointing sequential failures of the *Mars Climate Orbiter (MCO)* and *Mars Polar Lander (MPL)*, NASA officials sent the *Mars*

This artist's concept shows NASA's *Mars Odyssey 2001* spacecraft starting its multi-year scientific mission around the Red Planet in January 2002. Launched from Cape Canaveral on April 7, 2001, this robot orbiter spacecraft examined the surface distribution of minerals on Mars, especially those minerals that can form only in the presence of water. The spacecraft also measured the Martian radiation environment to determine the potential hazard to future human explorers. *(NASA/JPL)*

Odyssey 2001 spacecraft to the Red Planet on April 7, 2001. The scientific instruments onboard the orbiter spacecraft are designed to determine the composition of the planet's surface, to detect water and shallow buried ice, and to study the ionizing radiation environment in the vicinity of Mars.

The spacecraft arrived at the planet on October 24, 2001, and successfully entered orbit around it. After executing a series of aerobrake maneuvers that properly trimmed it into a near-circular polar orbit around Mars, the spacecraft began to make scientific measurements in January 2002. This space robot has examined the surface distribution of minerals on Mars, especially those minerals that can form only in the presence of water. The spacecraft also measured the Martian radiation environment to determine the potential hazard to future human explorers.

The orbiting spacecraft collected scientific data until the end of its primary scientific mission (late summer 2004). After its primary mission, the spacecraft has functioned as a communications relay, supporting information transfer from the *Mars Exploration Rover (MER)* spacecraft back to scientists on Earth.

✧ *Cassini* Spacecraft

The Cassini mission was successfully launched by a mighty Titan IV–Centaur vehicle on October 15, 1997, from Cape Canaveral Air Force Station, Florida. It remains a joint NASA and European Space Agency (ESA) project to conduct detailed exploration of Saturn, its major moon Titan, and its complex system of other moons. Following the example of the *Galileo* spacecraft, the *Cassini* spacecraft also took a gravity-assisted tour of the solar system. The spacecraft eventually reached Saturn following a Venus-Venus-Earth-Jupiter gravity assist (VVEJGA) trajectory. After a nearly seven-year journey through interplanetary space, covering 2.2 billion miles (3.5 billion km), the *Cassini* spacecraft arrived at Saturn on July 1, 2004.

The very large and complex robot spacecraft was named in honor of the Italian-born French astronomer Giovanni Domenico Cassini (1625–1712), who was the first director of the Royal Observatory in Paris and conducted extensive observations of Saturn.

The most critical phase of the mission after launch was Saturn orbit insertion (SOI). When *Cassini* arrived at Saturn, the sophisticated robot spacecraft fired its main engine for 96 minutes to reduce its speed and allow it to be captured as a satellite of Saturn. Passing through a gap between Saturn's F and G rings, the intrepid spacecraft successfully swung close to the planet and began the first of some six-dozen orbits that it will complete during its four-year primary mission.

This is an artist's concept of the *Cassini* spacecraft during the critical Saturn orbit insertion (SOI) maneuver, just after the main engines began firing on July 1, 2004. The SOI maneuver reduced the robot spacecraft's speed, allowing *Cassini* to be captured by Saturn's gravity and to enter orbit. Following the successful SOI maneuver, *Cassini* began a planned four-year exploration mission of Saturn, its mysterious moons, stunning rings, and complex magnetic environment. On December 25, 2004, *Cassini* released its hitchhiking companion, the *Huygens* probe—sending the robot on a historic one-way journey into the atmosphere of Saturn's largest moon, Titan. *(NASA/JPL)*

The arrival period provided a unique opportunity to observe Saturn's rings and the planet itself, since this was the closest approach the spacecraft will make to Saturn during the entire mission. As anticipated, the *Cassini* spacecraft went right to work upon arrival and provided scientific results.

Scientists examining Saturn's contorted F ring, which has baffled them since its discovery, have found one small body, possibly two, orbiting in the F ring region, and a ring of material associated with Saturn's moon, Atlas. *Cassini*'s close-up look at Saturn's rings revealed a small object moving near the outside edge of the F ring, interior to the orbit of Saturn's moon Pandora. This tiny object, which is about 3.1 miles (5 km) in diameter, has been provisionally assigned the name S/2004 S3. It may be a tiny moon that orbits Saturn at a distance of 87,600 miles (141,000 km) from Saturn's center. This object is located about 620 miles (1,000 km) from Saturn's F ring. A second object, provisionally called S/2004 S4, has also been observed in the initial imagery provided by the *Cassini* spacecraft. About the same size as S/2004 S3, this object appears to exhibit some strange dynamics, which take it across the F ring.

In the process of examining the F ring region, scientists also detected a previously unknown ring, now called S/2004 1R. This new ring is associated with Saturn's moon, Atlas. The ring is located 85,770 miles (138,000 km) from the center of Saturn in the orbit of the moon Atlas, between the A ring and the F ring. Scientists estimate the ring has a width of 185 miles (300 km).

Upon arrival at Saturn and the successful orbit insert burn (July 2004), the *Cassini* spacecraft began its extended tour of the Saturn system. This orbital tour involves at least 76 orbits around Saturn, including 52 close encounters with seven of Saturn's known moons. The *Cassini* spacecraft's orbits around Saturn are being shaped by gravity-assist flybys of Titan. Close flybys of Titan also permit high-resolution mapping of the intriguing, cloud-shrouded moon's surface. The *Cassini* orbiter spacecraft carries an instrument called the Titan imaging radar, which can see through the opaque haze covering that moon to produce vivid topographic maps of the surface.

The size of these orbits, their orientation relative to Saturn and the Sun, and their inclination to Saturn's equator, are dictated by various scientific requirements. These scientific requirements include: imaging radar coverage of Titan's surface; flybys of selected icy moons, Saturn, or Titan; occultations of Saturn's rings; and crossings of the ring plane.

The *Cassini* orbiter will make at least six close, targeted flybys of selected icy moons of greatest scientific interest—namely, Iapetus, Enceladus, Dione, and Rhea. Images taken with *Cassini*'s high-resolution telescopic cameras during these flybys will show surface features equivalent in spatial resolution to the size of a professional baseball diamond. At least two dozen more distant flybys (at altitudes of up to 62,000 miles [100,000 km]) will also be made of the major moons of Saturn—other than Titan. The varying inclination of the *Cassini* spacecraft's orbits around Saturn will allow the spacecraft to conduct studies of the planet's polar regions, as well as its equatorial zone.

In addition to the *Huygens* probe (discussed in the next section), Titan will be the subject of close scientific investigations by the *Cassini* orbiter. The spacecraft will execute 45 targeted, close flybys of Titan, Saturn's largest moon—some flybys as close as 590 miles (950 km) above the surface. Titan is the only Saturn moon large enough to enable significant gravity-assisted changes in *Cassini*'s orbit. Accurate navigation and targeting of the point at which the *Cassini* orbiter flies by Titan will be used to shape the orbital tour. This mission-planning approach is similar to the way in which the *Galileo* spacecraft used its encounters of Jupiter's large moons (the Galilean satellites) to shape its very successful scientific tour of the Jovian system.

As currently planned, the prime mission tour of the *Cassini* spacecraft will end on June 30, 2008. This date is four years after arrival at Saturn and 33 days after the last Titan flyby, which will occur on May 28, 2008. The aim point of the final flyby is being chosen (in advance) to position *Cassini* for an additional Titan flyby on July 31, 2008—providing mission controllers with the opportunity to proceed with more flybys during an extended mission, if resources (such as the supply of attitude-control propellant) allow. Nothing in the present design of the orbital tour of the Saturn system now precludes an extended mission.

The *Cassini* spacecraft, which originally included the orbiter and the *Huygens* probe, is the largest and most complex interplanetary spacecraft ever built. The orbiter spacecraft alone has a dry mass of 4,675 pounds (2,125 kg). When the 704-pound (320-kg) *Huygens* probe and a launch vehicle adapter were attached and 6,890 pounds (3,130 kg) of attitude-control and maneuvering propellants loaded, the assembled spacecraft acquired a total launch mass of 12,570 pounds (5,712 kg). At launch, the fully assembled *Cassini* spacecraft stood 22 feet (6.7 m) high and 12.9 feet (4 m) wide.

The Cassini mission involves a total of 18 science instruments, six of which are contained in the wok-shaped *Huygens* probe. This ESA-sponsored probe was detached from the *Cassini* orbiter spacecraft on December 25, 2004, and successfully conducted its own scientific investigations as it plunged into the atmosphere of Titan on January 14, 2005. The probe's science instruments included: the aerosol collector pyrolyzer, descent imager and spectral radiometer, Doppler wind experiment, gas chromatograph and mass spectrometer, atmospheric structure instrument, and surface science package.

The *Cassini* spacecraft's science instruments include a composite infrared spectrometer, imaging system, ultraviolet imaging spectrograph, visual and infrared mapping spectrometer, imaging radar, radio science, plasma spectrometer, cosmic dust analyzer, ion and neutral mass spectrometer, magnetometer, magnetospheric imaging instrument, and radio and plasma wave science. Telemetry from the spacecraft's communications antenna is also being used to make observations of the atmospheres of Titan and Saturn and to measure the gravity fields of the planet and its satellites.

Electricity to operate the *Cassini* spacecraft's instruments and computers is being provided by three long-lived radioisotope thermoelectric generators (RTGs). RTG power systems are lightweight, compact, and highly reliable. With no moving parts, an RTG provides the spacecraft with electric power by directly converting the heat (thermal energy) released by the natural decay of a radioisotope (here, plutonium-238, which decays by alpha-particle emission) into electricity through solid-state thermoelectric conversion devices. At launch (on October 15, 1997), *Cassini*'s three RTGs were providing a total of 885 watts of electrical power from 13,200 watts of nuclear-decay heat. By the end of the currently planned primary tour mission (June 30, 2008), the spacecraft's electrical power level will be approximately 633 watts. This power level is more than sufficient to support an extended exploration mission within the Saturn system, should other spacecraft conditions and resources permit.

The Cassini mission (including *Huygens* probe and orbiter spacecraft) is designed to perform a detailed scientific study of Saturn, its rings, its magnetosphere, its icy satellites, and its major moon Titan. The *Cassini* orbiter's scientific investigation of the planet Saturn includes cloud properties and atmospheric composition, winds and temperatures, internal structure and rotation, the characteristics of the ionosphere, and the origin and evolution of the planet. Scientific investigation of the Saturn ring system includes structure and composition, dynamic processes within the rings, the interrelation of rings and satellites, and the dust and micrometeoroid environment.

Saturn's magnetosphere involves the enormous magnetic bubble surrounding the planet that is generated by its internal magnet. The magnetosphere also consists of the electrically charged and neutral particles within this magnetic bubble. Scientific investigation of Saturn's magnetosphere includes its current configuration; particle composition, sources and sinks; dynamic processes; its interaction with the solar wind, satellites, and rings; and Titan's interaction with both the magnetosphere and the solar wind.

During the orbit tour phase of the mission (from July 1, 2004, to June 30, 2008), the *Cassini* orbiter spacecraft will perform many flyby encounters of all the known icy moons of Saturn. As a result of these numerous satellite flybys, the spacecraft's instruments will investigate the characteristics and geologic histories of the icy satellites, the mechanisms for surface modification, surface composition and distribution, bulk composition and internal structure, and interaction of the satellites with Saturn's magnetosphere.

The moons of Saturn are diverse, ranging from the planet-like Titan to tiny, irregular objects only tens of kilometers in diameter. Scientists currently believe that all of these bodies (except for perhaps Phoebe) hold not only water ice, but also other chemical components, such as methane, ammonia, and carbon dioxide. Before the advent of robotic spacecraft in space exploration, scientists believed the moons of the outer planets were relatively uninteresting and geologically dead. They assumed that (planetary) heat sources were not sufficient to have melted the mantles of these moons enough to provide a source of liquid, or even semi-liquid, ice or silicate slurries.

The *Voyager* and *Galileo* spacecraft have radically altered this view by revealing a wide range of geological processes on the moons of the outer planets. For example, Saturn's moon Enceladus may be feeding material into the planet's F ring—a circumstance that suggests the existence of current activity, such as geysers or volcanoes. Several of Saturn's medium-sized moons are large enough to have undergone internal melting with subsequent differentiation and resurfacing. The *Cassini* spacecraft is greatly increasing knowledge about Saturn's icy moons.

Finally, the *Cassini* mission (both orbiter and probe) involves a detailed investigation of the largest of Saturn's moons, Titan. This intriguing, planet-sized moon is the only one in the solar system with a dense atmosphere. Titan's hazy, opaque atmosphere prevents Earth-based astronomers from seeing its surface, however. One of the major objectives of the Cassini mission is to penetrate this natural veil of secrecy. Scientific objectives include a study of Titan's atmospheric composition; the distribution of trace gases and aerosols; winds and temperatures; the state (liquid or solid) and composition of the surface; and the conditions in the upper atmosphere of Titan.

An international team of planetary scientists is using data from the *Huygens* probe and remote sensing instruments on NASA's *Cassini* orbiter spacecraft to validate a new model of the evolution of Titan. Combinations of these data suggest that Titan's methane supply may be locked away in a kind of methane-rich ice, called clathrate hydrate, which forms a crust above a suspected subsurface ocean of liquid water mixed with ammonia that lies a few tens of miles (kilometers) below the moon's surface. Scientists now hypothesize that parts of the clathrate crust might be warned from time to time by cryovolcanic activity on Titan, causing the moon's crust to release some of its trapped methane into the atmosphere. (The phenomenon of cryovolcanism involves ice melting and ice degassing.) These outbursts might also produce temporary flows of liquid methane on the surface, accounting for the river-like features that the mission has detected on Titan's surface.

For example, as the *Huygens* probe descended below an altitude of 25 miles (40 km) on January 14, 2005, its onboard instruments obtained clear images of Titan's surface. These images revealed that this extraordinary world resembles primitive Earth in many respects—especially in meteorology, geomorphology, and fluvial activity, but with different ingredients. The *Huygens* images also provided strong evidence for erosion due to liquid flows, possibly methane.

Two-way communication with the *Cassini* spacecraft takes place through the large dish antennae of NASA's Deep Space Network (DSN). The spacecraft transmits and receives signals in the microwave X-band, using its own parabolic high-gain antenna. The orbiter spacecraft's high-gain antenna is also used for radio and radar experiments, and for receiving signals from the *Huygens* probe as it plunges into Titan's atmosphere.

Because of the enormous distances involved (on the average, Saturn is 890 million miles [1.43 billion km] away from Earth), real-time control of the *Cassini* spacecraft is not feasible. For example, when the spacecraft arrived at Saturn on July 1, 2004, the one-way speed-of-light time from Saturn to Earth was 84 minutes. During the four-year orbital tour mission, the one-way speed-of-light time from Saturn to Earth (and vice versa) will range between 67 and 85 minutes, depending on the relative position of the two planets in their journeys around the Sun. To overcome this problem, aerospace engineers have included a great deal of machine intelligence (using advanced computer hardware and software) to enable the sophisticated robot spacecraft to function with minimal human supervision.

Each of the *Cassini* spacecraft's science instruments is run by a microprocessor capable of controlling the instrument and formatting/packaging (packetizing) data. Ground controllers run the spacecraft at a distance by using a combination of centralized commands to control system-level resources and commands issued by the microprocessors of the individual science instruments. Packets of data are collected from each instrument on a schedule that can vary at different times in the orbit-tour phase of the mission. These data packets may be transmitted immediately to Earth, or else stored within *Cassini*'s onboard solid-state recorders for transmission at a later time.

Because the *Cassini* spacecraft's scientific instruments are fixed, the entire spacecraft must be turned to point them. So, the spacecraft is frequently reoriented, either through the use of its reaction wheels or by firing its set of small onboard thrusters. Most of the science observations are being made without a real-time communications link to Earth. The science instruments have different pointing requirements, however. These different requirements often conflict with each other and with the need to point the spacecraft toward Earth to transmit data home. Reprogrammable onboard software with embedded hierarchies that determine "who and what goes first" guides the onboard microprocessors and the spacecraft's main computer/clock subsystem, as they resolve scheduling conflicts. Mission designers have also carefully built into the design of the orbiter tour a sufficient number of periods during which the spacecraft's high-gain antenna points toward Earth.

✧ *Huygens* Spacecraft

The *Huygens* probe was carried to the Saturn system by the *Cassini* orbiter spacecraft. Bolted to the *Cassini* mother spacecraft and fed electrical power through an umbilical cable, *Huygens* rode along during the nearly seven-year journey largely in a sleep mode. Mission controllers did awaken the robot probe about every six months, however, for three-hour duration instrument and engineering checkups. *Huygens* was sponsored by the

Aerospace technicians examine the *Huygens* probe in the Payload Hazardous Servicing Facility (PHSF) at the Kennedy Space Center in 1997. The robot probe was launched on October 15, 1997, as part of the *Cassini/Huygens* spacecraft sent to Saturn. After a seven-year journey attached to the side of its *Cassini* mother spacecraft, *Huygens* was released toward Titan on December 25, 2004. The probe coasted for 20 days before plunging into the hazy atmosphere of Titan on January 14, 2005. *Huygens* sampled Titan's atmosphere from an altitude of 100 miles (160 km) all the way to the ground. The probe landed safely on the moon's frozen surface and continued to transmit data up to *Cassini* for about 70 minutes. Instruments on the probe indicated that Titan's surface resembles wet sand or clay with a thin solid crust and consists mostly of dirty water ice and hydrocarbon ice. *(NASA/Kennedy Space Center)*

European Space Agency and named after the Dutch physicist and astronomer Christiaan Huygens (1629–95), who first described the nature of Saturn's rings and discovered its major moon, Titan, in 1655.

The *Cassini* spacecraft's second flyby of Titan on December 13, 2004, left the spacecraft (which was still carrying the hitchhiking *Huygens* probe) on a trajectory that, if uncorrected, would lead to a subsequent flyby of Titan at an altitude of about 2,860 miles (4,600 km). To get the *Huygens* probe traveling into Titan's atmosphere at just the right angle, the *Cassini* mother spacecraft performed a targeting maneuver before it released its hitchhiking robot companion. On December 17, the *Cassini* spacecraft completed a precise targeting maneuver that shaped its course and pointed the cojoined robot spacecraft team on a direct impact trajectory to Titan.

On December 25, 2005 (at 02:00 universal time coordinated), the spin/eject device separated *Huygens* from *Cassini* at a relative speed of 1.1 feet per second (0.35 m/s) and a spin rate of 7.5 revolutions per minute. As a result of these successful maneuvers and actions, the spin-stabilized atmospheric probe was targeted for a southern-latitude landing site on the dayside of Titan. To support a variety of mission needs and parameters, the probe entry angle into Titan's atmosphere was set at a relatively steep 65 degrees. ESA mission controllers selected this entry angle to give the probe the best opportunity to reach the surface of Titan. Following probe separation, the *Cassini* orbiter performed some final maneuvers to avoid crashing into Titan and to position itself to collect data from *Huygens* as it descended into Titan's opaque, nitrogen-rich atmosphere.

On January 14, 2005, after coasting for 20 days, the *Huygens* probe reached the desired entry altitude of 790 miles (1,270 km) above Titan and started its parachute-assisted descent into the moon's atmosphere. Within five minutes, *Huygens* began transmitting its scientific data to the *Cassini* orbiter as the probe floated down through Titan's atmosphere.

During the first part of the *Huygens*'s atmospheric plunge, instruments on board the probe were controlled by a timer. For the final six to 12 miles (10 to 20 km) of descent, a radar altimeter on board the probe controlled its scientific instruments on the basis of altitude. About 138 minutes after starting its plunge into Titan's upper atmosphere, the *Huygens* probe came to rest on the moon's surface. Images of the site were collected just before landing. The probe survived impact on a squishy surface that was neither liquid nor frozen solid—the two candidate surface conditions most frequently postulated by planetary scientists for this cloud-enshrouded moon. As the *Cassini* orbiter disappeared over the horizon, the mother spacecraft stopped collecting data from its hardworking robot companion. The probe had been continuously transmitting data for about four and one-half hours.

✧ Pioneer Venus Mission

The Pioneer Venus mission consisted of two separate spacecraft launched by NASA to the planet Venus in 1978. The *Pioneer Venus Orbiter* spacecraft (also called *Pioneer 12*) was a 1,217-pound (553-kg) spacecraft that contained a 99-pound (45-kg) payload of scientific instruments. *Pioneer 12* was launched on May 20, 1978, and placed into a highly eccentric orbit around Venus on December 4, 1978. For 14 years (1978–92), the *Pioneer Venus Orbiter* spacecraft gathered a wealth of scientific data about the atmosphere and ionosphere of Venus and their interactions with the solar wind as well as details about the planet's surface. Then, in October 1992, NASA mission controllers sent this spacecraft on an intended final entry into Venus's atmosphere. The spacecraft gathered data up to its final fiery plunge, which dramatically ended the operations portion of the Pioneer Venus mission.

The *Pioneer Venus Multiprobe* spacecraft (also called *Pioneer 13*) consisted of a basic bus spacecraft, a large probe, and three identical smaller probes. The *Pioneer Venus Multiprobe* spacecraft was launched on August 8, 1978, and separated about three weeks before entry into the Venusian atmosphere. The four (now-separated) probes and their (spacecraft) bus successfully entered the Venusian atmosphere at widely dispersed locations on December 9, 1978, and returned important scientific data as they plunged toward the planet's surface. Although the probes were not designed to survive landing, one hardy probe did and transmitted data for about an hour after impact.

Collectively, *Pioneer Venus Orbiter* and *Multiprobe* spacecraft provided important scientific data about Venus, its surface, atmosphere, and interaction with the solar wind. For example, the orbiter spacecraft made an extensive radar map, covering about 90 percent of Venus's surface. Using its radar to peer through the planet's dense, opaque clouds, this spacecraft revealed that the surface of Venus is mostly gentle, rolling plains with two prominent plateaus: Ishtar Terra and Aphrodite Terra. This highly successful, two-spacecraft mission also provided important groundwork for NASA's subsequent *Magellan* mission to Venus.

✧ *Ulysses* Spacecraft

The Ulysses mission is an international space robot designed to study the poles of the Sun and the interstellar environment above and below these solar poles. The *Ulysses* spacecraft is named for the legendary Greek hero in Homer's epic saga of the Trojan War who wandered into many previously unexplored areas on his return home. The spacecraft's mission is a survey mission designed to examine the properties of the solar wind; the

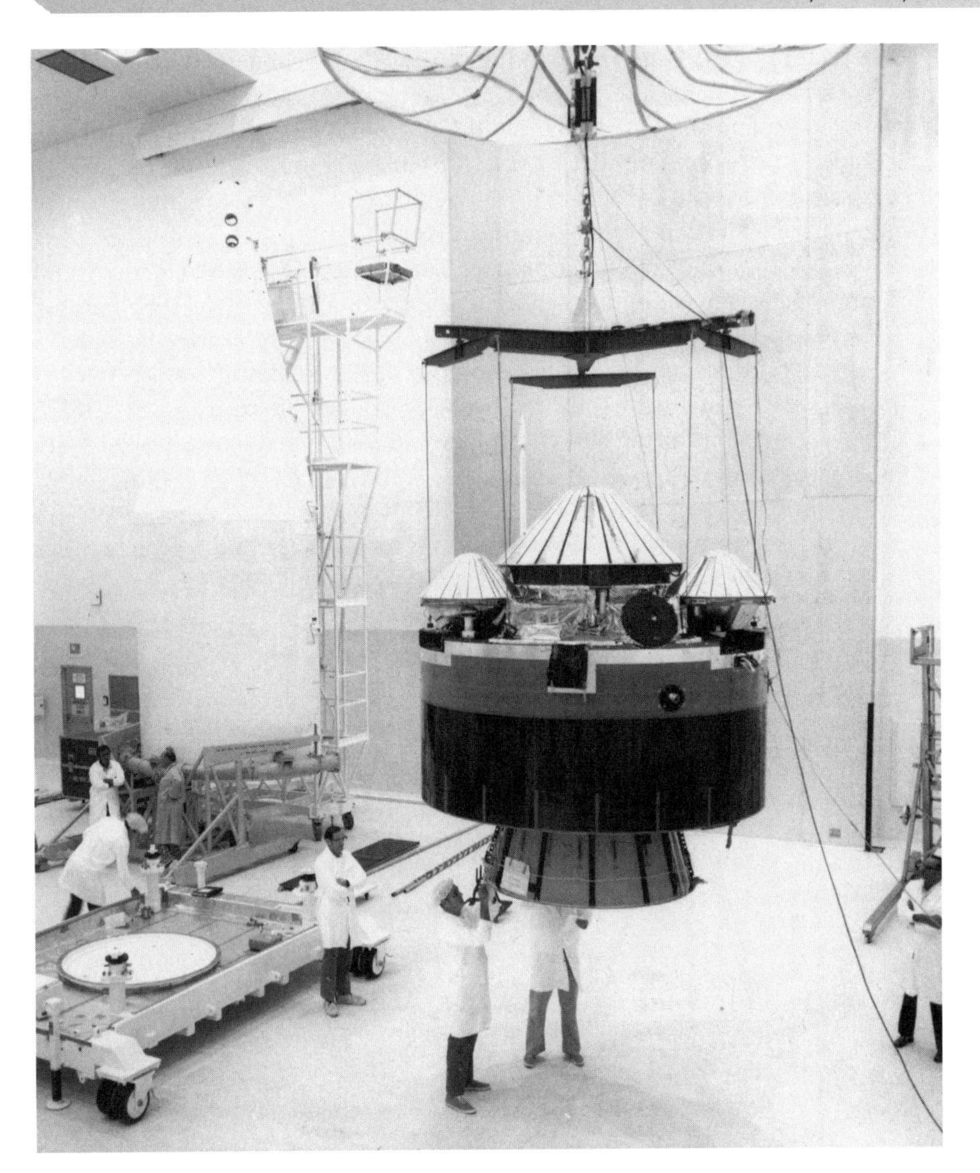

NASA's *Pioneer Venus Multiprobe* spacecraft is lifted for final inspection prior to encapsulation and launch in 1978 at Cape Canaveral, Florida. Also known as *Pioneer 13*, the multiprobe-carrying spacecraft was launched on August 8, 1978, traveled through interplanetary space, and then separated about three weeks before entry into the Venusian atmosphere. On December 9, 1978, the four (now-separated) robot probes successfully entered the atmosphere of the cloud-shrouded planet at widely dispersed locations. As each probe plunged through the dense atmosphere, it returned important scientific data. Although not designed to survive landing, one hardy probe did and transmitted data for about an hour after impact. Data from each probe were collected by a companion spacecraft, called the *Pioneer Venus Orbiter (Pioneer 12)*, and relayed back to Earth. *(NASA/Kennedy Space Center)*

structure of the Sun–solar wind interface; the heliospheric magnetic field, solar radio bursts, and plasma waves; solar and galactic cosmic rays; and the interplanetary/interstellar neutral gas and dust environment—all as a function of solar latitude. Dornier Systems of Germany built the *Ulysses* spacecraft for the European Space Agency (ESA), which is responsible for in-space operation of the scientific mission.

NASA provided launch support using the space shuttle *Discovery* and an upper-stage configuration. In addition, the United States, through the Department of Energy, provided the radioisotope thermoelectric generator (RTG) that supplies electric power to this spacecraft. *Ulysses* is tracked and its scientific data collected by NASA's Deep Space Network (DSN). Spacecraft monitoring and control, as well as data reduction and analysis, is performed at NASA's Jet Propulsion Laboratory (JPL) by a joint ESA/JPL team.

Ulysses is the first spacecraft to travel out of the ecliptic plane in order to study the unexplored region of space above the Sun's poles. To reach

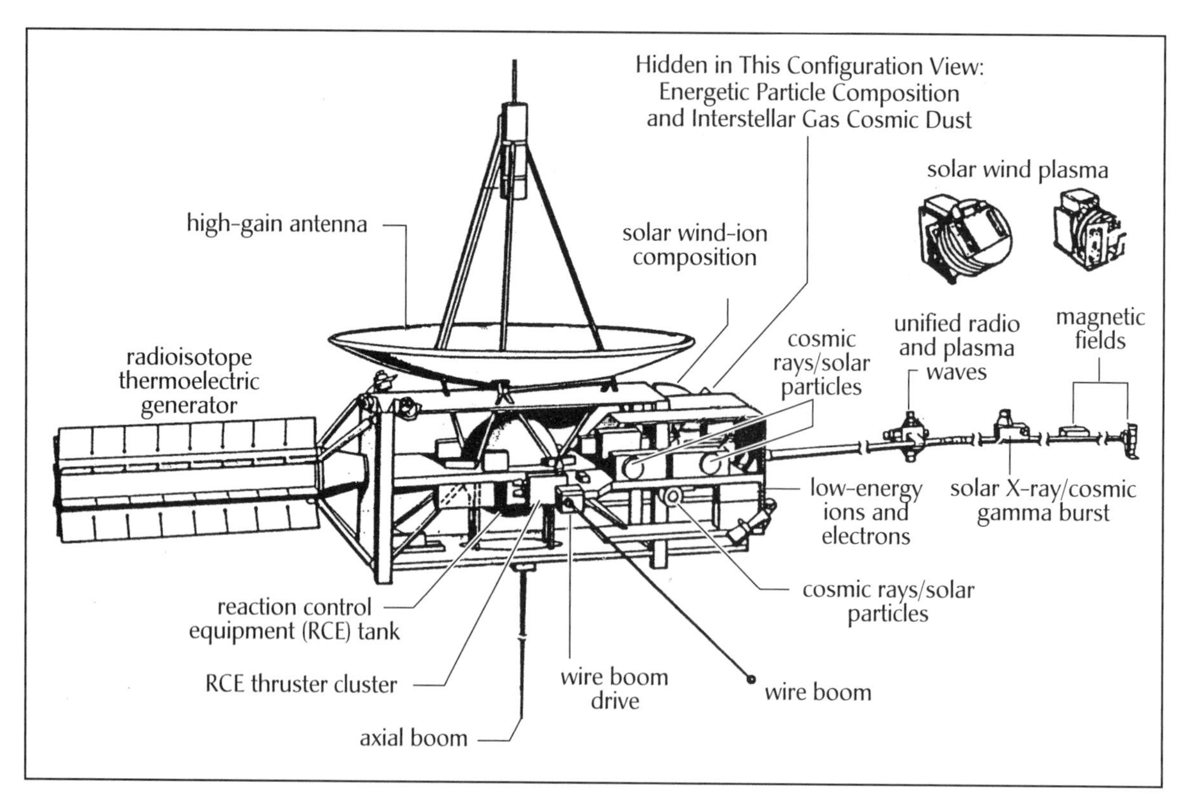

The compact *Ulysses* spacecraft and its array of scientific instruments *(NASA)*

the necessary high solar latitudes, *Ulysses* was initially aimed close to Jupiter so that the giant planet's large gravitational field would accelerate the spacecraft out of the ecliptic plane to high latitudes. The gravitational-assist encounter with Jupiter occurred on February 8, 1992. After the Jupiter encounter, *Ulysses* traveled to higher latitudes, with the maximum southern latitude of 80.2° being achieved on September 13, 1994 (South Polar Pass 1).

Because *Ulysses* was the first spacecraft to explore the third dimension of space over the poles of the Sun, space scientists experienced some surprising discoveries. For example, they learned that two clearly separate and distinct solar wind regimes exist, with fast wind emerging from the solar poles. Scientists also were surprised to observe how cosmic rays make their way into the solar system from galaxies beyond the Milky Way Galaxy. The magnetic field of the Sun over its poles turns out to be very different from what was expected, based on observations from Earth. Finally, *Ulysses* detected a beam of particles from interstellar space that was penetrating the solar system at a velocity of about 49,720 miles per hour (80,000 km/hr), or about 13.80 miles per second (22.22 km/s).

Ulysses then traveled through high northern latitudes from June through September 1995 (North Polar Pass 1). The spacecraft's high-latitude observations of the Sun occurred during the minimum portion of the 11-year solar cycle.

In order to fully understand the Sun, however, scientists also wanted to study our parent star at conditions of near-maximum activity during an 11-year cycle. The extended mission of the far-traveling, nuclear-powered scientific spacecraft provided the opportunity. During solar maximum conditions, *Ulysses* achieved high southern latitudes between September 2000 and January 2001 (South Polar Pass 2) and then traveled through high northern latitudes between September 2001 and December 2001 (North Polar Pass 2).

Now well into its extended mission, *Ulysses* continues to send back valuable scientific information on the inner workings of our parent star, especially concerning its magnetic field and how that magnetic field influences the solar system.

This mission was originally called the International Solar Polar Mission (ISPM). The original mission planned for two spacecraft, one built by NASA and the other by ESA. NASA canceled the spacecraft part of its original mission in 1981, however, and instead provided launch and tracking support for the single spacecraft built by ESA.

Lander and Rover Spacecraft

A lander is a robot spacecraft designed to safely reach the surface of a planet and to survive there long enough to send some useful scientific data back to Earth. Landers are generally fixed-in-place spacecraft, meaning once the robot touches down on a planetary body, it generally does not move from its original landing spot. To support investigation of the local environment, the lander may carry one or several robotic arms and perhaps a set of automated drilling equipment.

On some space-exploration missions to a planet's surface, the lander serves primarily as a surface-based mother spacecraft. It safely delivers one or several mini-rovers to the surface and then provides telecommunications services for the deployed mobile-robot system(s). Data from each mini-rover gets transmitted back to Earth through the lander spacecraft. Mission managers on Earth, in turn, use the lander's communications subsystem to provide commands and instructions to the mini-rover(s).

Robot rovers can assume a range of sizes from mini- to hefty (about the size of a small automobile). They can also display varying levels of machine intelligence, ranging from being totally dependent upon human supervision, to being semiautonomous, to being fully autonomous. This chapter describes some of the lander and rover spacecraft that have been successfully operated over the past four decades of solar system exploration. Chapters 7, 8, 9, and 10 provide additional glimpses and snapshots of some very exciting landers and rovers, which could be used in the next four decades of solar system exploration.

✧ Surveyor Project

NASA's highly successful Surveyor Project began in 1960. It consisted of seven robot lander spacecraft that were launched between May 1966 and

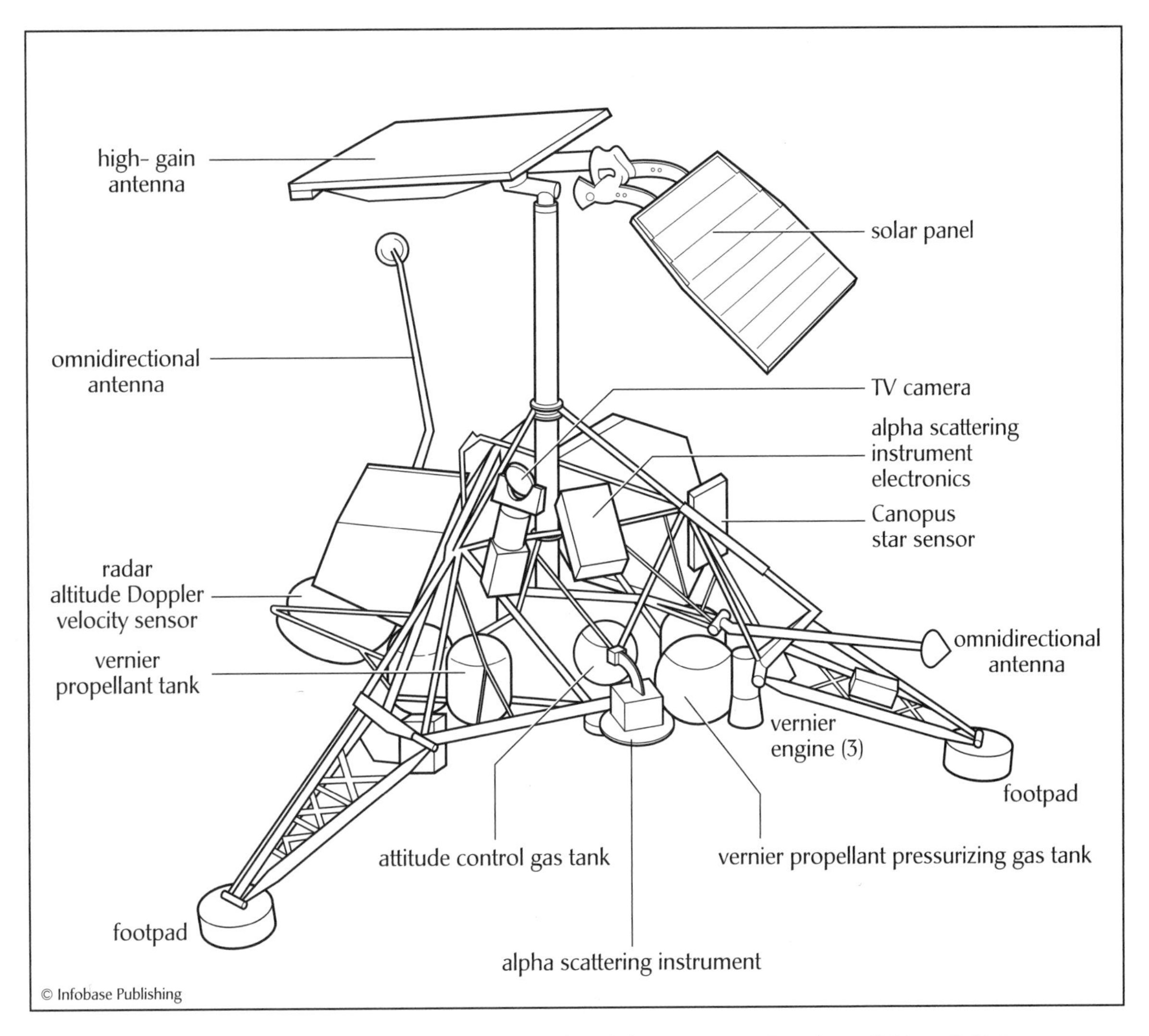

NASA's *Surveyor* spacecraft was a robot lander that explored the Moon's surface from 1966 to 1968, in preparation for the lunar landing missions of the Apollo astronauts (1969–72). *(NASA) / (Note: Drawing does not show main retro rocket.)*

January 1968, as a precursor to the human expeditions to the lunar surface in the Apollo Project. These robot lander craft were used to develop soft-landing techniques, to survey potential Apollo mission landing sites, and to improve scientific understanding of the Moon.

The *Surveyor 1* spacecraft was launched on May 30, 1966, and soft-landed in the Ocean of Storms region of the Moon. It found the bearing strength of the lunar soil was more than adequate to support the Apollo Project lander spacecraft (called the lunar module, or LM). This contradicted the then-prevalent hypothesis that the LM might sink out of sight in the fine lunar dust. The *Surveyor 1* spacecraft also telecast many pictures from the lunar surface.

The *Surveyor 3* spacecraft was launched on April 17, 1967, and soft-landed on the side of a small crater in another region of the Ocean of Storms. This robot spacecraft used a shovel attached to a mechanical arm to dig a trench and discovered that the load-bearing strength of the lunar soil increased with depth. It also transmitted many pictures from the lunar surface.

The *Surveyor 5* spacecraft was launched on September 8, 1967, and soft-landed in the Sea of Tranquility. An alpha particle–scattering device on board this craft examined the chemical composition of the lunar soil and revealed a similarity to basalt on Earth.

The *Surveyor 6* was launched on November 7, 1967, and soft-landed in the Sinus Medii (Central Bay) region of the Moon. In addition to performing soil-analysis experiments and taking many images of the lunar surface, this spacecraft also performed an extremely critical "hop experiment." NASA engineers back on Earth remotely fired the *Surveyor*'s vernier rockets to launch it briefly above the lunar surface. The spacecraft's launch did not create a dust cloud and resulted only in shallow cratering. This important demonstration indicated that the Apollo astronauts could safely lift off from the lunar surface with their rocket-propelled craft (upper portion of the lunar module [LM]), when their surface exploration mission was completed.

Finally, the *Surveyor 7* spacecraft was launched on January 7, 1968, and landed in a highland area of the Moon, near the crater Tycho. Its alpha particle-scattering device showed that the lunar highlands contained less iron than the soil found in the mare regions (lunar plains). Numerous images of the lunar surface also were returned.

Despite the fact that the *Surveyor 2* and *4* spacecraft crashed on the Moon (rather than soft-landed and functioned), the overall Surveyor Project was extremely successful and paved the way for the human-crewed Apollo surface expeditions that occurred between 1969 and 1972.

✧ *Lunokhod 1* and *2* Robot Rovers

Lost in the glare of the triumphant human landings on the Moon conducted by the United States between 1969 and 1972 were several highly successful Soviet robot spacecraft missions to the same celestial body.

Luna 16, launched on September 12, 1970, was the first successful automated (robotic) sample-return mission to the lunar surface. After landing on the Sea of Fertility, this robot spacecraft deployed a drill that bored 13.8 inches (35 cm) into the surface. The lunar soil sample, which

This 1971 postage stamp from the former German Democratic Republic (East Germany) depicts the Russian *Lunokhod 1* robot rover departing its lander. During the Russian *Luna 17* mission to the Moon in 1970, the mother spacecraft soft-landed on the lunar surface in the Sea of Rains and deployed the *Lunokhod 1* robot-rover vehicle. Controlled from Earth by radio signals, this eight-wheeled lunar rover vehicle traveled for months across the lunar surface, transmitted more than 20,000 television images of the surface, and performed more than 500 lunar soil tests at various locations. *(Author)*

had a mass of about 0.2 pound (0.1 kg), was transferred automatically to a return vehicle that then left the lunar surface and landed in the former Soviet Union on September 24, 1970. Of course, the U.S. *Apollo 11* and *12* lunar landing missions in July 1969 and November 1969, respectively, also returned relatively large lunar samples, which were collected by astronauts from the Moon's surface. So this interesting robot-sample-return mission was given little international notice at the time.

Luna 17 placed the first mobile robot, called *Lunokhod 1,* on the lunar surface. The lander spacecraft successfully touched down on the Sea of Rains and deployed the sophisticated *Lunokhod 1* rover. This eight-wheel vehicle was radio-controlled from Earth. The rover covered 6.5 miles (10.5 km) during a surface-exploration mission that lasted 10.5 months. The rover's cameras transmitted more than 20,000 images of the Moon's surface, and instruments of the vehicle analyzed properties of the lunar soil at many hundreds of locations.

Luna 20 (launched February 14, 1972) and *Luna 24* (launched August 9, 1976) were also successful robot-soil-sample-return missions. *Luna 21,* launched in January 1973, successfully deployed another robot rover, *Lunokhod 2,* in the Le Monnier crater in the Sea of Tranquility. This 1,848-pound (840-kg) rover vehicle traveled about 23 miles (37 km) during its four-month surface exploration mission. Numerous photographs were taken, and surface experiments conducted, by this robot rover, which was operated under radio control by Russian scientists and technicians on Earth.

✧ *Viking 1* and *2* Lander Spacecraft

The *Viking 1* lander spacecraft accomplished the first soft landing on Mars on July 20, 1976, on the western slope of Chryse Planitia (the Plains of Gold) at 22.46 degrees north latitude, 48.01 degrees west longitude. The *Viking 2* lander touched down successfully on September 3, 1976, at Utopia Planitia (Plains of Utopia) located at 47.96 degrees north latitude, 225.77 degrees west longitude.

NASA's *Viking 1* lander took this image of Mars on August 8, 1978–730 days after landing at Chryse Planitia (the Plains of Gold). Parts of the robot lander are visible in the foreground. The square structure on the left is the top of a landing leg; to its right are the wind and temperature sensor and a brush used for cleaning off the scoop that collected soil samples. On the surface can be seen a field of dust accumulation on the left and a rocky plain, which extends to the horizon a few miles (km) away. Most of the rocks measure around 19.5 inches (50 cm) across; the large rock on the left, about eight feet (2.5 m) wide and 26 feet (8 m) away from the lander spacecraft, was nicknamed "Big Joe." *(NASA/JPL)*

Each 1,258-pound (572-kg) lander carried instruments to achieve the primary objectives of the lander portion of NASA's Viking Project. These objectives were to study the biology, chemical composition (organic and inorganic), meteorology, seismology, magnetic properties, appearance, and physical properties of the Martian surface. Power was provided to the lander spacecraft by two radioisotope thermoelectric generator (RTG) units containing plutonium-238.

The Viking landers carried many instruments carefully selected and designed to help exobiologists answer the intriguing question about the existence of life on Mars. Despite many experiments, tests, and surface activities, which were performed well by both robot landers, the exobiology results ranged from negative to indeterminate—leaving the question wide open for resolution by robot (and possibly human) explorers this century.

The Viking-lander biology instrument, consisted of three separate experiments designed to detect evidence of microbial life in the Martian soil. There was a gas chromatograph/mass spectrometer (GCMS) that searched the Martian soil for complex organic molecules. The robot lander also carried an X-ray fluorescence spectrometer that analyzed samples of the Martian soil to determine its elemental composition. A meteorology boom, as well as holding-temperature, wind-direction, and wind-velocity sensors, extended out and up from the top of one of the lander legs. The robot spacecraft also had a pair of slow-scan cameras that were mounted about three feet (1 m) apart on the top of each lander. These cameras provided black-and-white, color, and stereoscopic photographs of the Martian surface. Finally, scientists also designed a seismometer to record any Mars quakes that might occur on the Red Planet. Such information would help planetary scientists determine the nature of Mars's internal structure. Unfortunately, the seismometer on *Lander 1* did not function after landing and the instrument on *Lander 2* observed no clear signs of internal (tectonic) activity.

Each Viking lander also had a surface sampler boom that employed its collector head to scoop up small quantities of Martian soil to feed the biology, organic-chemistry, and inorganic-chemistry instruments. This articulating robot arm also provided clues to the soil's physical properties. Magnets attached to the sampler, for example, provided information about the soil's iron content.

Even the lander radios were used to conduct interesting scientific experiments. Physicists were able to refine their estimates of Mars's orbit by measuring the time for radio signals to travel between Mars and Earth. The great accuracy of these radio-wave measurements also allowed scientists to confirm portions of Albert Einstein's General Theory of Relativity.

NASA scientists received their last data from the *Viking 2* lander on April 11, 1980. In January 1982, NASA renamed the *Viking 1* lander the *Thomas Mutch Memorial Station.* The *Viking 1* lander made its final transmission on November 11, 1982. After over six months of effort to regain contact with the *Viking 1* lander, the Viking mission came to an end on May 23, 1983.

With the single exception of the seismic instruments, the entire complement of scientific instruments of the Viking Project acquired far more data about Mars than had ever been anticipated. The primary objective of the landers was to determine whether (microbial) life currently exists on Mars. The evidence provided by the landers is still subject to debate, although most scientists feel these results are strongly indicative that life does not now exist on Mars. Recent analyses of Martian meteorites, however, have renewed interest in this very important question, and Mars is once again the target of intense scientific investigation by even more sophisticated scientific spacecraft.

✧ Mars Pathfinder Mission

NASA launched the *Mars Pathfinder* to the Red Planet using a Delta II expendable launch vehicle on December 4, 1996. This mission, previously called the Mars Environmental Survey (or MESUR) Pathfinder, had the primary objective of demonstrating innovative technology for delivering an instrumented lander and free-ranging robotic rover to the Martian surface. The *Mars Pathfinder* not only accomplished this primary mission but also returned an unprecedented amount of data, operating well beyond its anticipated design life.

Mars Pathfinder used an innovative landing method that involved a direct entry into the Martian atmosphere, assisted by a parachute to slow its descent through the planet's atmosphere and then a system of large airbags to cushion the impact of landing. From its airbag-protected bounce-and-roll landing on July 4, 1997, until the final data transmission on September 27, the robotic lander/rover team returned numerous close-up images of Mars and chemical analyses of various rocks and soil found in the vicinity of the landing site.

The landing site was at 19.33 N, 33.55 W, in the Ares Vallis region of Mars, a large outwash plain near Chryse Planitia (the Plains of Gold), where the *Viking 1* lander had successfully touched down on July 20, 1976. Planetary geologists speculate that this region is one of the largest outflow channels on Mars—the result of a huge ancient flood that occurred over a short period of time and flowed into the Martian northern lowlands.

The lander, renamed by NASA the *Carl Sagan Memorial Station,* first transmitted engineering and science data collected during atmospheric

NASA's *Mars Pathfinder* lander and mini-rover on the surface of the Red Planet (July 4, 1997). The view is to the west. A great flood of water washed over this region long ago, passing from left to right across this portion of the landscape. The Twin Peaks on the horizon are about 0.6 mile (1 km) away. In this scene, the robot rover is still crouched on one of the lander petals. Airbag material billows out from beneath the petal. Rolled into tight cylinders at either end of the robot rover are the (as yet) undeployed rover ramps. *(NASA/JPL)*

entry and landing. The American astronomer Carl Edward Sagan (1934–96) popularized astronomy and astrophysics and wrote extensively about the possibility of extraterrestrial life.

Just after arrival on the surface, the lander's imaging system (which was on a pop-up mast) obtained views of the rover and the immediate surroundings. These images were transmitted back to Earth to assist the human flight team in planning the robot rover's operations on the surface of Mars. After some initial maneuvering to clear an airbag out of the way, the lander deployed the ramps for the rover. The 23.3-pound (10.6-kg) mini-rover had been stowed against one of the lander's petals. On a command from Earth, the tiny robot explorer came to life and rolled onto the Martian surface. Following rover deployment, the bulk of the lander's remaining tasks were to support the rover by imaging rover operations and relaying data from the rover back to Earth. Solar cells on the lander's three petals, in combination with rechargeable batteries, powered the lander, which also was equipped with a meteorology station.

The rover, renamed *Sojourner* (after the American civil rights crusader Sojourner Truth), was a six-wheeled vehicle that was teleoperated (that is, driven over great distances by remote control) by personnel at the Jet Propulsion Laboratory. The rover's human controllers used images obtained by both the rover and the lander systems. Teleoperation at interplanetary distances required that the rover be capable of some semiautonomous operation, since the time delay of the signals averaged between 10 and 15 minutes depending up on the relative positions of Earth and Mars.

For example, the rover had a hazard-avoidance system, and surface movement was performed very slowly. The small rover was 11 inches (28 cm) high, 24.8 inches (63 cm) long, and 18.9 inches (48 cm) wide, with a ground clearance of 5 inches (13 cm). While stowed in the lander, the rover had a height of just 7.1 inches (18 cm). After deployment on the Martian surface, however, the rover extended to its full height and rolled down a deployment ramp. The relatively far-traveling little rover received its supply of electrical energy from its 2.2-square-foot (0.2-m^2) array of solar cells. Several nonrechargeable batteries provided backup power.

The rover was equipped with a black-and-white imaging system. This system provided views of the lander, the surrounding Martian terrain, and

Mars Polar Lander (MPL)—Another Martian Mystery

Originally designated as the lander portion of the *Mars Surveyor '98* mission, NASA launched the *Mars Polar Lander (MPL)* robot spacecraft from Cape Canaveral, Florida on January 3, 1999, using a Delta II expendable launch vehicle. *MPL* was an ambitious mission to land a robot spacecraft on the frigid surface of Mars near the edge of the planet's southern polar cap. Two small penetrator probes (called *Deep Space 2*) hitchhiked along with the lander spacecraft on the trip to Mars. After an uneventful interplanetary journey, all contact with the *MPL* and the *Deep Space 2* experiments was lost as the spacecraft arrived at the planet on December 3, 1999. The missing lander was equipped with cameras, a robotic arm, and instruments to measure the composition of the Martian soil. The two tiny penetrators were to be released as the lander spacecraft approached Mars and then were to follow independent ballistic trajectories, making impact on the surface and then plunging below it in search of water ice.

The exact fate of the lander and its two tiny microprobes remains a mystery. Some NASA engineers believe that the *MPL* may have tumbled down into a steep canyon, while others speculate that the *MPL* may have experienced too rough a landing and become disassembled. A third hypothesis suggests the *MPL* may have suffered a fatal failure during its descent through the Martian atmosphere. No firm conclusions could be drawn, because the NASA mission controllers were completely unable to communicate with the missing lander or either of its hitchhiking planetary penetrators.

even the rover's own wheel tracks, which helped scientists estimate soil properties. An alpha particle X-ray spectrometer (APXS) on board the rover was used to assess the composition of Martian rocks and soil.

Both the lander and the rover outlived their design lives—the lander by nearly three times and the rover by 12 times. Data from this very successful lander/rover surface mission suggest that ancient Mars was once warm and wet, stimulating further scientific and popular interest in the intriguing question of whether life could have emerged on the planet when it had liquid water on the surface and a thicker atmosphere.

✧ Mars Exploration Rover (MER) 2003 Mission

In summer 2003, NASA launched identical twin Mars rovers that were to operate on the surface of the Red Planet during 2004. *Spirit* (MER-A) was launched by a Delta II rocket from Cape Canaveral on June 10, 2003, and successfully landed on Mars on January 4, 2004. *Opportunity* (MER-B) was launched from Cape Canaveral on July 7, 2003, by a Delta II rocket and successfully landed on the surface of Mars on January 25, 2004. Both landings resembled the successful airbag bounce-and-roll arrival demonstrated during the *Mars Pathfinder* mission.

This artist's concept shows NASA's *Mars Exploration Rover (MER)* operating on the surface of the Red Planet (circa 2004–05). *(NASA/JPL)*

This interesting mosaic image was taken by the navigation camera on NASA's *Mars Exploration Rover Spirit* on January 4, 2004. NASA scientists have reprocessed the image to project a clear overhead view of the robot rover and its lander mother spacecraft on the surface of Mars. *(NASA/JPL)*

Following arrival on the surface of the Red Planet, each rover drove off and began its surface exploration mission in a decidedly different location on Mars. *Spirit* (MER-A) landed in Gusev Crater, which is roughly 15 degrees south of the Martian equator. NASA mission planners selected Gusev Crater because it had the appearance of a crater lakebed. *Opportunity* (MER-B) landed at Terra Meridiani—a region of Mars that is also known as the Hematite Site because this location displayed evidence of coarse-grained hematite, an iron-rich mineral that typically forms in water. Among this mission's principal scientific goals is the search for and characterization of a wide range of rocks and soils that hold clues to past water activity on Mars. At the end of July 2006, both rovers were continuing to function and move across Mars far beyond the primary mission goal of 90 days.

With much greater mobility than the *Mars Pathfinder* minirover, each of these powerful new robot explorers has successfully traveled up to 330 feet (100 m) per Martian day across the surface of the planet. Each rover carries a complement of sophisticated instruments that allows it to search for evidence that liquid water was present on the surface of Mars in ancient times. Each rover has visited a different region of the planet. Immediately after landing, each rover performed reconnaissance of the particular landing site by taking panoramic (360°-degree) visible (color) and infrared images. Then, using images and spectra taken daily by the rovers, NASA scientists at the Jet Propulsion Laboratory used telecommunications and teleoperations to supervise the overall scientific program. With intermittent human guidance, the pair of mechanical explorers functioned like robot prospectors, examining particular rocks and soil targets and evaluating composition and texture at the microscopic level.

Each rover has a set of five instruments with which to analyze rocks and soil samples. The instruments include a panoramic camera (Pancam), a miniature thermal emission spectrometer (Mini-TES), a Mössbauer spectrometer (MB), an alpha particle X-ray spectrometer (APXS), magnets, and

The tracks created by NASA's *Spirit* rover at the end of a drive on Mars (May 5, 2005). The *Mars Exploration Rover (MER) Spirit* collected this mosaic image with its navigation camera. *Spirit* previously had to abandon climbing hills (on April 14, 2005) because of steep slopes. The backtracking was fortuitous, allowing the science team to discover layered outcrops of rock that had been overlooked on the first drive past this area. Since then, *Spirit* had been examining the so-called Methuselah outcrops in the Columbia Hills region for several weeks. This mosaic image looks back at the tracks *Spirit* left while backtracking and heading to the Methuselah outcrop. *(NASA/JPL)*

a microscopic imager (MI). There is also a special rock abrasion tool (or RAT) that allowed each rover to expose fresh rock surfaces for additional study of interesting targets.

Both *Spirit* and *Opportunity* has a mass of 407 pounds (185 kg) and a range of up to 330 feet (100 m) per sol (Martian day). Surface operations have lasted longer than the goal of 90 sols. Communication back to Earth is accomplished primarily with Mars-orbiting spacecraft, like the *Mars Odyssey 2001*, serving as data relays.

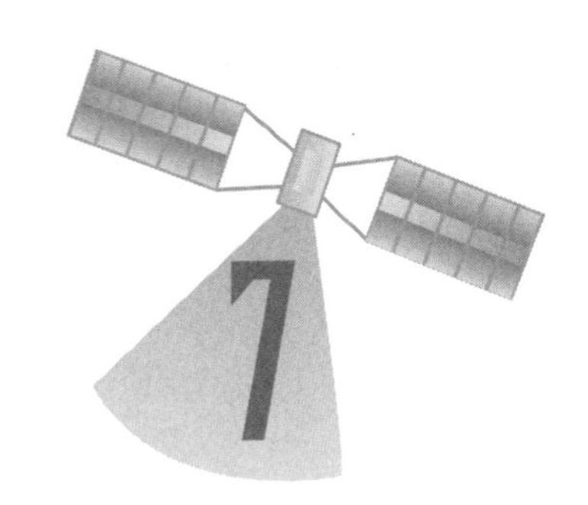

Sample Return Missions

This chapter describes some of NASA's current and future programs to use robot spacecraft to collect samples of extraterrestrial materials and return them to Earth for study by scientists.

There are three fundamental approaches toward handling extraterrestrial samples to avoid possibly contaminating Earth with alien microorganisms. (This undesirable consequence of space exploration is called back contamination.) First, scientists can sterilize a sample while it is en route to Earth from its native world. Second, the mission supervisors can place the alien material in quarantine in a remotely located, maximum-confinement facility on Earth, were scientists can examine it closely. Finally, exobiologists and astronauts may perform a preliminary hazard analysis (called the extraterrestrial protocol test) on the alien sample in an orbiting quarantine facility before the samples are allowed to enter the terrestrial biosphere. To be adequate, a quarantine facility must be capable of (1) containing all alien organisms present in a sample of extraterrestrial material; (2) detecting these alien organisms during protocol testing; and (3) controlling these organisms after detection, until scientists can dispose of them in a safe manner.

How likely is it that Earth will get contaminated by alien microorganisms brought back during a sample-return mission? There is no direct evidence to suggest an immediate threat. But to err on the side of extreme caution is the logical and prudent approach. From a historic perspective, the human-crewed U.S. Apollo Project missions to the Moon (1969–72) stimulated a great deal of scientific debate about the potential issue of forward and back contamination. Early in the 1960s, scientists began speculating in earnest about whether there was life on the Moon. Some of the bitterest technical exchanges during the Apollo Project concerned this particular question. If there was life, no matter how primitive or microscopic, scientists wanted to examine it carefully and compare it with life forms of terrestrial origin. This careful search for microscopic lunar life would, however, be very difficult and expensive because of the forward-contamination problem. For example, all equipment and materials landed on the Moon would need rigorous sterilization and decontamination procedures.

There was also the glaring uncertainty about back contamination. If microscopic life did indeed exist on the Moon, some scientists openly wondered whether the (hypothetical) alien life would represent a serious hazard to life on Earth. Because of this potential extraterrestrial-contamination problem, some members of the scientific community pressed for time-consuming and expensive sterilization and quarantine procedures.

On the other side of this early 1960s contamination argument were the exobiologists, who emphasized the apparently extremely harsh lunar conditions: virtually no atmosphere, probably no water, extremes of temperature, and unrelenting exposure to lethal doses of ultraviolet, charged-particle and X-ray radiations from the Sun. These scientists argued that no life-form possibly exists under such extremely hostile conditions.

And so the great extraterrestrial-contamination debate raged back and forth, until finally the *Apollo 11* expedition departed on the first lunar-landing mission. As a compromise, the *Apollo 11* mission flew to the Moon with careful precautions against back contamination but with only a very limited effort to protect the Moon from forward contamination by terrestrial organisms.

The Lunar Receiving Laboratory (LRL) at the Johnson Space Center in Houston provided quarantine facilities for two years after the first lunar landing. During the Apollo Project, however, no evidence was discovered that native alien life was then present or had ever existed on the Moon. Scientists at the Lunar Receiving Laboratory performed a careful search for carbon, because terrestrial life is carbon-based. One hundred to 200 parts per million of carbon were found in the lunar samples. Of this amount, only a few tens of parts per million are considered indigenous to the lunar material, with the bulk of carbon having been deposited by the solar wind. Exobiologists and lunar scientists concluded that none of this carbon appeared to be derived from biological activity. In fact, after the first few Apollo expeditions to the Moon, even the back-contamination quarantine procedures of isolating the Apollo astronauts for a period of time were dropped.

Nevertheless, what scientists learned during these Apollo Project–era quarantine and sample-analysis operations serves as a useful starting point for planning new quarantine activities, whether Earth-based or space-based. In the future, a well-isolated quarantine facility will most likely be needed to accept, handle, and test extraterrestrial materials from Mars, Europa, and other solar-system bodies of interest in the expanded search for alien life-forms (extant or extinct) in this century.

✧ *Genesis* Solar-Wind Sample Return Mission

This artist's concept shows NASA's *Genesis* spacecraft in collection mode. This means that the robot spacecraft has its science canister opened up to collect and store samples of solar-wind particles. *(NASA)*

The primary mission of NASA's *Genesis* spacecraft was to collect samples of solar-wind particles and to return these samples of extraterrestrial material safely to Earth for detailed analysis. The mission's specific scientific objectives were to obtain precise solar isotopic and elemental abundances and to provide a reservoir of solar matter for future investigation. A detailed study of captured solar wind materials would allow scientists to test various theories of solar system formation. Access to these materials would also help them resolve lingering issues about the evolution of the solar system and the composition of the ancient solar nebula.

The basic robot spacecraft was 7.5 feet (2.3 m) long and 6.6 feet (2 m) wide. The spacecraft's solar array had a wingspan of 25.9 feet (7.9 m). At launch, the spacecraft payload had a total mass of 1,400 pounds (636 kg), composed of the 1,087-pound (494-kg) dry spacecraft and 312 pounds (142 kg) of onboard propellant. The scientific instruments included the solar-wind collector arrays, ion concentrator, and solar-wind (ion and electron) monitors. A combined solar-cell array with a nickel-hydrogen storage battery provided up to 254 watts of electric power just after launch.

The mission started on August 8, 2001, when an expendable Delta II rocket successfully launched the *Genesis* spacecraft from Cape Canaveral Air Force Station, Florida. Following launch, the cruise phase of the mission lasted slightly more than three months. During this period, the spacecraft traveled 932,000 miles (1.5 million km) from Earth to the Lagrange libration point 1 (L1). The *Genesis* spacecraft entered a *halo orbit* around the L1 point on November 16, 2001. Upon arrival, the spacecraft's large thrusters fired, putting *Genesis* into a looping, elliptical orbit around the L1 Lagrangian point. The *Genesis* spacecraft then completed five orbits around L1; nearly 80 percent of the mission's total time was spent collecting particles from the Sun.

On December 3, 2001, *Genesis* opened its collector arrays and began accepting particles of solar wind. A total of 850 days were logged, exposing the special collector arrays to the solar wind. These collector arrays are circular trays composed of palm-sized hexagonal tiles made of various high-quality materials, such as silicon, gold, sapphire, and diamond-like carbon. After the sample-return capsule opened, the lid of the science canister opened as well, exposing a collector for the bulk solar wind. As long as the science canister's lid was opened, this bulk collector array was exposed to different types of solar wind that flowed past the spacecraft.

The ion and electron monitors of the *Genesis* spacecraft were located on the equipment deck outside the science canister and sample-return capsule. These instruments looked for changes in the solar wind and then relayed information about these changes to the main spacecraft computer, which would then command the collector array to expose the appropriate collector. By recognizing characteristic values of temperature, velocity, density, and composition, the spacecraft's monitors were able to distinguish among three types of solar wind—fast, slow, and coronal mass ejections. The versatile robot-sampling spacecraft would then fold out and extend one of three different collector arrays when a certain type of solar wind passed by.

The other dedicated science instrument of the *Genesis* spacecraft was the solar-wind collector. As its name implies, this instrument would concentrate the solar wind onto a set of small collector tiles, made of diamond, silicon carbide, and diamond-like carbon. As long as the lid of the science canister was opened, the concentrator was exposed to the solar wind throughout the collection period.

On April 1, 2004, ground controllers ordered the robot spacecraft to stow the collectors, and so its collection of pristine particles from the Sun ended. The closeout process was completed on April 2, when the *Genesis* spacecraft closed and sealed its sample-return capsule. Then, on April 22, the spacecraft began its journey back toward Earth. However, because of the position of the landing site—the United States Air Force's Utah Testing and Training Range in the northwestern corner of that state—and the unique geometry of the *Genesis* spacecraft's flight path, the robot sampling craft could not make a direct approach and still make a daytime landing. In order to allow the *Genesis* chase-helicopter crews an opportunity to capture the return capsule in midair in daylight, the Genesis mission controllers designed an orbital detour toward another Lagrange point, L2, located on the other side of Earth from the Sun. After successfully completing one loop around the L2 point, the *Genesis* spacecraft was prepared for its return to Earth on September 8. On September 8, the spacecraft approached Earth and performed a number of key maneuvers prior to releasing the sample-return capsule. Sample capsule release took place when the spacecraft flew past Earth at an altitude of about 41,000 miles (66,000 km). As planned, the *Genesis* return capsule successfully reentered Earth's atmosphere at a velocity of 6.8 miles per second (11 km/s) over northern Oregon.

Unfortunately, during reentry on September 8, the parachute on the *Genesis* sample-return capsule failed to deploy (apparently because of an improperly installed gravity switch) and the returning capsule smashed into the Utah desert at a speed of 193 miles per hour (311 km/hr). The high-speed impact crushed the sample return capsule and breached the sample collection capsule—possibly exposing some of collected of pristine solar materials to potential contamination by the terrestrial environment.

Mission scientists worked diligently, however, to recover as many samples as possible from the spacecraft wreckage and then to ship the recovered materials in early October to the Johnson Space Center in Houston, Texas for evaluation and analysis. One of the cornerstones of the recovery process was the discovery that the gold-foil collector was undamaged by the hard landing. Another post-impact milestone was the recovery of the *Genesis* spacecraft's four separate segments of the concentrator target. Designed to measure the isotopic ratios of oxygen and nitrogen, the segments contain within their structure the samples that are the mission's most important science goals.

The United States is a signatory to the Outer Space Treaty of 1966. This document states, in part, that exploration of the Moon and other celestial bodies shall be conducted "so as to avoid their harmful contamination and also adverse changes in the environment of the Earth resulting from the introduction of extraterrestrial matter." The *Genesis* sample consists of atoms collected from the Sun. NASA's planetary protection officer has categorized the Genesis mission as a mission "safe for unrestricted Earth return." This declaration means that exobiologists and other safety experts have concluded that there is no chance of extraterrestrial biological contamination during sample collection at the L1 point. The U.S. National Research Council's Space Studies Board has also concurred on a planetary protection designation of "unrestricted Earth return" for the Genesis mission. The board determined that the sample had no potential for containing life. Consequently, there is no significant issue of extraterrestrial contamination of Earth because of the aborted sample-return operation of the *Genesis* capsule. The issue of planetary contamination remains of concern, however, when robot-collected sample capsules return from potentially life-bearing celestial bodies, such as Mars and Europa.

✧ Stardust Mission

Stardust is the first U.S. space mission dedicated solely to the exploration of a comet and the first robotic spacecraft mission designed to return extraterrestrial material from outside the orbit of the Moon. Launched on February 7, 1999, from Cape Canaveral, the spacecraft traveled through

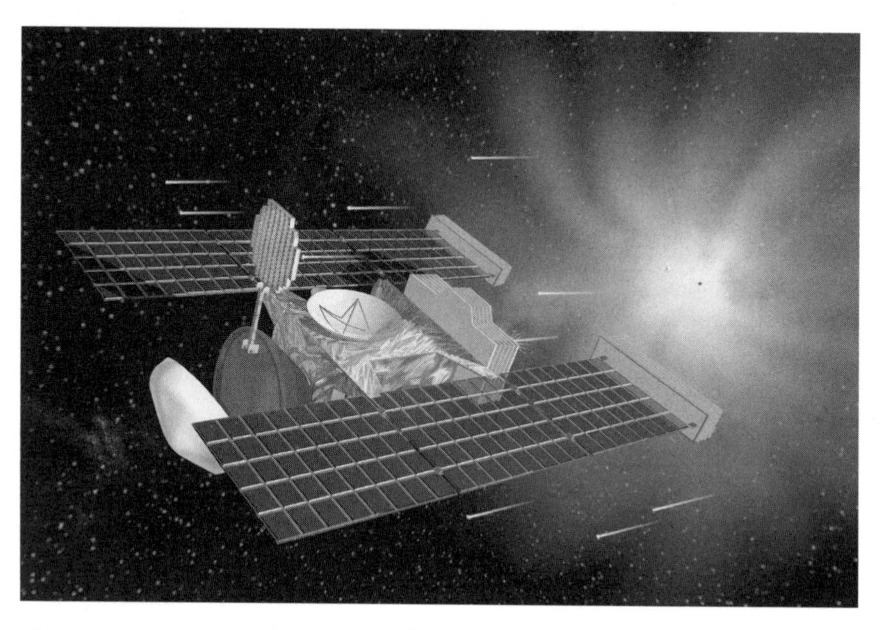

This artist's concept shows NASA's *Stardust* spacecraft encountering Comet Wild 2 (circa January 2004) and collecting dust and volatile material samples in the coma of the comet. The robot spacecraft collected, stowed, and sealed the comet material samples in the special storage vault of an Earth-return reentry capsule, which was also carried onboard *Stardust. (NASA/JPL)*

interplanetary space and successfully flew by the nucleus of Comet Wild 2 on January 2, 2004. When the *Stardust* craft flew past the comet's nucleus, it did so at an approximate relative velocity of 3.8 miles per second (6.1 km/s). At its closest approach during this encounter, the spacecraft came within 155 miles (250 km) of the comet's nucleus and returned images of the nucleus. The spacecraft's dust-monitor data indicate that many particle samples were collected. Mission scientists therefore believe that, during the close encounter, *Stardust* captured thousands of particles and volatiles of cometary material comet.

The spacecraft has also collected samples of interstellar dust, including recently discovered dust streaming into our solar system from the direction of Sagittarius. These materials are believed to consist of ancient pre-solar interstellar grains and nebular remnants that date back to the formation of the solar system. Scientists anticipate their analysis of such fascinating celestial specks will yield important insights into the evolution of the Sun, the planets, and possibly life itself.

In January 2006, *Stardust* returned to the vicinity of Earth in order to drop off its cargo of extraterrestrial material. The spacecraft delivered these materials by precisely ejecting a specially designed, 132-pound (60-kg) reentry capsule. The capsule passed at high speed through Earth's atmosphere and then parachuted to the planet's surface.

Specifically, the *Stardust* capsule successfully returned to Earth on January 15, 2006. Soon after the spacecraft made its final trajectory maneuver at an altitude of about 69,000 miles (111,000 km), it released the sample return capsule. Following capsule release, the main spacecraft performed a maneuver in order to avoid entering Earth's atmosphere. Upon completion of this avoidance maneuver, *Stardust* went into orbit around the Sun.

Once rejected, the sample return capsule entered Earth's atmosphere at a velocity of approximately 8.0 miles per second (12.8 km/s). Aerospace engineers had given the capsule an aerodynamic shape and center of gravity similar to that of a badminton shuttlecock. Because of this special design, the capsule automatically oriented itself with its nose down as it enters the atmosphere.

As the capsule descended, atmospheric friction on the heat shield reduced its speed. When the capsule reached an altitude of about 18.6 miles (30 km), it was slowed down to about 1.4 times the speed of sound. At that point, a small pyrotechnic charge fired, releasing a drogue parachute. After descending to an altitude of about 1.9 miles (3 km), the line holding the drogue chute was cut, allowing the drogue parachute to pull out a larger parachute that carried the capsule to its soft landing. At touchdown, the capsule was traveling at approximately 14.8 feet per second (4.5 m/s), or 9.9 miles per hour (16 km/hr). About 10 minutes elapsed between the start of the capsule's entry into Earth's atmosphere and the time when the main parachute deployed.

Scientists chose the landing site at the Utah Test and Training Range near Salt Lake City because the area is a vast and desolate salt flat controlled by the U.S. Air Force in conjunction with the U.S. Army. The landing footprint for the sample return capsule was about 18.6 by 52.2 miles (30 by 84 km)—an ample space to allow for aerodynamic uncertainties and winds that might affect the direction that the capsule traveled in the atmosphere. The sample return capsule approached the landing zone on a heading of approximately 122 degrees on a northwest to southeast trajectory. The landing took place on early Sunday morning.

A UHF radio beacon on the capsule transmitted a signal as the capsule descended to Earth. Scientists also tracked the parachute and capsule by radar. As soon as the capsule successfully "soft-landed," the retrieval crew flew by helicopter to the site and recovered it. By January 17, the recovered sample capsule was safely delivered to NASA's Johnson Space Center in Houston, Texas. NASA officials then began the process of distributing the collected samples to the 150 or so scientists around the world, who are participating in the program. Preliminary analyses of the samples indicate that comets may not be as simple as the clouds of ice, dust, and gases they were once though to comprise. As suggested by some of the very tiny gemstone-like particles collected from the coma of Comet Wild 2, these intriguing celestial objects may actually be quite diverse with complex and varied histories.

✧ Mars Sample Return Mission

The purpose of a Mars Sample Return Mission (MSRM) is, as the name implies, to use a combination of robot spacecraft and lander systems to collect soil and rock samples from Mars and then return them to Earth for detailed laboratory analysis. A wide variety of options for this type of advanced robot spacecraft mission are being explored.

For example, one or several small robot rover vehicles could be carried and deployed by the lander vehicle. These rovers (under some level of supervision and control by mission managers) would travel away from the original landing site and collect a wider range of interesting rock and soil samples for return to Earth. Another option is to design a nonstationary, or mobile, lander that could travel (again guided by human beings on Earth) to various surface locations and collect interesting specimens. After the soil collection mission was completed, the upper portion of the lander vehicle would lift off from the Martian surface and rendezvous in orbit around the planet with a special carrier spacecraft.

While avoiding direct contact with the ascent vehicle (to prevent back contamination), this automated rendezvous/return carrier spacecraft would carefully remove the soil sample canisters from the ascent portion of the lander vehicle and then depart Mars's orbit on a trajectory that would bring the samples back to Earth. After an interplanetary journey of about one year, the automated carrier spacecraft, with its precious cargo of Martian soil and rocks, would achieve orbit around Earth.

To avoid any potential problems of extraterrestrial contamination of Earth's biosphere by alien microorganisms that might be contained in the Martian soil or rocks, the sample canisters might first be analyzed in a special human-tended orbiting quarantine facility. An alternate return mission scenario would be to bypass an Earth-orbiting quarantine process altogether and use a direct reentry vehicle operation to bring the encapsulated Martian soil samples to Earth's surface in an isolated area.

Whatever sample-return mission profile ultimately is selected, contemporary analysis of Martian meteorites (that have fallen to Earth) has stimulated a great scientific interest in obtaining well-documented and well-controlled virgin samples of Martian soil and rocks. Carefully analyzed in laboratories on Earth, these samples will provide a wealth of

This is an artist's concept of a Mars Sample Return Mission. The sample-return spacecraft is shown departing the surface of the Red Planet after soil and rock samples, previously gathered by robot rovers, have been stored on board in a specially sealed capsule. To support planetary protection protocols, once in rendezvous orbit around Mars, the sample return spacecraft would use a mechanical device to transfer the sealed capsule(s) of Martian soil samples to an orbiting Earth-return mother spacecraft. This craft would then take the samples back to Earth for detailed study by scientists. *(NASA/JPL; artist, Pat Rawlings)*

Extraterrestrial Contamination

Extraterrestrial contamination may be defined as the contamination of one world by life-forms, especially microorganisms, from another world. Using the Earth and its biosphere as a reference, this planetary-contamination process is called forward contamination, if an extraterrestrial soil sample or the alien world itself is contaminated by contact with terrestrial organisms, and back contamination if alien organisms are released into Earth's biosphere.

An alien species will usually not survive when introduced into a new ecological system, because it is unable to compete with native species that are better adapted to the environment. Once in a while, however, alien species actually thrive, because the new environment is very suitable, and indigenous life-forms are unable to defend themselves successfully against these alien invaders. When this war of biological worlds occurs, the result might very well be a permanent disruption of the host ecosphere, with severe biological, environmental, and possibly economic consequences.

Of course, the introduction of an alien species into an ecosystem is not always undesirable. Many European and Asian vegetables and fruits, for example, have been successfully and profitably introduced into the North American environment. Any time a new organism is released in an existing ecosystem, however a finite amount of risk is also introduced.

Frequently, alien organisms that destroy resident species are microbiological life-forms. Such microorganisms may have been nonfatal in their native habitat, but once released in the new ecosystem, they become unrelenting killers of native life-forms that are not resistant to them. In past centuries on Earth, entire human societies fell victim to alien organisms against which they were defenseless, as, for example, the rapid spread of diseases that were transmitted to native Polynesians and American Indians by European explorers.

But an alien organism does not have to directly infect humans to be devastating. Who can easily ignore the consequences of the potato fungus that swept through Europe and the British Isles in the 19th century, causing a million people to starve to death in Ireland alone?

In the Space Age, it is obviously of extreme importance to recognize the potential hazard of extraterrestrial contamination (forward or back). Before any species is intentionally introduced into another planet's environment, scientists must carefully determine not only whether the organism is pathogenic (disease-causing) to any indigenous species but also whether the new organism will be able to force out native species—with destructive impact on the original ecosystem. The introduction of rabbits into the Australian continent is a classic terrestrial example of a nonpathogenic life-form creating immense problems when introduced into a new ecosystem. The rabbit population in Australia simply exploded in size because of their high reproduction rate, which was essentially unchecked by native predators.

At the start of the space age, scientists were already aware of the potential extraterrestrial-contamination problem—in either direction. Quarantine protocols (procedures) were established to avoid the forward-contamination of alien worlds by outbound unmanned spacecraft, as well as the back-contamination of the terrestrial biosphere when lunar samples were returned to Earth as part of the Apollo program. For example, the United States is a signatory to a 1967 international agreement, monitored by the Committee on Space Research (COSPAR) of the International Council of Scientific Unions, which requires the avoidance of forward and back contamination of planetary bodies during space exploration.

A quarantine is a forced isolation to prevent the movement or spread of a contagious disease. Historically, quarantine was the period during which ships suspected of carrying persons or cargo (for example, produce or livestock) infected with contagious diseases were detained at their port of arrival. The length of the quarantine, generally 40 days, was considered sufficient to cover the incubation period of most highly infectious terrestrial diseases. If no symptoms appeared at the end of the quarantine, then the travelers were permitted to disembark. In modern times, the term *quarantine* has obtained a new meaning, namely, that of holding a suspect organism or infected person in strict isolation until it is no longer capable of transmitting the disease. With the Apollo Project and the advent of the lunar quarantine, the term now has elements of both meanings. Of special interest in future space missions to the planets and their major moons is how we avoid the potential hazard of back contamination of Earth's environment when robot spacecraft and human explorers bring back samples for more detailed examination in laboratories on Earth.

important and unique information about the Red Planet. These samples might even provide further clarification of the most intriguing question of all: Is there (or, at least, has there been) life on Mars? A successful Mars Sample Return Mission is also considered a significant and necessary step toward eventual human expeditions to Mars this century.

Mobile Robots as Scientific Laboratories

NASA engineers are planning to add a strong dose of artificial intelligence (AI) to planetary landers and rovers to make these robot spacecraft much more self-reliant and capable of making basic decisions during a mission without human control or supervision. In the past, robot rovers contained very simple AI systems, which allowed them to make a limited number of basic, noncomplicated decisions. In the future, however, mobile robots will possess much higher levels of AI or machine intelligence and be able to make decisions now being made by human mission controllers on Earth.

One of the technical challenges that robot engineers face is how to encapsulate the process by which human beings make decisions in response to changes in their surroundings into a robot rover or complex lander spacecraft sitting on a planet millions of miles (kilometers) away. To make the detailed exploration of the Moon and Mars by mobile robots practical over the next two decades, future robot rovers will have to be intelligent enough to navigate the surface of the Moon or Mars without a continuous stream of detailed instructions from, and decision-making by, scientists on Earth.

Large teams of human beings on Earth are needed to direct the Mars Exploration Rovers (MER) *Spirit* and *Opportunity* as two robot rovers roll across the terrain of Mars looking for evidence of water. (See chapter 6). In a very slow and deliberate process, it takes human-robot teams on two worlds millions of miles (kilometers) apart several days to achieve each of many individual mission milestones and objectives. Specifically, it takes about three (Earth) days for the *Spirit* or *Opportunity* robot rover to visualize a nearby target, get to the target, and do some contact science. Mission controllers currently measure a great day of robot exploring on Mars in terms of travel up to 330 feet (100 m) per Martian day (sol) across the surface of the planet. (A sol is a Martian day and is about 24 hours, 37 minutes, 23 seconds in duration, using Earth-based time units.) Imagine trying to explore an entire continent here on Earth using a system that travels a maximum distance each day equivalent to the length of just one football or soccer field.

This chapter examines how in future a mobile robot with more onboard machine intelligence (or AI) will collect data about its environment, and then make an on-the-spot evaluation of appropriate tasks and actions without being dependent upon decisions made by humans. Advanced AI systems on board such smart future mobile robots will eventually allow them to mimic human thought processes and perform tasks that a human explorer would do. For example, such smart rovers might pause to make an on-the-spot soil analysis of an interesting sample, communicate with an orbiting robot spacecraft for additional data about the immediate location, or even signal other robot rovers to gather (swarm) at the location in order to perform a collective evaluation of the unusual discovery.

Within the next two decades, teams of smart robots, interacting with each other, should be able to map and evaluate large tracts on the surface of the Moon or Mars. An interactive team of smart robot rovers would provide much better coverage of a large area of land and perhaps even exhibit a level of collective intelligence while performing tasks too difficult or complex for a single robot system. With a team of robots, the mission objectives can be accomplished, even if one robot fails to perform or is severely damaged in an accident.

✧ Prospecting for Lunar Water with Smart Robots

The Moon is nearby and accessible, so it is a great place to try out many of the new space technologies, including advanced robot spacecraft, which will prove critical in the detailed scientific study and eventual human exploration of more distant alien worlds, such as Mars. Whether a permanent lunar base turns out to be feasible depends on the issue of logistics, especially the availability of water in the form of water ice. The logistics problem is quite simple. Water is dense and rather heavy, so shipping large amounts of water from Earth's surface to sustain a permanent human presence on the Moon this century could be prohibitively expensive. Establishing a permanent human base on the Moon becomes much easier and far more practical if large amounts of water (frozen in water ice deposits) are already there.

This unusual resource condition is possible, because scientists now hypothesize that comets and asteroids smashing into the lunar surface eons ago left behind some water. Of course, water on the Moon's surface does not last very long. It evaporates in the intense sunlight and quickly departs this airless world by drifting off into space. Only in the frigid recesses of permanently shadowed craters do scientists expect to find any of the water that might have been carried to the Moon and scattered across the lunar surface by ancient comet or asteroid impacts. In the 1990s, two spacecrafts, *Clementine* and *Lunar Prospector,* collected tantalizing data suggesting that the shadowed craters at the lunar poles may contain significant quantities of water ice.

NASA plans to resolve this very important question by using smart robots as scouts. First into action will be the *Lunar Reconnaissance Orbiter (LRO)*—a robot spacecraft mission planned for launch by late 2008. The LRO mission emphasizes the overall objective of collecting science data that will facilitate a human return to the Moon. As part of NASA's strategic plan for solar-system exploration, a return to the Moon by human beings is considered a critical step in field-testing the equipment necessary for a successful human expedition to Mars later in this century.

The *LRO* will orbit the Moon for at least one year using an 18.6–31.1-mile- (30–50-km-) altitude polar orbit to map the lunar environment in greater detail than ever before. The six instruments planned for the *Lunar Reconnaissance Orbiter* will do many things: they will map and photograph the Moon in great detail, paying special attention to the permanently shadowed polar regions. The *LRO*'s instruments will also measure the Moon's ionizing-radiation environment and conduct a very detailed search for signs of water-ice deposits. No single spacecraft-borne instrument can provide definitive evidence of ice on the Moon, but if all the data from the *LRO*'s collection of water-hunting instruments point to suspected ice in the same area, those data would be most compelling.

Within NASA's current strategic vision for robot-human partnership in space exploration, the *LRO* is just the first in a string of smart robots with missions to the Moon over the next two decades. Once compelling evidence for the presence of water ice is obtained by the LRO, then the next logical step is to send a smart scout robot to that location to scratch and sniff the site and to perform on the spot (in situ) analyses. The rover robot's detailed investigations will confirm the existence of any water ice. The semiautonomous mobile robot may expand investigations of the area to provide a first-order estimate of the total quantity of the water available.

Finally, if suitable water resources are located and inventoried, teams of smart robot prospectors would be sent to the Moon to harvest the particular site or sites in preparation for the return of human beings to the lunar surface. Supervised and teleoperated by humans from Earth, a team of semiautonomous water-harvesting robots would make the construction

This is an artist's concept of an advanced, semiautonomous robot rover making remote sample collections at the Moon's south pole. With minimal supervision and teleoperation by controllers on Earth, this type of advanced robot sample collector would help validate the presence of water ice and quantify any promising resource data collected by lunar-orbiting resource reconnaissance spacecraft. The presence of ample quantities of water ice in the permanently shadowed polar regions of the Moon would be a major stimulus in the development of human bases. Robot-assisted lunar-ice mining could become the major industry on the Moon later in this century. *(NASA/Johnson Space Center)*

and operation of a permanent human base practical (from a logistics perspective) and prepare the way for an eventual human expedition to Mars.

✧ Smarter Robots to the Red Planet

NASA's planned *Phoenix Mars Scout* will land in icy soils near the north polar permanent ice cap of the Red Planet and explore the history of water in these soils and any associated rocks. This sophisticated space robot serves as NASA's first exploration of a potential modern habitat on Mars and opens the door to a renewed search for carbon-bearing compounds, last attempted with the *Viking 1* and *2* lander spacecraft missions in the 1970s.

The *Phoenix* spacecraft is currently in development and will launch in August 2007. The robot explorer will land in May 2008 at a candidate site in the Martian polar region, previously identified by the *Mars Odyssey* orbiter spacecraft as having high concentrations of ice just beneath the top layer of soil. *Phoenix* is a fixed-in-place lander spacecraft, which means it cannot move from one location to another on the surface of Mars. Rather, once the spacecraft has safely landed on the surface, it will stay there and use its robotic arm to dig the ice layer and bring samples to its suite of on-deck science instruments. These instruments will analyze samples directly on the Martian surface, sending scientific data back to Earth via radio signals, which will be collected by NASA's Deep Space Network.

The *Phoenix* spacecraft's stereo color camera and a weather station will study the surrounding environment, while its other instruments check excavated soil samples for water, organic chemicals, and conditions that could indicate whether the site was ever hospitable to life. Of special interest to exobiologists, the spacecraft's microscopes would reveal features as small as one one-thousandth the width of a human hair.

The *Phoenix* lander's science goals of learning about ice history and climate cycles on Mars complements the robot spacecraft's most exciting task—to evaluate whether an environment hospitable to microbial life may exist at the ice-soil boundary. One tantalizing question is whether cycles on Mars, either short-term or long-term, can produce conditions in which even small amounts of near-surface water may stay melted. As

This artist's concept shows NASA's planned *Phoenix* robot lander spacecraft deployed on the surface of Mars (circa 2008). The lander would use its robotic arm to dig into a spot in the water ice–rich northern polar region of Mars for clues concerning the Red Planet's history of water. The robot explorer would also search for environments suitable for microscopic organisms (microbes). *(NASA)*

This artist's concept shows NASA's planned *Mars Science Laboratory (MSL)* traveling near a canyon on the Red Planet (circa 2010). With a greater range than any previous robot rover used on Mars, the *MSL* will be able to analyze dozens of samples scooped up from the soil and cored from rocks at scientifically interesting locations on the planet. One of the primary objectives of this sophisticated robot explorer is to investigate the past or present ability of Mars to support life. *(NASA/JPL)*

studies of arctic environments on Earth have indicated, if water remains liquid only—even just for short periods during long intervals—life can persist, if other factors are right.

Building upon the success of the two Mars Exploration Rover (MER) spacecraft, *Spirit* and *Opportunity*—which arrived on the surface of the Red Planet in January 2004—NASA's next mobile rover mission to Mars is being planned for arrival on the planet in late 2010. Called the *Mars Science Laboratory (MSL)*, this mobile robot will be twice as long and three times as massive as either *Spirit* or *Opportunity.* The *Mars Science Laboratory* will collect Martian soil samples and rock cores and analyze them on the spot for organic compounds and environmental conditions that could have supported microbial life in the past, or possibly even now in the present.

An *aerobot* is an autonomous robotic aerovehicle (that is, a free-flying balloon or a specially designed robot airplane), which is capable of flying in the atmospheres of Venus, Mars, Titan, or any of the outer planets. (A Mars robot airplane is discussed chapter 10.)

The Mars balloon is a specially designed balloon package (or *aerobot*) that could be deployed into the planet's atmosphere and then used to explore the surface. During the Martian daytime, the balloon would become buoyant enough, owing to solar heating, to lift its instrumented guide-rope off the surface. Then at night, the balloon would sink when cooled and the instrumented guide-rope would again come in contact with the Martian surface, allowing various surface scientific measurements to be made. A typical balloon exploration system might operate for 10 to 50 sols (that is, from 10 to 50 Martian days) and provide surface and (*in situ*) atmospheric data from many locations. Data would be relayed back to Earth via a Mars orbiting spacecraft.

One proposed NASA mission, called the Mars Geoscience Aerobots mission, involves high spatial-resolution spectral mapping of the Martian surface from aerobot platforms. One or more aerobots would be deployed at an altitude between 2.5 miles (4 km) and 3.7 miles (6 km) and operate for up to 50 days. Onboard instruments would perform high-resolution mineralogy and geochemistry measurements in support of future exobiologic sample-return missions.

To be useful as robot exploration systems, aerobots should be capable of one or more of the following activities: autonomous state determination, periodic altitude variations, altitude control and the ability to follow a designated flight path within a planetary atmosphere using prevailing planetary winds, and landing at a designated surface location.

Recent advances in microelectronic technology and mobile robotics have made it possible for engineers to consider the creation and use of extremely small automated or remote-controlled vehicles, called *nanorovers,* in planetary surface exploration missions. For convenience, engineers often define a nanorover as a robot system with a mass of between 0.35 ounce (10 g) and 1.77 ounces (50 g). One or several of these tiny robots could be used to survey areas around a lander and to look for a particular substance, such as water ice or microfossils. The nanorover would then communicate its scientific findings back to Earth via the lander spacecraft, possibly in conjunction with an orbiting mother spacecraft or communications hub—such as the proposed *Mars Telecommunications Orbiter (MTO).*

A cluster of nanorovers endowed with some degree of collective intelligence could perform detailed analysis of an interesting Martian surface or subsurface site suspected of harboring microbial life. How do the nanorovers get to that interesting site? In one exploration scenario, a large surface rover, serving as a mother spacecraft and mobile base camp, carries several populations of these nanorovers, releasing or injecting them as part of its own test protocol in the search for suspected life-sites (extinct or existent) on the Red Planet.

NASA engineers expect to launch the *MTO* in September 2009, have the spacecraft arrive at Mars in August 2010, and then start performing its mission for six to 10 years from a high-altitude orbit around the Red Planet. This future spacecraft's mission is to serve as the Mars hub for interplanetary telecommunications. By providing reliable and more available communications channels to Earth for rovers and stationary landers working on the surface of Mars, the *MTO* greatly increases the overall information payoff from all future robot missions.

Eventually, mobile space robots will achieve higher levels of artificial intelligence, autonomy, and dexterity, so that servicing and exploration operations will become less and less dependent on a human operator's being present in the control loop. These robots would be capable of interpreting very high-level commands and executing them without human intervention. Erroneous command structures, incomplete task operations, and the resolution of differences between the robot's built-in world model and the real-world environment it is encountering would be handled autonomously. This is important because when more sophisticated robots are sent deeper into the outer solar system, telecommunications delays of minutes become hours.

Collective intelligence is another interesting concept for future robots. Just as human beings can self-organize into groups or teams to achieve complicated goals, collections of smart robots will learn to self-organize into teams (or swarms) to perform more complicated missions. For example, a team of robot rovers could gather at a particularly interesting surface site to harvest all of the scientific data available; or else several mobile robots might rush to the assistance of a stranded robot. Such collective actions and group behavior will allow teams of future space robots to exceed the performance capabilities and artificial-intelligence levels of any individual machine. Collective machine intelligence would open up entirely new avenues for the use of robot systems on Mars and elsewhere in the solar system.

The development of higher levels of autonomy, and the demonstration of collective machine intelligence by teams of robots, are very important technology milestones in robotics. Once attained, these capabilities will also result in the effective use of robots in the construction and operation of permanent lunar or Martian-surface bases. In such complex undertakings on alien worlds, teams of smart machines will serve as scouts, mobile science platforms, and eventually construction workers, who set about their tasks with little or no direct human supervision. As future space robots learn to think more like humans, these machines will anticipate the needs of their human partners in space exploration, and

Using surveying instruments, teams of robot rovers equipped with robust levels of artificial intelligence will be able to map large tracts of the surface of Mars in about 2020 without detailed instructions from scientists on Earth. This artist's concept shows a pair of robot rovers exploring an interesting surface site on Mars. Smart robot rovers, operating in teams, would also be available to lend assistance to one another, whenever a difficult situation was encountered. Some robot engineers suggest that a team of smart rovers might possess the limited "collective intelligence" necessary to repair a damaged member of their mechanical clan or to rescue a fellow robot in distress. *(NASA)*

simply perform the necessary tasks with little or no human supervision. If a human explorer shows strong interest in a particular outcropping on Mars, his or her mobile robot companion will also focus its sensors and attention on the site. When an astronaut drops a tool on Mars during the construction of a surface base, his or her companion construction robot will immediately fetch the tool with its mechanical arm and "hand" it back to the astronaut, without blinking an electronic eye.

Robot Spacecraft Visiting Small Bodies in the Solar System

Just two decades ago, scientists did not have very much specific information about the small bodies in the solar system, such as comets and asteroids. There was a great deal of speculation about the true nature of a comet's nucleus, and no one had ever seen the surface of an asteroid up close. All that changed very quickly, when robot spacecraft missions flew past, imaged, sampled, probed, and even landed on several of these interesting celestial objects. This chapter discusses some of the most significant small-body missions that have taken place or will soon do so.

Asteroids and comets are believed to be the ancient remnants of the earliest years of the formation of the solar system, which took place more than four billion years ago. From the beginning of life on Earth to the spectacular collision of Comet Shoemaker-Levy 9 with Jupiter (in July 1994), these so-called small bodies influence many of the fundamental processes that have shaped the planetary neighborhood in which Earth resides.

A comet is a dirty ice rock consisting of dust, frozen water, and gases that orbits the Sun. As a comet approaches the inner solar system from deep space, solar radiation causes its frozen materials to vaporize (sublime), creating a coma and a long tail of dust and ions. Scientists think these icy planetesimals are the remainders of the primordial material from which the outer planets were formed billions of years ago. As confirmed by spacecraft missions, a comet's nucleus is a type of dirty ice ball, consisting of frozen gases and dust. While the accompanying coma and tail may be very large, comet nuclei generally have diameters of only a few tens of miles (kilometers) or less.

As a comet approaches the Sun from the frigid regions of deep space, the Sun's radiation causes the frozen (surface) materials to vaporize (sublime). The resultant vapors form an atmosphere, or coma, with a diameter that may reach 62,150 miles (100,000 km). Measurements suggest that an enormous cloud of hydrogen atoms also surrounds the visible coma. This hydrogen was first detected in comets in the 1960s.

Ions produced in the coma are affected by the charged particles in the solar wind, while dust particles liberated from the comet's nucleus are impelled in a direction away from the Sun by the pressure of the solar wind. The results are the formation of the plasma (Type I) and dust (Type II) comet tails, which can extend for up to 62 million miles (100 million km). The Type I tail, composed of ionized gas molecules, is straight and extends radially outward from the Sun as far as 62 million miles (100 million km). The Type II tail, consisting of dust particles, is shorter, generally not exceeding 6 million miles (10 million km) in length. The Type II tail curves in the opposite direction to the orbital movement of the comet around the Sun.

Throughout history, no astronomical object, other than perhaps the Sun or the Moon, has attracted more attention or interest. The ancient Greeks called these objects "hairy stars" (κομετες). Since ancient times, comets have generally been regarded as harbingers of momentous human events that are evil or harmful. For example, the famous English playwright William Shakespeare, wrote in his play *Julius Caesar:* "When beggars die, there are no comets seen; but the heavens themselves blaze forth the death of princes." Shakespeare used this phrase to emphasize the significance of the murder of Julius Caesar by Brutus, Cassius, and other conspirators within the Roman Senate.

Many scientists think that comets are icy planetesimals, which represent the cosmic leftovers when the rest of the solar system formed over four billion years ago. In 1950, the Dutch astronomer Jan Hendrik Oort (1900–92) suggested that most comets reside far from the Sun in a giant cloud (now called the *Oort Cloud*). The comet-rich region is thought to extend to the limits of the Sun's gravitational attraction, creating a giant sphere with a radius of between 50,000 and 80,000 astronomical units (AU). (An astronomical unit is the distance from Earth to the Sun, a distance of about 93 million miles [150 million km]). Billions of comets may reside in this distant region of the solar system, and their total mass is estimated to be roughly equal to the mass of Earth. Every once a while, an Oort-Cloud comet enters the planetary regions of the solar system, possibly through the gravitational perturbations caused by neighboring stars or some other chaotic phenomenon.

Oort's suggestion was followed quickly by an additional hypothesis concerning the location and origin of the periodic comets seen passing through the solar system. In 1951, the Dutch-American astronomer Gerard Kuiper (1905–73) proposed the existence of another, somewhat nearer, region populated with cometary nuclei and icy planetesimals. Unlike the very distant Oort Cloud, this region—now called the Kuiper belt—lies roughly in the plane of the planets at a distance of 30 astronomical units

(Neptune's orbit) out to about 1,000 astronomical units from the Sun. (See chapter 10 for additional discussion about icy objects in the Kuiper belt.)

Once a comet approaches the planetary regions of the solar system, it is also subject to the gravitational influences of the major planets, especially Jupiter, and the comet may eventually achieve a quasi-stable orbit within the solar system. By convention, comet orbital periods are often divided into two classes: long-period comets (which have orbital periods of more than 200 years) and short-period comets (which have periods of less than 200 years). Astronomers sometimes use the term *periodic comet* for short-period comet.

During the space age, robot spacecraft have explored several comets and greatly improved what scientists know about these very interesting celestial objects. Several of the most interesting robot spacecraft, like *Giotto* and *Deep Impact,* are discussed in this chapter.

An asteroid is a small, solid rocky body without atmosphere that orbits the Sun independent of a planet. The vast majority of asteroids, which are also called minor planets, have orbits that congregate in the asteroid belt or the main belt: a vast doughnut-shaped region of heliocentric space located between the orbits of Mars and Jupiter. The asteroid belt extends from approximately two to four astronomical units (AU) (or a distance of about 186 million miles [300 million km] to 373 million miles [600 million km]) from the Sun and probably contains millions of asteroids ranging widely in size from Ceres—which is 584 miles (940 km) in diameter, making it about one-quarter the diameter of the Moon—to numerous solid bodies that are less than 0.6 mile (1 km) across. There are more than 20,000 numbered asteroids.

NASA's *Galileo* spacecraft was the first to observe an asteroid close up, flying past main belt asteroids Gaspra and Ida in 1991 and 1993, respectively. Gaspra and Ida proved to be irregularly shaped objects, rather like potatoes, riddled with craters and fractures. The *Galileo* spacecraft also discovered that Ida had its own moon—a tiny body called Dactyl, in orbit around its parent asteroid. Astronomers suggest Dactyl, may be a fragment from past collisions in the asteroid belt.

NASA's *Near-Earth Asteroid Rendezvous (NEAR)* spacecraft was the first scientific mission dedicated to the exploration of an asteroid. As discussed in this chapter, the *NEAR-Shoemaker* spacecraft caught up with the asteroid Eros in February 2000 and orbited the small body for a year, studying its surface, orbit, mass, composition, and magnetic field. Then, in February 2001, mission controllers guided the spacecraft to the first-ever landing on an asteroid.

Scientists currently believe that asteroids are the primordial material that was prevented by Jupiter's strong gravity from accreting (accumulating) into a planet-size body when the solar system was born about 4.6 billion years ago. It is estimated that the total mass of all the asteroids (if assembled together) would make up a celestial body about 932 miles (1,500 km) in diameter—an object less than half the size (diameter) of the Moon.

Known asteroids range in size from about 584 miles (940 km) in diameter (Ceres, the first asteroid discovered) down to pebbles a few inches (centimeters) in diameter. Sixteen asteroids have diameters of 149 miles (240 km) or more. The majority of main-belt asteroids follow slightly elliptical, stable orbits, revolving around the Sun in the same direction as Earth and the other planets, and taking from three to six years to complete a full circuit of the Sun.

Future robot exploration of the solar system's small bodies will require technological developments in several areas. Landing and surface operations in the very low-gravity environment of small interplanetary bodies (where the acceleration due to gravity is typically 0.0003 foot per second per second (0.0001 m/s^2) to 0.03 ft/s^2 (0.01 m/s^2) is an extremely challenging engineering problem. The robot explorer for the surface of an asteroid or comet must have mechanisms and autonomous control algorithms to perform landing, anchoring, surface/subsurface sampling, and sample manipulation for a suite of scientific instruments. A robot lander might use crushable material on the underside of a base-plate design to absorb almost all of the landing kinetic energy. An anchoring, or attachment, system would then be used to secure the lander and compensate for the reaction forces and moments generated by the sample acquisition mechanisms. The European Space Agency's *Rosetta* robot spacecraft mission is now traveling through interplanetary space and is scheduled to rendezvous with Comet 67P/Churyumov-Gerasimenko, drop a probe on the surface of the comet's nucleus, and study the comet from orbit.

✧ *Giotto* Spacecraft

The European Space Agency (ESA)'s scientific spacecraft, called *Giotto,* was launched on July 2, 1985, from the agency's Kourou, French Guiana, launch site by an Ariane 1 rocket. The spacecraft was designed to encounter and investigate Comet Halley during its 1986 return to the inner solar system. Following this successful encounter, *Giotto* studied Comet Grigg-Skjellerup during an extended mission in 1992.

The major objectives of the Giotto mission were to obtain color photographs of the Comet Halley's nucleus; determine the elemental and isotopic composition of volatile components in the cometary coma, particularly parent molecules; characterize the physical and chemical processes that occur in the cometary atmosphere and ionosphere; determine the elemental and isotopic composition of dust particles; measure the total gas-production rate and dust flux and size/mass distribution and derive the dust-to-gas ratio; and investigate the macroscopic systems of plasma flows resulting from the cometary-solar wind interaction.

The *Giotto* spacecraft encountered Comet Halley on March 13, 1986, at a distance of 0.89 AU from the Sun and 0.98 AU from Earth and at an angle of 107 degrees from the comet-Sun line. A design goal of this mission was to have the spacecraft come within 311 miles (500 km) of the comet's nucleus at closest encounter. The actual closest-approach distance was measured at 370 miles (596 km). To protect itself during the encounter, *Giotto* had a dust shield designed to withstand impacts of particles up to 0.0035 ounce (0.1 g). The scientific payload consisted of 10 hardware experiments: a narrow-angle camera; three mass spectrometers for neutral particles, ions, and dust particles; various dust detectors; a photopolarimeter; and a set of plasma experiments. All experiments performed well and produced an enormous quantity of important scientific results. Perhaps the most significant accomplishment was the clear identification of the comet's nucleus and confirmation of the hypothesized dirty-snowball (that is, rock-and-ice) model.

Fourteen seconds before closest approach, the *Giotto* spacecraft was hit by a large dust particle. The impact caused a significant shift (about 0.9 degree) of the spacecraft's angular momentum vector. Following this bump, scientific data were received intermittently for the next 32 minutes. Some experiment sensors suffered damage during this 32-minute interval. Other experiments (the camera baffle and deflecting mirror, the dust detector sensors on the front sheet of the bumper shield, and most experiment apertures) were exposed to dust particles regardless of the accident, and also suffered damage.

Giotto's cameras recorded numerous images and gave scientists a unique opportunity to determine the shape and material composition of Comet Halley's nucleus—an irregular, peanut-shaped object somewhat larger (about 9.32 miles [15 km] by 6.2 miles [10 km]) than they had estimated. The nucleus was dark and surrounded by a cloud of dust. Though damaged by multiple-particle impacts, *Giotto* successfully conducted the Halley encounter. Upon completion of the historic flyby, ESA mission controllers put the spacecraft in a hibernation (or quiet) mode, as it continued to travel in deep space. In February 1990, ESA controllers reactivated the hibernating spacecraft for a new task—to observe Comet Grigg-Skjellerup, a short period comet with an orbital period 5.09 years as it experienced perihelion in July 1992.

During the Giotto Extended Mission (GEM), the hardy and far-traveled robot spacecraft successfully encountered Comet Grigg-Skjellerup on July 10, 1992. The closest approach was approximately 125 miles (200 km). At the time of the encounter, the heliocentric distance of the spacecraft was 1.01 AU, and the geocentric distance 1.43 AU. ESA controllers had switched on the scientific payload in the previous evening (July 9). On July 23, ESA controllers terminated operations for the Comet Grigg-Skjellerup encounter. The again-hibernating *Giotto* spacecraft flew by Earth on July 1, 1999, at a closest approach of about 136,110 miles (219,000 km). During its home-planet flyby, the spacecraft was moving at about 2.2 miles per second (3.5 km/s) relative to Earth.

The European Space Agency named the *Giotto* spacecraft after the Italian painter Giotto di Bondone (1266–1337), who apparently witnessed the 1301 passage of Comet Halley. The Renaissance artist then included the first scientific representation of this famous comet in his renowned fresco *Adoration of the Magi,* which can be found in the Scrovegni Chapel in Padua, Italy.

✧ *Deep Space 1 (DS1)* Spacecraft

The *Deep Space 1 (DS1)* mission was the first of a series of technology demonstration spacecraft and probes developed by NASA's New Millennium Program. It was primarily a mission to test advanced spacecraft technologies that had never been flown in space. In the process of testing its solar electric-propulsion system, autonomous navigation system, advanced microelectronics and telecommunications devices, and other cutting-edge aerospace technologies, *Deep Space 1* encountered the Mars-crossing near-Earth asteroid Braille (formerly known as 1992 KD) on July 20, 1999. Then, in September 2001, the robot spacecraft encountered Comet Borrelly and, despite the failure of the system that helped determine its orientation, returned useful images of a comet's nucleus.

As part of its mission to flight-demonstrate new space technologies, the robot spacecraft carried the miniature integrated camera-spectrometer (MICAS), an instrument combining two visible imaging channels with ultraviolet (UV) and infrared (IR) spectrometers. MICAS allowed scientists to study the chemical composition, geomorphology, size, spin-state, and atmosphere of the target celestial objects. *Deep Space 1* also carried the plasma experiment for planetary exploration (PEPE), an ion and electron spectrometer that measures the solar wind during cruise, the interaction of the solar wind with target bodies during encounters, and the composition of a comet's coma.

Aerospace engineers built the *Deep Space 1* spacecraft on an octagonal aluminum frame bus that measured 3.6 feet (1.1 m) by 3.6 feet (1.1 m) by 4.9 feet (1.5 m) in size. With instruments and systems attached, the spacecraft measured 8.2 feet (2.5 m) high, 6.9 feet (2.1 m) deep, and 5.6 feet (1.7 m) wide. The launch mass of the spacecraft was about 1,070 pounds (486 kg), including 68.4 pounds (31.1 kg) of hydrazine and 179 pounds (81.5 kg) of xenon gas. The robot probe received its electricity from a system of batteries and two solar panel wings attached to the sides of the frame. When deployed, the solar array spanned roughly 38.6 feet (11.75 m) across.

The solar panels represented one of the major technology demonstration tests on the spacecraft. A cylindrical lens concentrated sunlight on a strip of photovoltaic cells and also protected the cells. Each solar array consisted of four 5.25-foot- (1.60-m-) by-3.71-foot (1.13-m) panels. At the beginning of the mission, the solar array provided 2,500 watts (W) of electric power. As the spacecraft moved away from the Sun and as the solar cells aged, the array provided less electricity to the spacecraft. Mission controllers used the spacecraft's high-gain antenna, three low-gain antennae, and a Ka-band antenna (all mounted on top of the spacecraft except one low-gain antenna, which was mounted on the bottom) to communicate with the spacecraft as it traveled through interplanetary space.

This artist's concept shows NASA's *Deep Space 1* spacecraft approaching Comet Borrelly in September 2001. With its primary mission to serve as a space technology demonstrator—especially the flight-testing of an advanced xenon electric propulsion system—successfully completed by September 1999, *Deep Space 1* headed for a risky, exciting rendezvous with a comet. On September 22, 2001, this robot spacecraft entered the coma of Comet Borrelly and successfully made its closest approach to the nucleus at a distance of about 1,370 miles (2,200 km). At the time of the encounter, *Deep Space 1* was traveling at approximately 10.3 miles per second (16.5 km/s) relative to the comet's nucleus. *(NASA/JPL)*

A xenon-ion engine mounted in the propulsion unit on the bottom of the frame provided the spacecraft with the thrust it needed to accomplish its journey. The 11.8-inch- (30-cm-) diameter electric rocket engine consisted of an ionization chamber into which xenon gas was injected. Electrons emitted by a cathode traversed the discharge tube and collided with the xenon gas, stripping off electrons and creating positive ions in the process. The ions were then accelerated through a 1,280-volt grid and reached a velocity of 19.6 miles per second (31.5 km/s) before ejecting from the spacecraft as an ion beam. The demonstration xenon ion engine produced a thrust of 0.02 pounds of force (0.09 newton [N]) under conditions of maximum electric power (namely 2,300 W) and a thrust of about 0.0045 pounds of force (0.02 N) at the minimum operational electrical-power level of 500 W. To neutralize electric charge, excess electrons were collected and injected into the ion beam as it left the spacecraft. Of the 179 pounds (81.5 kg) of xenon propellant, approximately 37 pounds (17 kg) were consumed during the robot spacecraft's primary mission.

Deep Space 1 was launched from Cape Canaveral on October 24, 1998. A successful third-stage rocket burn placed the *Deep Space 1* spacecraft into its solar orbit trajectory. The spacecraft then separated from the upper stage of the launch vehicle about 342 miles (550 km) above the Indian Ocean. About 97 minutes after launch, mission controllers received telemetry from the spacecraft through the NASA Deep Space Network. The telemetry indicated that all critical spacecraft systems were performing well. Following this set of communications, the space robot went about its scientific business of encountering asteroids and comets.

DS1 flew by the near-Earth asteroid Braille on July 29, 1999, at a distance of about 16 miles (26 km) and a relative velocity of approximately 9.6 miles per second (15.5 km/s). Just prior to this encounter, a software problem caused the spacecraft to go into a safing mode. The problem was soon solved, however, and the spacecraft returned to normal operations. Up to six minor trajectory correction maneuvers were scheduled in the 48 hours prior to the flyby. The spacecraft made its final pre-encounter transmission about seven hours before its closest approach to the asteroid, after which it turned its high-gain antenna away from Earth to point the MICAS camera/spectrometer camera toward the asteroid. Unfortunately, the spacecraft experienced a target-tracking problem and the MICAS instrument was not pointed toward the asteroid as it approached, so the probe could not obtain any close-up images or spectra. MICAS was turned off about 25 seconds before its closest approach at a distance of about 218 miles (350 km) and measurements were taken with the PEPE plasma instrument.

The spacecraft then turned after the encounter to obtain images and spectra of the opposite side of the asteroid, as it receded from view. Once again an equipment anomaly surfaced. Because of the target-tracking problem, only two black-and-white images and a dozen spectra were obtained. The two images were captured at 915 and 932 seconds after closest approach from a distance of approximately 8,700 miles (14,000 km). The spectra were taken about 3 minutes later. Over the next few days, *Deep Space 1* transmitted these data back to Earth. The diameter of Braille is estimated at 1.4 miles (2.2 km)at its longest and 0.6 mile (1 km) at its shortest. The spectra indicated that the asteroid was similar to the large minor planet Vesta. The primary mission of *Deep Space 1* lasted until September 18, 1999.

By the end of 1999, the ion engine had used approximately 48 pounds (22 kg) of xenon to impart a total delta V (velocity change) of 4,265 feet per second (1,300 m/s) to the spacecraft. The original plan was to fly by the dormant Comet Wilson-Harrington in January 2001 and then past the Comet Borrelly in September 2001. But the spacecraft's star tracker failed on November 11, 1999, so mission planners drew upon techniques developed to operate the spacecraft without the star tracker and came up with a new extended mission to fly by Comet Borrelly. As a result of these innovative actions, on September 22, 2001, *Deep Space 1* entered the coma of Comet Borrelly and successfully made its closest approach (a distance of about 1,367 miles [2,200 km]) to the nucleus. At the time of this encounter, *DS1* was traveling at 10.3 miles per second (16.5 km/s) relative to the nucleus. The PEPE instrument was active throughout the encounter. As planned, MICAS started making measurements and imaging 80 minutes before encounter and operated until a few minutes before encounter. Both instruments successfully returned data and images.

On December 18, 2001, NASA mission controllers commanded the spacecraft to shut down its ion engines. This action ended the *Deep Space 1* mission. Mission managers decided to leave the spacecraft's radio receiver on, however, just in case they desired to make future contact with *Deep Space 1*.

✧ *Deep Impact* Spacecraft

In early July 2005, NASA's *Deep Impact* robot spacecraft performed a complex experiment in space that probed beneath the surface of a comet and helped reveal some of the secrets of its interior. As a larger flyby spacecraft released a smaller impactor spacecraft into the path of Comet Tempel 1, the experiment became one of a cometary bullet chasing down a spacecraft bullet (the penetrator), while a third spacecraft bullet (the flyby robot) sped along to watch.

The greatest challenge for the engineers who created the *Deep Impact* flight system and its collection of science instruments was to target and successfully hit the 3.7-mile- (6-km-) diameter nucleus of Comet Tempel 1. Traveling at a relative velocity of 6.2 miles per second (10 km/s) and released by the flyby spacecraft from a distance of about 537,000 miles (864,000 km), the self-guided impactor had to strike in an area on the sunlit side of the nucleus. This allowed the flyby spacecraft's science instruments to take images of the collision and its aftermath.

The *Deep Impact* flight system consisted of two robot spacecraft: the flyby spacecraft and the impactor. Each spacecraft had its own instruments and capabilities to receive and transmit data. The flyby spacecraft carried the primary imaging instruments and hauled the impactor to the vicinity of the comet's nucleus. Serving as a mother spacecraft for the mission, the *Deep Impact* flyby spacecraft released the impactor about 24 hours prior to comet impact. The flyby spacecraft then received data from the impactor as it traveled to the target, used its on-board instruments to record images of the impactor-comet collision, observed post-collision phenomena

This artist's concept provides a look at the moment of impact and the formation of a human-generated crater on Comet Temple 1. This event occurred on July 4, 2005, when the 820-pound (372-kg) copper projectile launched by NASA's *Deep Impact* spacecraft smashed into Comet Temple 1. For billions of years, Earth has been bombarded by comets and asteroids. So, from a nonscientific perspective, the *Deep Impact* mission was payback time—the first time in history that a comet got whacked by Earth, or, more correctly, by a robot probe sent from Earth. *(NASA)*

(including the resultant crater and ejected materials), and then transmitted all of the scientific data back to Earth.

After releasing the impactor, the flyby spacecraft had to slow down and carefully align itself in order to observe the impact, ejecta, crater development, and crater interior. All this occurred rather quickly, as the flyby spacecraft approached to within 310 miles (500 km) of the comet's nucleus and then zipped past.

The primary task of the impactor was to guide itself to the comet's nucleus and strike on the sunlit side. The impactor also needed to have sufficient kinetic energy when it smashed into the comet that the violent collision would excavate a crater approximately 330 feet (100 m) wide and 92 feet (28 m) deep.

The *Deep Impact* flyby spacecraft is a three-axis stabilized spacecraft that uses a fixed solar array and a small rechargeable nickel-hydrogen (NiH_2) rechargeable battery to generate 620 watts of electric power. The spacecraft is approximately 10.5 feet (3.2 m) long, 5.7 feet (1.7 m) wide, and 7.7 feet (2.3 m) high. The *Deep Impact* flyby has a mass of 1,430 pounds (650 kg), a structure made of aluminum and aluminum honeycomb material, and a propulsion system that uses hydrazine to provide a total velocity change (delta V) capability of 623 feet per second (190 m/s). The spacecraft's thermal control system used blankets, surface radiators, special finishes, and heaters to keep temperatures with proper limits.

During the encounter phase with Comet Tempel 1 in July 2005, the *Deep Impact* flyby spacecraft used its high-gain antenna to transmit near–real time images of the impact back to Earth. Engineers also gave special attention to the problem of dust and debris in the comet's coma and used debris shielding in strategic locations on the spacecraft to protect potentially vulnerable instruments and subsystems.

The impactor spacecraft was cylindrical in shape, had a mass of 816 pounds (370 kg), and consisted of mostly copper (about 49 percent), and aluminum (about 24 percent), to minimize corruption of spectral emission lines from materials excavated from inside the comet's nucleus. Mission planners wanted the impactor to strike the comet's nucleus at a relative velocity of 6.3 miles per second (10.2 km/s), since scientists estimated the high-speed collision would have an explosive energy equivalent of 4.8 tons of trinitrotoluene (TNT) (19 gigajoules [GJ]). The impactor was mechanically and electrically attached to the flyby spacecraft for all but the last 24 hours of the mission. Only during the last 24 hours did the impactor operate on internal battery power. The flyby spacecraft released the impactor 24 hours before collision at a distance of about 536,865 miles (864,000 km) from the target. The impactor then used its high-precision star tracker and autonavigation software to guide itself to the sunlit side of the comet's nucleus. The impactor also employed a small hydrazine propulsion system for attitude control and minor trajectory corrections during its terminal flight to ground zero on the comet's nucleus.

This unusual and interesting mission began with a successful launch from Cape Canaveral on January 12, 2005. Following launch, the *Deep Impact* flyby spacecraft and the hitchhiking (co-joined) impactor transferred into a heliocentric orbit and rendezvoused with Comet Tempel 1 in early July. On July 3, the flyby spacecraft released the impactor and executed a velocity decrease and re-alignment maneuver to better observe the impact. On July 4 (about 24 hours later), the impactor smashed into the comet's nucleus.

As planned, *Deep Impact*'s impactor successfully collided with Comet Tempel 1 on the sunlit side. A camera on the impactor spacecraft captured and relayed images of the comet's nucleus until just seconds before the collision. Upon impact, there was a brilliant and rapid release of dust that momentarily saturated the cameras onboard the *Deep Impact* flyby spacecraft. Audiences around the world watched this spectacular event, as dramatic images were shown in near-real-time on NASA TV and provided over the Internet. All available NASA orbiting telescopes, including the *Hubble Space Telescope (HST)*, the *Chandra X-ray Observatory (CXO)*, and the *Spitzer Space Telescope (SST)* observed the unique event.

The flyby spacecraft was approximately 6,200 miles (10,000 km) away at the time of impact and began collecting images about 60 seconds before impact. At approximately 600 seconds after impact, the flyby spacecraft was about 2,500 miles (4,000 km) from the nucleus; and observations of the crater started and continued until closest approach to the nucleus at a distance of about 310 miles (500 km). Sixteen minutes after impact, imaging ended as the flyby spacecraft aligned itself to cross the inner coma. Within 21 minutes, the crossing of the inner coma was complete and the *Deep Impact* flyby spacecraft once again aligned itself—this time to look back at the comet. After 50 minutes, the flyby spacecraft began to playback all of the accumulated scientific data to scientists on Earth. The *Deep Impact* flyby spacecraft is now in a hibernation (or sleep) mode, awaiting a possible wake-up call for further scientific investigations.

The impact, while powerful, was not forceful enough to make an appreciable change in the comet's orbital path around the Sun. Ice, heated by the energy of the impact, vaporized, and dust and debris was ejected from the crater. As scientists sort through the data gathered by the flyby spacecraft, they will be able to learn a little more about the structure of a comet's interior and whether it is substantially different from its surface.

✧ *Rosetta* Spacecraft

Rosetta is a European Space Agency (ESA) robot spacecraft mission to rendezvous with Comet 67P/Churyumov-Gerasimenko, drop a probe on the surface of the comet's nucleus, study the comet from orbit, and fly by at least one asteroid while traveling to the target comet. The major goals of this mission are to study the origin of comets, the relationship between cometary and interstellar material, and the implications of the latter in regard to the origin of the solar system.

An Ariane 5 rocket successfully launched the *Rosetta* spacecraft on March 2, 2004, from the Kourou launch complex in French Guiana. As of July 31, 2006, *Rosetta* continues in the interplanetary cruise phase of its mission. The spacecraft has a complex trajectory, including three Earth and one Mars gravity-assist maneuvers, before it finally reaches the target comet in late 2014. Upon arrival at Comet 67P/Churyumov-Gerasimenko, the spacecraft will study the comet remotely as well as by means of a sophisticated instrument probe, called the *Philae* lander, which will land on the surface of the nucleus. Following a successful landing, *Philae* will transmit data from the comet's surface to the *Rosetta* spacecraft, which will then relay these data back to scientists on Earth. The *Rosetta* spacecraft will orbit around the comet as it passes through perihelion (which occurs in August 2015) and continue to remain in orbit around the comet until the nominal end of the mission (which is scheduled to take place in December 2015).

The *Rosetta* spacecraft carries the following scientific instruments: an imager, infrared and ultraviolet spectrometers, a plasma package, a magnetometer, particle analysis instruments, and a radio-frequency sounder to study any subsurface layering of materials in the comet's nucleus. Instruments carried by the lander probe include an imager, a magnetometer, and an alpha particle/proton/X-ray spectrometer to determine the chemical composition of materials on the surface of the comet's nucleus.

Comet 67P/Churyumov-Gerasimenko was discovered in 1969 by astronomers Klim "Comet" Churyumov and Svetlana Gerasimenko from Kiev (Ukraine) as they were conducting at survey of comets at the Alma-Ata Astrophysical Institute. This comet has been observed from Earth on six approaches to the Sun—1969 (discovery), 1976, 1982, 1989, 1996, and 2002. These observations have revealed that the comet has a small nucleus (about 1.9 miles [3 km] by 3.1 miles [5 km]), which rotates in a period of approximately 12 hours. The comet travels around the Sun in a highly eccentric (0.632) orbit, characterized by an orbital period of 6.57 years, a perihelion distance from the Sun of 1.29 astronomical units (AU) and an aphelion distance from the Sun of 5.74 AU.

The *Rosetta* spacecraft was originally going to rendezvous with and examine Comet 46P/Wirtanen. Because the previously planned launch date, was postponed, however, the original target comet was dropped and a new target comet (Comet 67P/Churyumov-Gerasimenko) was selected by ESA mission planners.

✧ *Near Earth Asteroid Rendezvous (NEAR)* Spacecraft

NASA's *Near Earth Asteroid Rendezvous (NEAR)* spacecraft was launched on February 17, 1996, from Cape Canaveral by a Delta II expendable launch vehicle. The *NEAR* spacecraft was equipped with an X-ray/gamma ray spectrometer, a near-infrared imaging spectrograph, a multispectral camera fitted with a charge-coupled device (CCD) imaging detector, a laser altimeter, and a magnetometer. The primary goal of this mission was

An artist's rendering of the *Near Earth Asteroid Rendezvous (NEAR)* spacecraft's encounter with the asteroid Eros, which began on February 14, 2000. After going into orbit around Eros and examining the asteroid for a year, the mission ended when the *NEAR-Shoemaker* spacecraft touched down in the saddle region of the minor planet on February 12, 2001. NASA renamed this robot spacecraft *NEAR-Shoemaker* in honor of the American geologist and astronomer Eugene M. Shoemaker (1928–97). *(NASA/Johnson Space Center; artist, Pat Rawlings)*

to rendezvous with and achieve orbit around the near-Earth asteroid Eros (also called 433 Eros).

Eros is an irregularly shaped S-class asteroid about 8 by 8 by 20.5 miles (13 by 13 by 33 km) in size. This asteroid, the first near Earth asteroid to be found, was discovered on August 13, 1898, by the German astronomer Gustav Witt (1866-1946). In Greek mythology, Eros (Roman name:

Cupid) was the son of Hermes (Roman name: Mercury) and Aphrodite (Roman name: Venus) and served as the god of love.

As a member of the Amor group of asteroids, Eros has an orbit that crosses the orbital path of Mars, but does not intersect the orbital path of Earth around the Sun. The asteroid follows a slightly elliptical orbit, circling the Sun in 1.76 years at an inclination of 10.8 degrees to the ecliptic. Eros has a perihelion of 1.13 astronomical units (AU) and an aphelion of 1.78 AU. The closest approach of Eros to Earth in the 20th century occurred on January 23, 1975, when the asteroid came within 0.15 AU (about 13.7 million miles [22 million km]) of Earth.

After launch and departure from Earth orbit, *NEAR* entered the first part of its cruise phase. The robot spacecraft spent most of this phase in a minimal activity (hibernation) state that ended a few days before the successful flyby of the asteroid Mathilde on June 27, 1997. During that encounter, the spacecraft flew within 745 miles (1,200 km) of Mathilde at

This is a mosaic of four images of Eros taken by the *NEAR-Shoemaker* robot spacecraft on September 5, 2000, from a distance of about 62 miles (100 km) above the asteroid. The knobs sticking out of the surface near the top of the mosaic image surround a boulder-strewn area and are probably remnants of ancient impact craters. *(NASA/JPL/Johns Hopkins University Applied Physics Laboratory)*

a relative velocity of 6.2 miles per second (9.93 km/s). Imagery and other scientific data were collected.

On July 3, 1997, *NEAR* executed its first major deep-space maneuver, a two-part propulsive burn of its main 100-pound-force (450-newton) thruster—a rocket engine that used hydrazine and nitrogen tetroxide as its propellants. This maneuver successfully decreased the spacecraft's velocity by 915 feet per second (279 m/s) and lowered the perihelion from 0.99 AU to 0.95 AU. Then, on January 23, 1998, the spacecraft performed an Earth-gravity-assist flyby—a critical maneuver that altered its orbital inclination from .5 to 10.2 degrees and its aphelion distance from 2.17 AU to 1.77 AU. This gravity-assist maneuver gave *NEAR* orbital parameters that nearly matched those of Eros, the target asteroid.

The original mission plan was to rendezvous with and achieve orbit around Eros in January 1999, and then to study the asteroid for approximately one year. A software problem caused an abort of the first encounter-rocket engine-burn, however; and NASA revised the mission plan to include a flyby of Eros on December 23, 1998. This flyby was then followed by an encounter and orbit on February 14, 2000.

The radius of the spacecraft's orbit around Eros was brought down in stages to a 31-by-31-mile (50-by-50-km) orbit on April 30, 2000, and decreased to a 21.8-by-21.8-mile (35-by-35-km) orbit on July 14, 2000. The orbit was then raised over the succeeding months to 125-by-125-mile (200-by-200-km) and next slowly decreased and altered to a 21.8-by-21.8-mile (35-by-35-km) retrograde orbit around the asteroid on December 13, 2000. The mission ended with a touchdown in the saddle region of Eros on February 12, 2001. NASA renamed the spacecraft *NEAR-Shoemaker* in honor of American astronomer and geologist Eugene M. Shoemaker (1928–97), following his untimely death in an automobile accident on July 18, 1997.

✧ *Dawn* Spacecraft

The *Dawn* spacecraft, which is scheduled to launch in July 2007, is part of NASA's Discovery Program—an initiative for highly focused, rapid-development scientific spacecraft. The goal is to understand the conditions and processes during the earliest history of the solar system. To accomplish its scientific mission, the *Dawn* robot spacecraft will investigate the structure and composition of the two minor planets Ceres and Vesta. These large main-belt asteroids have many contrasting characteristics and appear to have remained intact since their formation more than 4.6 billion years ago.

Ceres and Vesta reside in the extensive zone between Mars and Jupiter with many other smaller bodies, in a region of the solar system that astronomers call the main asteroid belt. The objectives of the mission are to characterize these two large asteroids, with particular emphasis being placed on their internal structure, density, shape, size, composition, and mass. The spacecraft will also return data on surface morphology, the extent of craters, and magnetism. These measurements will help scientists determine the thermal history, size of the core, the role of water in asteroid evolution, and what meteorites found on Earth originate from these bodies. For both asteroids, the data returned will include full surface imagery, full surface spectrometric mapping, elemental abundances, topographic profiles, gravity fields, and mapping of remnant magnetism, if any.

The top-level question that the mission addresses is the role of size and water in determining the evolution of the planets. Scientists consider that Ceres and Vesta are the right two celestial bodies with which to address this important question. Both asteroids are the most massive of the protoplanets in the asteroid belt—miniature planets whose growth was interrupted by the formation of Jupiter. Ceres is very primitive and wet, while Vesta is evolved and dry. Planetary scientists suggest that Ceres may have active hydrological processes leading to seasonal polar caps of water frost. Ceres may also have a thin, permanent atmosphere—an interesting physical condition that would set it apart from the other minor planets. Vesta may have rocks more strongly magnetized than those on Mars—a discovery that would alter current ideas of how and when planetary dynamos arise.

As currently planned, the spacecraft will launch from Cape Canaveral on a Delta 7925H expendable rocket in July 2007. After a four-year heliocentric cruise, *Dawn* will reach Vesta in 2011 and then go into orbit around the minor planet for 11 months. One high-orbit period at 435-mile (700-km) altitude is planned, followed by a low orbit at an altitude of 75 miles (120 km). Upon completion of its rendezvous and orbital reconnaissance of Vesta, the Dawn spacecraft will depart this minor planet in 2012 and fly on to Ceres. The spacecraft will reach Ceres in 2015, where it will again go into 11 months of orbital reconnaissance mission. Both a high-orbit scientific investigation of Ceres at an altitude of 553 miles (890 km) and a low-orbit study at an altitude of 87 miles (140 km) are currently planned. The Ceres orbital reconnaissance phase of this mission will end sometime in 2016.

Aerospace mission planners anticipate that 634 pounds (288 kg) of xenon propellant will be required to reach Vesta and 196 pounds (89 kg) to reach Ceres. The spacecraft's hydrazine thrusters will be used for orbit capture. Depending on the remaining supply of onboard propellants and the general health of the *Dawn* spacecraft and its complement of scientific instruments, mission controllers at the Jet Propulsion Laboratory (JPL) could elect to continue this robot spacecraft's exploration of the asteroid belt beyond 2016.

The *Dawn* spacecraft structure is made of aluminum and is box-shaped with two solar-panel wings mounted on opposite sides. A parabolic fixed high-gain dish antenna (4.6 feet [1.4 m] in diameter) is mounted on one side of the spacecraft in the same plane as the solar arrays. A medium-gain fan-beam antenna is also mounted on the same side. A 16.4-foot- (5-m-) long magnetometer boom extends from the top panel of the spacecraft. Also mounted on the top panel is the instrument bench, holding the cameras, mapping spectrometer, laser altimeter, and star trackers. In addition, there is a gamma ray/neutron spectrometer mounted on the top panel. The solar arrays provide 7,500 watts to drive the spacecraft and the solar electric ion propulsion system.

The *Dawn* spacecraft is the first purely scientific NASA space exploration mission to be powered by ion propulsion. The ion-propulsion technology for this mission is based on the *Deep Space 1* spacecraft ion drive, and uses xenon as the propellant that is ionized and accelerated by electrodes. The xenon ion engines have a thrust of .02 pounds of force (90 millinewtons [mN]) and a specific impulse of 3,100 seconds. The spacecraft maintains attitude control through the use of twelve strategically positioned .2-pound-force (.9 N-thrust) hydrazine engines. The spacecraft communicates with scientists and mission controllers on Earth by means of high- and medium-gain antennae, as well as a low-gain omnidirectional antenna that uses a 135-watt traveling wave-tube amplifier.

In summary, the goal of the *Dawn* spacecraft mission is to help scientists better understand the conditions and processes that took place during the formation of the solar system. Ceres and Vesta represent two of the few large protoplanets that have not been heavily damaged by collisions with other bodies. Ceres is the largest asteroid in the solar system and Vesta is the brightest asteroid—the only one visible with the unaided eye. What makes the Dawn mission especially significant is the fact that Ceres and Vesta possess striking contrasts in composition. Planetary scientists speculate that many of these differences arise from the conditions under which Ceres and Vesta formed during the early history of the solar system. In particular, Ceres formed wet and Vesta formed dry. Water kept Ceres cool throughout its evolution, and there is some evidence to indicate that water is still present on Ceres, as either frost or vapor on the surface or possibly even liquid water under the surface. In sharp contrast, Vesta was hot, melted internally, and became volcanic early in its development. The two large protoplanets followed distinctly different evolutionary pathways. Ceres remains in its primordial state, while Vesta has evolved and changed over millions of years.

Future Generations of Robot Explorers

Over the next three decades, a variety of interesting spacecraft, representing several new generations of space robots, will explore the outermost reaches of the solar system, travel into the inner realm of the solar system, and make contact with several of the potential life-bearing alien worlds that lie between these extreme locations. One common technical characteristic that each new generation of space robot will have is an improved level of machine intelligence. The robot spacecraft mentioned in this chapter will enjoy higher levels of autonomy, fault management, self-assessment and repair, and possibly even exhibit a primitive form of learning.

The great distances involved in exploring the extrema of the solar system—distances of light-minutes with respect to the Sun or light-hours with respect to Pluto and the Kuiper belt beyond—preclude any possibility of real-time human control or interaction with the spacecraft during the prime data-collecting phase of its mission. The far-traveling robot must be able to take care of itself, especially if it encounters an unanticipated anomaly in its environment or a glitch in one of its onboard instruments.

Other future automated missions will require a level of machine intelligence capable of prompt decision-making to ensure the robot's survival. One example is that of a robot airplane flying low in the Martian atmosphere as it skims across the surface of Red Planet in search of interesting water-related sites. Future robots with suitable design and instrumentation will also allow a human controller to perform hazardous planetary exploration from the comfort of a permanent lunar base or Mars surface base. Virtual reality might turn out to be a better experience for a human explorer than physical presence in a cumbersome space suit. When supported by a technically rich, virtual-reality environment (that is, a computer-simulated environment based on real, at-the-scene data) and a properly "wired" robot, teleoperation and telepresence should work very well. The major limitation of using this technique in space exploration will be the speed-of-light distance between the human participant and the collaborative robot that mimics human behaviors. This distance should not exceed a few light-seconds, or else the human being will not be able to respond properly. In situations with more than five-second time delays in the communications loop, the brain of the human controller might not have time to recognize a serious problem and respond before the at-risk collaborative robot would have become toast—that is, have injured itself or destroyed itself.

Today, when human controllers at NASA's Jet Propulsion Laboratory interact with either the *Spirit* or *Opportunity* robot rovers on Mars, the resultant travel by either rover across the surface takes place at an extremely slow, but prudently cautious, pace. Progress is typically measured in feet (meters) traveled per day, not in miles (kilometers) per day.

Collaborative control and advance levels of human/robot interaction will form the operational basis for many important activities on the Moon or Mars performed by future generations of robots. While some of tomorrow's lunar robots may be controlled from Earth (it takes about 2.6 seconds for a radio wave to travel to the Moon and return to Earth), the most flexible use of teleoperated planetary rover robots will occur when human beings can exercise collaborative control over these advanced robots while remaining in the shirtsleeve comfort of a sheltered habitat on the surface of the Moon or Mars. The space robot takes the heat (or cold), the dust, and the risk, while its human partner makes the discovery, gathers targeted resources, or performs/supervises various construction tasks in and around the surface base.

In the future, semi- or fully autonomous robots will perform detailed exploration of the icy regions on Mars and search for life within the suspected subsurface liquid-water ocean of Europa. The other icy moons of Jupiter (Ganymede and Callisto) may also provide some interesting scientific surprises as a planned robot mission called the *Jupiter Icy Moons Orbiter (JIMO)* pays them a visit. Finally, a truly rugged and robust space robot, called the *Star Probe,* will go where no robot has dared to go before. Traveling through extreme environments characteristic of the innermost regions of the solar system, *Star Probe* will allow scientists to make their first really close-up measurements of a star's outer regions—the target star being humans' parent star, the Sun.

✧ New Horizons Pluto–Kuiper Belt Flyby Mission

Originally conceived as the *Pluto Fast Flyby (PFF),* NASA's New Horizons Pluto–Kuiper Belt Flyby mission was launched on January 19, 2006, from Cape Canaveral. This reconnaissance-type exploration mission will help

This is an artist's concept of NASA's *New Horizons* spacecraft during its planned encounter with the tiny planet Pluto (foreground), and its relatively large moon, Charon—possibly as early as July 2016. [Astronomical observations in 2005 suggest that Pluto may also have two smaller moons, which do not appear in this rendering.] A long-lived, plutonium-238-fueled, radioisotope thermoelectric generator (RTG) system (cylinder on lower left portion of the spacecraft) provides electric power to the robot spacecraft as it flies past these distant icy worlds billions of miles (km) from the Sun. As depicted here, one of the spacecraft's most prominent features is a 6.9-foot- (2.1-m-) diameter dish antenna, through which *New Horizons* can communicate with scientists on Earth from as far as 4.7 billion miles (7.5 billion km) away. Following its encounter with Pluto, the *New Horizons* spacecraft hopes to explore one or several icy planetoid targets of opportunity in the Kuiper belt. *(NASA/JPL)*

scientists understand the interesting, yet poorly understood, worlds at the edge of the solar system. The first robot spacecraft flyby of Pluto and Charon, the frigid double-planet system, could take place as early as 2015.

The mission will then continue beyond Pluto and visit one or more Kuiper belt objects of opportunity by 2026. The spacecraft's long journey will help to resolve some basic questions about the surface features and properties of these icy bodies, as well as to examine their geology, interior makeup, and atmospheres.

With respect to the Pluto-Charon system, some of the major scientific objectives include the characterization of the global geology and geomorphology of Pluto and Charon, the mapping of the composition of Pluto's surface, and the determination of the composition and structure of Pluto's transitory atmosphere. The spacecraft is intended to reach Pluto before the tenuous Plutonian atmosphere can refreeze onto

Kuiper Belt

The Kuiper belt is a vast region of billions of solid, icy planetesimals, or cometary nuclei, lying in the far outer regions of the solar system. This belt is believed to extend from the orbit of Neptune (about 30 astronomical units [AU]) out to a distance of 1,000 AU from the Sun. The existence of this region was first suggested in 1951 by the Dutch-American astronomer, Gerard P. Kuiper (1905–73), for whom it is now named.

The first Kuiper belt object, called 1992 QB, was discovered in 1992. This icy planetesimal has a diameter of approximately 125 miles (200 km), an orbital period of some 296 years, and an average distance from the Sun of about 44 astronomical units. 1992 QB is about the size of a major asteroid, with the suspected icy composition of a cometary nucleus. It is, therefore, similar to an interesting group of icy bodies called the Centaurs that are found in the outer solar system between the orbits of Neptune and Saturn.

Quaoar is a large object in the Kuiper belt that was first observed in June 2002. Quaoar is an icy world with a diameter of about 775 miles (1,250 km)—making it about half the size of Pluto. Located some 1 billion miles (1.6 billion km) beyond Pluto and about 4 billion miles (6.4 billion km) away from Earth, Quaoar takes 285 years to go around the Sun. The icy planetesimal travels in a nearly circular orbit around the Sun. The name Quaoar (pronounced kwah-o-wahr) comes from the creation mythology of the Tongva—a Native American people who inhabited the Los Angeles, California, area before the arrival of European explorers and settlers.

Like the planet Pluto, Quaoar dwells in the Kuiper belt—an icy debris field of comet-like bodies extending 5 billion miles (8 billion km) beyond the orbit of Neptune. While astronomers generally treat Pluto as both a planet and a member of the Kuiper belt, they regard Quaoar, however, as simply a Kuiper belt object (KBO). In addition, despite its size, astronomers do not consider Quaoar to be the long-sought, hypothesized tenth planet. Nevertheless, it remains an intriguing and impressive new world, most likely consisting of equal portions of rock and various ices, including water ice, methane ice (frozen natural gas), methanol ice (alcohol ice), carbon-dioxide ice (dry ice), and carbon-monoxide ice. Measurements made at the Keck Telescope indicate the presence of water ice on Quaoar.

The Kuiper belt is thought to be the source of the short-period comets that visit the inner solar system. Scientists now believe that the icy objects found in this region are remnants of the primordial materials from which the solar system formed.

the surface as the planet recedes from its 1989 perihelion. Studies of the double-planet system will actually begin some 12 to 18 months before the spacecraft's closest approach to Pluto in about 2015. The modest-sized spacecraft will have no deployable structures and will receive all of its electric power from long-lived radioisotope thermoelectric generators (RTGs) that are similar in design to those used on the *Cassini* spacecraft now orbiting Saturn.

This important mission will complete the initial scientific reconnaissance of the solar system with robot spacecraft. At present, Pluto is the most poorly understood planet in the solar system. As some scientists speculate, the tiny planet may even be considered the largest member of the family of primitive icy objects that reside in the Kuiper belt. In addition to the first close-up view of Pluto's surface and atmosphere, the spacecraft will obtain gross physical and chemical surface properties of Pluto, Charon, and (possibly) several Kuiper belt objects.

✧ Telepresence, Virtual Reality, and Robots with Human Traits

Telepresence, or virtual residency, makes use of telecommunications, interactive displays, and a collection of sensor systems on a robot (which is at some distant location) to provide the human operator with a sense of actually being present where the robot system is located. Depending on the level of sophistication in the operator's workplace as well as upon the robot system, this telepresence experience can vary from a simple stereoscopic view of the scene to a complete virtual-reality activity, in which sight, sound, touch, and motion are provided.

Telepresence actually combines the technologies of virtual reality with robotics. Some day in the not-too-distant future, human controllers (on Earth or at the permanent lunar base), will wear sensor-laden bodysuits and three-dimensional viewer helmets so that they can use telepresence to actually to walk and work on the Moon. Human explorers will also use telepresence to direct their machine surrogates across the surface of Mars. Properly designed future robots will allow human beings to efficiently explore remote regions of the Red Planet from the comfort and safety of the first Mars surface base, or perhaps from an orbiting, human-crewed spacecraft.

Virtual reality (VR) is a computer-generated artificial reality that captures and displays in varying degrees of detail the essence or effect of physical reality (that is, the real-world scene, event, or process) being modeled or studied. With the aid of a data glove, headphones, and/or

This is an artist's concept of a robot field geologist, called the TeleProspector. This advanced mobile robot would be capable of allowing human geologists comfortably located at a permanent lunar base (or back on Earth) to extend their visual and tactile senses to a remote location on the Moon through telepresence and virtual-reality technologies. Enabled by the robot's stereovision, motion sensors, and ability to duplicate human movements and provide tactile sensations, the human operator is surrounded by a virtual experience that mimics much of the environment the robot is physically experiencing in the field. Here, for example, both the robot and the human geologist (through virtual reality and telepresence) have just discovered a cluster of interesting crystals carried up to the Moon's surface from many miles (km) below by an ancient lava flow. *(NASA/Johnson Space Center; artist, Pat Rawlings)*

head-mounted stereoscopic display, a person is projected into the three-dimensional world created by the computer.

A virtual-reality system generally has several integral parts. There is always a computerized description (that is, the database) of the scene or event to be studied or manipulated. It can be a physical place, such as a planet's surface made from digitized images sent back by space robots. It can even be more abstract, such as a description of the ozone levels at various heights in Earth's atmosphere or the astrophysical processes occurring inside a pulsar or a black hole.

VR systems also use a special helmet or headset (goggles) to supply the sights and sounds of the artificial, computer-generated environment. Video displays are coordinated to produce a three-dimensional effect. Headphones make sounds appears to come from any direction. Special sensors track head motions, so that the visual and audio images shift in response.

Most VR systems also include a glove with special electronic sensors. The data glove lets a person interact with the virtual world through hand gestures. He (or she) can move or touch objects in the computer-generated visual display, and these objects then respond as they would in the physical world. Advanced versions of such gloves also provide artificial tactile sensations, so that an object feels like the real thing being touched or manipulated (for example, smooth or rough, hard or soft, cold or warm, light or heavy, flexible or stiff, etc.).

The field of virtual reality is quite new, and rapid advances should be anticipated over the next decade, as computer techniques, visual displays, and sensory feedback systems (for example, advanced data gloves) continue to improve in their ability to project and model the real world. VR systems have many potential roles in the aerospace industry and space exploration. For example, sophisticated virtual-reality systems will let scientists walk on another world while working safely here on Earth or at a safe and secure planetary surface base.

Future space-mission planners will use virtual reality to identify the best routes (based on safety, resource consumption, and mission objectives) for both robots and humans to explore the surface of the Moon and Mars, before the new missions are even launched. Astronauts will use VR training systems regularly to try out space maintenance and repair tasks and to perfect their skills long before they lift off on an actual mission. Aerospace engineers will use VR systems as an indispensable design tool to fully examine and test new aerospace hardware, long before any metal is bent in building even a prototype model of the item.

NASA researchers envision future robots that mimic people, so as to enable these advanced machines to work more efficiently with astronauts in outer space or on planetary surfaces. The immediate focus of this research is not to develop robots that have the same thought processes as humans do, but rather to have the robots act and respond "more naturally"—that is, to behave in ways similar to the ways in which human beings interact with each humans. In collaborative control, a human being and advanced robot speak to one another and work as partners. The robot does not have to be anthropomorphic, but this characteristic could be very useful in some circumstances. A key benefit of collaborative control is that the robot would be able to ask its human partner questions, in order to compensate for its machine-intelligence limitations.

To further accelerate the development of robots with human traits, NASA researchers are also building robots that have reasoning mechanisms (machine intelligence) that work in a fashion similar to human reasoning. The more a robot begins to think like a human being, the greater the probability that the human-robot team will understand each other in performing a specific task or executing some exploration protocol. The overarching goal of these efforts is to make the human-robot interaction more natural and humanlike.

Within NASA's basic space-exploration mission, there are many specific tasks on which robots and humans can collaborate. Some of these human-robot interaction areas are shelter construction on a planet's surface, the assembly and inspection of pipes, pressure vessel assembly, habitat inspection, and the collection and transport of resources found on a planet's surface. The robots will help assemble buildings, test equipment, dig with small tools, and weld structures. Human-robot teams will use a checklist and a plan to guide their collaborative efforts. Conventional human-robot dialogue has been limited to "master-slave" commanding and supervising (monitoring), that is, the way most robots have been controlled up to now. This limits the performance of the robotic system to the skill of the human operator and the quality of the interface (usually

Androids and Cyborgs

An android is an anthropomorphic machine—that is, a robot with near-human form, features, and/or behavior. Although it originated in science fiction, engineers and scientists now use the term android to describe robot systems being developed with advanced levels of machine intelligence and electromechanical mechanisms, so that the machines can "act" like people. A future human-form field geologist robot, able to communicate with its human partner (as the team explored the surface of the Moon) by using a radio-frequency transmitter, as well as by turning its head and gesturing with its arms, would be an example of an android.

The term *cyborg* is a contraction of the expression: *cyb*ernetic *org*anism. Cybernetics is the branch of information science dealing with the control of biological, mechanical, and/or electronic systems. While the word *cyborg* is quite common in contemporary science fiction—for example, the frightening "Borg collective" in the popular *Star Trek: The Next Generation* motion picture and television series—the concept was actually first proposed in the early 1960s by several scientists who were then exploring alternative ways of overcoming the harsh environment of space. The overall strategy they suggested was simply to adapt a human being to space by developing appropriate technical devices that could be incorporated into an astronaut's body. With these implanted or embedded devices, astronauts would become cybernetic organisms, or cyborgs.

computer-based) between the human and the robot. One real advantage of the human-robot relationship within the collaborative effort environment is that the human operator does not have to continuously engage in robot teleoperation or supervision.

In field demonstration tests on Earth, robots have worked as field geologists in partnership with human scientists. In the future, these human-like machines will also do a great deal of nonscientific work. They will not only search for resources on the surface of a planetary body, but they will also be equipped to process the materials they discover. Other future jobs for robots that are developed with human behavior traits are scouting, surveying, carrying equipment, inspecting and maintaining resource-harvesting machines, and inspecting, monitoring, and repairing human habitats on the surface of a planet. Human-like machines would autonomously perform their tasks and seek help from human beings only when they encounter problems they cannot resolve themselves.

Designing a human-like robot, or android, that can work autonomously and ask for human help and knowledge only when really necessary is a very challenging task in machine intelligence. But the prospect exists for intelligent androids to assist human astronauts in space exploration

Instead of simply protecting an astronaut's body from the harsh space environment by enclosing the person in some type of spacesuit, space capsule, or artificial habitat (the technical approach actually chosen), the scientists who advocated the cyborg approach boldly asked, Why not create cybernetic organisms that could function in the harsh environment of space without special protective equipment? For a variety of technical, social, and political reasons, this proposed line of research quickly ended, but the term cyborg has survived.

Today, the term is usually applied to any human being (whether on Earth, under the sea, or in outer space) using a technology-based, body-enhancing device. For example, a person with a pacemaker, hearing aid, or an artificial knee could be considered a cyborg. When a person straps on wearable, computer-interactive components, such as the special vision and glove devices that are used in a virtual-reality system, that person has (in fact) become a temporary cyborg.

By further extension, the term cyborg is sometimes used to describe fictional artificial humans or very sophisticated robots with near-human (or super-human) qualities. The Golem (a mythical clay creature in medieval Jewish folklore) and the Frankenstein monster (from Mary Shelley's classic 1818 novel *Frankenstein: The Modern Prometheus*) are examples of the former, while Arnold Schwarzenegger's portrayal of the superhuman Terminator robot (in the *Terminator* motion picture trilogy) is an example of the latter usage.

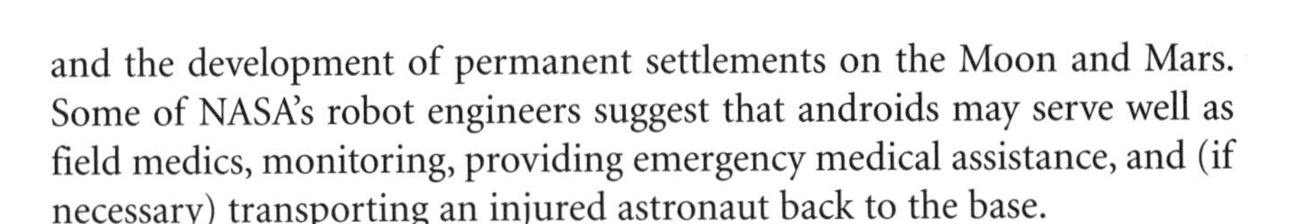

and the development of permanent settlements on the Moon and Mars. Some of NASA's robot engineers suggest that androids may serve well as field medics, monitoring, providing emergency medical assistance, and (if necessary) transporting an injured astronaut back to the base.

✧ Mars Airplane

The Mars airplane is a conceptual low-mass robotic (uncrewed) aerial platform that can deploy experiment packages or conduct detailed reconnaissance operations on Mars. In some mission scenarios, the Mars airplane would be used to deploy a network of science stations, such as seismometers or meteorology stations, at selected Martian sites, with an accuracy of a few miles (kilometers).

When designed with a payload capacity of about 110 pounds (50 kg), the robot flying platform could collect high-resolution images or conduct detailed geochemical surveys of candidate surface sites of great interest in exobiology. The ultralight aerial robot would be capable of flying at altitudes between 1,640 feet (500 m) and 9.3 miles (15 km), with corresponding ranges of 15.5 miles (25 km) to 4,165 miles (6,700 km). Scientists

This artist's concept shows a robot airplane exploring Mars (circa 2020). The hydrazine-powered Mars aircraft would have a 65-foot (19.8-m) wingspan and would be deployed from a package parachuted to the surface of the Red Planet by an orbiting mother spacecraft. This conceptual Mars aircraft would be capable of cruising 2,500 miles (4,000 km) with a 100-pound (45-kg) scientific-instrument payload, including imaging capability. *(NASA)*

might deploy a robot airplane on Mars to perform aerial reconnaissance up long valleys and canyons. Flying in a giant canyon, the robot airplane would cover a large amount of interesting territory and gather very high-resolution images. Such scouting missions would identify specific sites worthy of more detailed study by surface rovers and/or human explorers.

NASA strategic planners have entertained two basic design approaches for a Mars airplane. In the first approach, the airplane is designed as a one-way disposable aerial platform. After descending into the thin Martian atmosphere from a mother spacecraft, the robot airplane would automatically deploy its large wings and perform aerial surveys, atmospheric soundings, and other scientific investigations, finally crashing when its hydrazine fuel supply is exhausted.

In the second design scenario, engineers equip the Mars airplane with a small, variable-thrust rocket motor and land gear, so that it can make a soft (survivable) landing on the surface of the Red Planet, conduct some scientific investigation, and then take off. Because the Martian atmosphere is so thin, taking off from the ground requires an aircraft with very big wings and a power plant that supports a very fast takeoff. A rocket-assisted takeoff represents one viable engineering approach. This type of robot aircraft would have the ability to make in situ measurements and to gather samples at several widely separated sites on the Red Planet. The soil specimens could be examined on the spot or else delivered to a lander/ascent vehicle robot spacecraft, as part of a Mars sample-return mission. (Chapter 7 provides a discussion of sample-return missions.)

Mars mission planners recognize that a fleet of robot aircraft would provide a great deal of exploration flexibility and support to a human expedition to Mars. These aerial platforms could help the astronauts evaluate candidate-landing sites, deploy special sensors in support of network science projects, or collect soil and rock specimens from remote locations. Should several of the astronauts get stranded or lost while exploring the surface, Mars airplanes could effectively perform wide-area search operations. Finally, a Mars airplane, equipped with radio frequency transmitter/receiver hardware, could loiter in a fixed high-altitude holding pattern and serve as a temporary telecommunications relay station between astronaut explorers and their base camp or between astronauts at the base camp and a team of robot rovers, automated science stations, or other robot aircraft.

✧ Robots Exploring Icy Regions

The icy, northern polar region of Mars is interesting to exobiologists, because that is where the (frozen) water is—and where there is water (even in the form of ice), there may be life, extant or extinct. Similarly, close-up study of Europa by NASA's *Galileo* spacecraft has provided tantalizing hints that this major moon of Jupiter may possess a liquid water ocean beneath its icy crust. Once again, where there is liquid water, there is the intriguing possibility that alien life-forms (ALFs) may be found. So NASA planners are examining (on a conceptual and limited experimental basis) several future robot missions to quite literally "break the ice" in the search for alien life beyond Earth.

After scientists have identified a geophysical signature for water, from orbit around Mars or from surface rover mission investigations, the next important step is to drill in that location. The goal of exobiologists is to "follow the water" in search of signs of life on Mars. Although scientists do not know much about subsurface conditions on the Red Planet, data from previous missions to Mars have provided abundant evidence that in ancient times the planet once had surface water, including streams and possibly shallow seas or even oceans. While there is photographic evidence for recent gullies, possibly cut by flowing water, there is no evidence for liquid water currently at the surface.

Drawing upon experience with some of the colder parts of Earth, for example, Iceland or Antarctica, scientists know that water can be stored as a mixture of frozen mud and ice in a layer of permafrost, and, beneath a permafrost layer, as liquid groundwater. So even if the ancient surface

This is an artist's concept of a cryobot—a robot probe for penetrating into the icy surface of a planet or moon. The cryobot moves through ice by melting the surface directly in front of it, while allowing liquid to flow around the torpedo-shaped robot probe and refreeze behind it. As it makes its mole-like passage into an alien world, the cryobot's instruments take measurements of the encountered environment and send collected data back to the surface lander. On Mars, it appears that a communications cable could be used for penetration of shallower depths. On Europa, the thicker ice would require use of a network of mini–radio wave transceiver relays embedded in the ice. The use of semiautonomous steering and levels of artificial intelligence that promote fault management will help reduce the risk of the robot probe's getting trapped by subsurface obstructions, such as large rocks. *(NASA/JPL)*

water on Mars has evaporated, there may still be substantial reservoirs of water, in either liquid or frozen form, beneath the planet's surface.

To get to the zone where frozen water, and possibly dormant life, may be present on Mars, scientists anticipate that they will have to drill or penetrate to a depth of about 660 feet (200 m). In all likelihood, any liquid water (if present) will be even deeper below the surface. Deep subsurface access on Mars presents unique engineering challenges. One approach to reach below the polar-region surface on Mars is to use a system called the cryobot ice-penetrating robot probe.

After a successful soft landing in the treacherous northern polar regions of Mars, the lander spacecraft activates the ice-penetrating probe it carries. By heating the torpedo-shaped nose of the cryobot, the device is pulled by gravity down through a tunnel that the probe melts in the ice. Instruments carried within the body of the cryobot automatically take measurements and perform analyses of the gases and other materials encountered. As a smart robot, the cryobot probe will use innovative heating and steering features to adroitly maneuver around subsurface obstacles (primarily large rocks). Endowed with a high level of machine intelligence, the cryobot probe even has the ability to alter its downward course slightly to adjust for subsurface conditions or to exploit unexpected scientific opportunities, such as an encounter with a liquid-water aquifer deep beneath the frozen surface.

The cryobot moves through ice by melting the surface directly in front of it and allowing the liquid to flow around the robot probe and refreeze behind it. The probe takes measurements that characterize the encountered environment and then relays the scientific data up to the lander craft through a thin cable spooled out from its aft section as the robot probe descends into the frozen material. For ice layers more than a mile (1.6 km) or so thick, it may be more practical to have the robot-probe communicate back to the lander at the surface through a series of mini–radio wave transceiver relays that the probe deposits in the resolidified material as it descends.

The cryobot probe represents an innovative combination of active and passive melting systems. The cryobot method of subsurface penetration is more effective than conventional drilling techniques because it uses less power than mechanical cutting. Furthermore, since the cryobot travels downward in a self-sealing pathway through the ice, there is no deeply drilled hole that must be encased with massive steel tubes to prevent cave-in or collapse. Finally, the use of semiautonomous steering and fault management allow the probe to reduce the risk of its getting trapped by unanticipated subsurface conditions or obstructions. As will be discussed very shortly, the cryobot is also an ideal robot system for penetrating the ice crust of Europa.

EUROPA

Europa is the smooth, ice-covered moon of Jupiter, discovered in 1610 by the Italian scientist, Galileo Galilei (1564–1642). Flyby visits by robot spacecraft lead scientists to think that this intriguing moon has a liquid-water ocean beneath its frozen surface. Europa has a diameter of 1,942 miles (3,124 km) and a mass of 10.6 × 10^{22} pounds (4.84 × 10^{22} kg). The moon is in synchronous orbit around Jupiter at a distance of 416,970 miles (670,900 km). An eccentricity of 0.009, an inclination of 0.47 degree, and a period of 3.551 (Earth) days further characterize the moon's orbit around Jupiter. The acceleration of gravity on the surface of Europa is 4.33 feet per second per second (1.32 m^2/s) and the icy moon has an average density of 188 pounds per cubic foot (3,020 kg/m^3). Next to Mars, exobiologists favor Europa as a world within the solar system that may possibly harbor some form of alien life.

NASA's proposed *Jupiter Icy Moons Orbiter (JIMO)* mission involves an advanced-technology robot spacecraft that would orbit three of Jupiter's most intriguing moons—Callisto, Ganymede, and Europa. All three planet-sized moons may have liquid-water oceans beneath their icy surfaces. Following up on the historic discoveries made by the *Galileo* spacecraft, the *JIMO* mission would make detailed studies of the makeup, history, and potential for sustaining life of each of these three large icy moons. The mission's proposed scientific goals include: scouting for potential life on the moons, investigating the origin and evolution of the moons, exploring the radiation environment around each moon, and determining how frequently each moon is battered by space debris.

The *JIMO* spacecraft would pioneer the use of electric propulsion powered by a nuclear fission reactor. Contemporary electric-propulsion technology—successfully tested on the NASA's *Deep Space 1* spacecraft—would allow the planned *JIMO* spacecraft to orbit three different moons during a single mission. Current spacecraft, like *Cassini,* have enough onboard propulsive-thrust capability (upon arrival at a target planet) to orbit that single planet and then use various orbits to fly by any moons or other objects of interest, such as ring systems. In contrast, the *JIMO*'s proposed nuclear-electric-propulsion system would have the necessary long-term thrust capability to gently maneuver through the Jovian system and allow the spacecraft to successfully orbit each of the three icy moons of interest.

One very interesting robot space mission to search for life on Europa involves an orbiting mother spacecraft, a cryobot and a hydrobot. The

This is an artist's concept of NASA's proposed Project Prometheus nuclear reactor–powered, ion-propelled spacecraft entering the Jovian system, circa 2015. The Jupiter Icy Moons Orbiter (JIMO) mission would perform detailed scientific studies of Callisto, Ganymede, and Europa (in that order), searching for liquid-water oceans beneath their frozen surfaces. Europa is of special interest to the scientific community, because its suspected ocean of liquid water may contain alien life-forms. *(NASA/JPL)*

Europa Orbiter would serve as the robot command post for the entire mission. Once in orbit around Europa, this robot spacecraft would release a lander robot to a special location on the surface, identified as being of great scientific interest to exobiologists and other investigators. After it soft-lands up on the ice-covered surface of Europa, the lander deploys a large cryobot probe, which also contains a hydrobot. The cryobot probe melts its way down through the icy crust of Europa until it reaches the suspected subsurface ocean. The cryobot has left a trail of radio transponders behind

This artist's concept shows a proposed ice-penetrating cryobot (background) and a submersible *hydrobot* (foreground)—an intriguing advanced robot combination that could be used to explore the suspected ice-covered ocean on Jupiter's moon Europa. In this scenario, a lander robot would arrive on Europa's surface and deploy the cryobot/hydrobot package, remaining on the surface to function as a communications relay station. The cryobot would melt its way through the ice cover and then deploy the hydrobot into the ice-covered ocean. The hydrobot is a self-propelled underwater vehicle that can analyze the chemical composition of the subsurface ocean and search for signs of alien life. The artwork here shows the autonomous robot submarine (hydrobot) examining a *hypothesized* underwater thermal vent and various alien aquatic life-forms gathered around his life-sustaining phenomemon. *(NASA/JPL)*

in the resolidified ice in order to communicate with the lander, which in turn relays data to the orbiting mother spacecraft in burst-transmission mode. The mother spacecraft (*Europa Orbiter*) keeps scientists back on Earth informed of the mission's progress, but this is a totally automated operation without any direct human supervision of the orbiter spacecraft, the cryobot, or the hydrobot.

Once the cryobot has penetrated Europa's thick icy crust and found the currently suspected subsurface ocean, the torpedo-shaped robot probe releases an autonomous, self-propelled underwater robot, called the *hydrobot.* The hydrobot scoots off and starts making scientific measurements of its aquatic environment. The robot submarine also diligently investigates the waters of Europa for signs of alien life. Data from the hydrobot are relayed back to Earth via the cyrobot, surface lander, and orbiting mother spacecraft. The accompanying figure is an artist's rendering of what would be one of the great scientific discoveries of this century. This *hypothesized* scene shows the hydrobot examining an underwater thermal vent and various alien life-forms (ALFs) in Europa's subsurface ocean. The team of very smart robots (orbiter, lander, cryobot, and hydrobot) in this postulated scenario has allowed their human creators to discover life on another world in the solar system. Whether or not life actually exists on Europa, this type of advanced space-exploration mission, with a team of future robots exercising collective machine intelligence, will be remarkable and will serve as a precursor to even more exciting missions.

✧ Star Probe Mission

Star Probe is a conceptual robot spacecraft that can survive an approach to within about 1 million miles (1.6 million km) of the Sun's surface (photosphere). This close encounter with the nearest star will give scientists their first direct (in situ) measurements of the physical conditions in the corona (the Sun's outer atmosphere). The challenging mission requires advanced robot-spacecraft technologies, including superior thermal protection, specialized instrumentation, guidance and control, communications, and propulsion. NASA's advanced-mission planners suggest that this type of robot-spacecraft mission might be flown sometime between the years 2020 and 2030, with 2030 being the more conservative projection.

The science objects of *Star Probe* include a determination of where and what physical processes heat the Sun's corona and accelerate the solar wind to its supersonic velocity. As now envisioned, the robot probe will combine remote sensing of the corona with in situ sampling within the corona, to produce a unique set of measurements not collected by any other spacecraft. Because of the extreme thermal environment in which the probe will have to operate, radioisotope thermoelectric generator units

This artist's concept shows a solar probe traveling within a million miles of the Sun. The robot spacecraft would use this visit to the nearest star to perform first-hand investigations of the physical conditions in the solar corona. The primary science objective of the solar probe is to help physicists to understand the processes that heat the solar corona and produce the solar wind. *(NASA)*

will be used as a reliable source of electric power throughout the mission. Gravity-assist from the planet Jupiter will be used to give the three-axis stabilized spacecraft the final velocity it needs to fly on a trajectory very close to the Sun and sample the corona.

Star Probe will operate in hostile space environments over interplanetary distances ranging from about .2 astronomical unit (AU) to 5 AU from the Sun. One astronomical unit corresponds to a distance of 93 million miles (149.6 million km). When the probe is at a distance of about .02 AU, the robot spacecraft will be just 1 million miles (1.6 million km) from the visible surface of the Sun (the photosphere) and approximately 1.4 million miles (2.3 million km) from the center of its thermonuclear reacting core. When the probe is at 5 AU away from the Sun, it will be traveling in close proximity to the planet Jupiter.

✧ Space Nuclear Power

Through the cooperative efforts of the U.S. Department of Energy (DOE), formerly called the Atomic Energy Commission, and NASA, the United States has used nuclear energy in its space program to provide electrical power for many missions, including science stations on the Moon, extensive exploration missions to the outer planets—Jupiter, Saturn, Uranus, Neptune, and beyond—and even to search for life on the surface of Mars.

For example, when the *Apollo 12* mission astronauts departed from the lunar surface on their return trip to Earth (November 1969), they left behind a nuclear-powered science station that sent information back to scientists on Earth for several years. That science station, as well as similar stations left on the Moon by the *Apollo 14* through *17* missions, operated on electrical power supplied by plutonium-238-fueled, radioisotope thermoelectric generators (RTGs). Since 1961, nuclear-power systems have helped assure the success of many space missions, including the *Pioneer 10* and *11* missions to Jupiter and Saturn; the *Viking 1* and *2* landers on Mars; the spectacular *Voyager 1* and *2* missions to Jupiter, Saturn, Uranus, Neptune, and beyond; the *Ulysses* mission to the Sun's polar regions; the *Galileo* mission to Jupiter, and the *Cassini* mission to Saturn.

Energy supplies that are reliable, transportable, and abundant represent a very important technology in the development of solar-system civilization. Space nuclear-power systems will play an ever-expanding role in supporting more ambitious deep space–exploration missions by robots and in supporting human spaceflight beyond Earth orbit, when astronauts return to the Moon to build a permanent settlement and then visit Mars to establish a surface base.

Space nuclear-power supplies offer several distinct advantages over the more traditional solar and chemical space-power systems. These advantages include compact size, modest mass requirements, very long operating lifetimes, the ability to operate in extremely hostile environments (such as intense trapped-radiation belts, the surface of Mars, the moons of the outer planets, and even interstellar space), and the ability to operate independent of distance from, or orientation to, the Sun.

Space nuclear-power systems use the thermal energy or heat released by nuclear processes. These processes include the spontaneous (but predictable) decay of radioisotopes, the controlled fission or splitting of heavy atomic nuclei (such as fissile uranium-235) in a self-sustained neutron chain reaction, and (eventually) the joining together, or fusing, of light atomic nuclei (such as deuterium and tritium) in a controlled thermonuclear reaction. This nuclear-reaction heat is converted directly or through a variety of thermodynamic (heat-engine) cycles into electric power. Until controlled thermonuclear fusion capabilities are achieved, space nuclear-power applications will be based on the use of either radioisotope decay or nuclear fission reactors.

The radioisotope thermoelectric generator consists of two main functional components: the thermoelectric converter and the nuclear heat source. The radioisotope plutonium-238 has been used as the heat source in all U.S. space missions involving radioisotope power supplies. Plutonium-238 has a half-life of about 87.7 years and therefore supports a long operational life. (The half-life is the time required for one-half the number of unstable nuclei present at a given time to undergo radioactive decay.) In the nuclear decay process, plutonium-238 emits primarily alpha radiation that has very low penetrating power. Consequently, only a small amount of shielding is required to protect the spacecraft from its nuclear emissions. A thermoelectric converter uses the thermocouple principle to directly convert a portion of the nuclear (decay) heat into electricity.

A space fission-power system is a device designed and engineered to generate power for space applications using a nuclear reactor to fission (or split) uranium atoms. During the fission process, a neutron strikes a uranium atom, causing it to release energy as it splits into smaller atoms, called fission products. The released thermal energy (heat) is then converted into electricity through a conversion system to power the spacecraft. This fission process can be sustained and controlled to provide power at needed levels in a continuous manner in a reactor system.

Space reactors are designed differently from terrestrial reactors. The space reactors are much smaller, typically about the size of a 5-gallon can of paint. Aerospace safety engineers also design space reactors to remain in a cold, inactive state until arriving at a designated startup location in space. Once at this designated location the reactor receives the command signal to initiate operation. This design feature enhances system launch and operations safety.

Although the design of a space fission-power system is quite complicated, the basic theory on which it operates is fairly simple. To generate electric power, there are only three basic subsystems: a controlled fission reactor core to produce heat, a cooling loop or mechanism that removes heat from the core, and a power conversion subsystem that receives the heat from the cooling loop and converts a portion of the input heat into electric power. The principles of thermodynamics govern that not all of the input heat can be converted into useful electric energy, so some of the input heat must be rejected to the environment (outer space). Engineers use radiators to remove this excess (or waste) thermal energy from the space power system and reject it to outer space.

Different power-conversion technologies can be used to convert heat from the reactor into electricity. The final choice of a power-conversion technology depends on the requirements of the mission and compatibility with the rest of the spacecraft, including scientific payload. Engineers also use a radiation shield to protect electronic components and other sensitive equipment from the radiation emitted from the reactor during operation.

The Russian space program has flown several space nuclear reactors (most recently a system called Topaz). The United States has flown only one space nuclear reactor, an experimental system called the SNAP-10A, which was launched and operated on-orbit in 1965. The objective of the SNAP-10A program was to develop a space nuclear-reactor power unit capable of producing a minimum of 500 watts-electric for a period of one year, while operating in space. The SNAP-10A reactor was a small (about the size of a garden pail) zirconium hydride (ZrH) thermal reactor fueled by uranium-235. The SNAP-10A orbital test was successful, although the mission was prematurely (and safely) terminated on-orbit by the failure of an electronic component outside the reactor.

Since the United States first used nuclear power in space, great emphasis has been placed on the safety of people and the protection of the terrestrial environment. A continuing major objective in any new space nuclear-power program is to avoid undue risks. In the case of radioisotope power supplies, this means designing the system to contain the radioisotope fuel under all normal and potential accident conditions. For space nuclear reactors, such as the SNAP-10A and more advanced systems, this means launching the reactor in a "cold" (non-operating) configuration and starting up the reactor only after a safe, stable Earth orbit or interplanetary trajectory has been achieved.

✧ The Need for High Levels of Machine Intelligence

Tomorrow's advanced space robots promise to take over much of the data-processing and information-sorting activities that are now performed by human mission controllers here on Earth. In the past, the amount of data made available by space missions has been considerably larger than what scientists could comfortably sift through. For example, the Viking missions to Mars returned image data of the Red Planet that were transferred unto approximately 75,000 reels of magnetic tape. Smart space robots (orbiters, landers, and rovers) with advanced onboard computers will be capable of deciding what information gathered by an orbiting spacecraft or surface robot is worth relaying back to Earth and what information should be stored or discarded. Robots with an inherent ability to selectively filter data will relieve stress on interplanetary telecommunications links and avoid the creation of an avalanche of data that simply buries scientists back on Earth.

Space robots with advanced machine intelligence capable of making these kinds of decisions would have a large number of pattern classification templates or "world models" stored in their computer memories. These templates would represent the characteristics of objects or features of interest in a particular mission. The robot explorers would compare patterns or objects they "see" with their machine-vision systems to patterns or objects stored in their computer memories. Only the objects or patterns that match data in the exploration protocol-classification template would be stored (in computer memory) and then tagged as interesting when these data are sent back to Earth.

Unlike current spacecraft, the advanced future robots would discard any unnecessary or unusable data. As soon as something unusual appeared, however, the smart machine explorer would carefully examine the object or phenomenon more closely. The robot would then report the unusual findings and alert its human mission managers on Earth (or at a permanent lunar settlement or Martian surface base) to the potential significance of the data. This special alert would be the robot's version of the "Eureka" (I've found it) exclamation by the ancient Greek engineer Archimedes (ca. 287 B.C.E.–212 B.C.E.). According to legend, Archimedes was working a particularly challenging problem for the king. He suddenly solved the problem by discovering the principle of buoyancy while taking a bath. Overwhelmed with excitement, he dashed naked out of the bath through the streets of Syracuse to the palace, shouting "eureka" as he ran. Through automated data-selection and data-filtering operations, smart-robot exploring machines will free human experts for more demanding intellectual activities.

The advanced machine intelligence (or artificial intelligence) requirements for a general-purpose space robot used as an exploring machine can be summarized in terms of two fundamental tasks: (1) the smart robot must be capable of learning about new environments; and (2) it must be able to formulate hypotheses about these new environments. Hypothesis formation and learning represent the key problems in the successful development of advanced machine intelligence. Deep space interplanetary and interstellar robotic space systems will need a machine-intelligence system capable of autonomously conducting intense studies of alien world objects. For interstellar probes, the machine-intelligence system may even need to identify, or at least properly catalog, artifacts from an intelligent alien civilization.

Simply stated, the machine-intelligence levels necessary to support these deep space missions must be capable of producing scientific knowledge concerning previously unknown objects. Since the production of scientific knowledge is a high-level intelligence capability, the machine-intelligence requirements for smart autonomous space-robot missions are often called advanced-intelligence machine intelligence, or just advanced machine intelligence.

For a really autonomous deep-space exploration system to undertake knowing and learning tasks, it must have the ability to mechanically or artificially formulate hypotheses using all three of the logical patterns of inference: analytic, inductive, and abductive.

Analytic inference is needed by the advanced space robot to process raw data and to identify, describe, predict, and explain extraterrestrial objects, phenomena, and processes in terms of existing knowledge structures. Inductive inference is needed so that the robot explorer can formulate quantitative generalizations and abstract the common features of objects, phenomena, or processes occurring on alien worlds. Such logic activities amount to the creation of new knowledge structures. Finally, abductive inference is needed by the truly smart robot exploring machines to formulate hypotheses about new scientific laws, theories, concepts, models, etc. The formulation of this type of hypothesis is really the key to the ability of a smart robot to create a full range of new knowledge structures. These new knowledge structures, in turn, are needed if human beings are to successfully use advanced space robots to explore and scientifically investigate all the interesting planetary bodies in this solar system (including moons, asteroids, and comets) and eventually alien worlds around distant stars.

Although the three patterns of inference just described are distinct and independent, they can be ranked by order of difficulty or complexity. Analytical inference is at the low end of the new knowledge-creation scale. An automated system that performs only this type of logic could probably successfully undertake only extraterrestrial reconnaissance missions. A machine capable of performing both analytic and inductive inference could most likely successfully perform space missions that combine reconnaissance and exploration. This assumes, however, that the celestial object being visited is represented well enough by the world models with which the smart robot has been preprogrammed. If the target alien world cannot be well represented by such fundamental world models, however, then automated-exploration missions will also require an ability to perform abductive inference. This logical pattern is the most difficult to perform and lies at the heart of knowledge creation. An automated space-robot system capable of abductive reasoning could successfully undertake missions combining reconnaissance, exploration, and intensive study—all with little or no direct human supervision.

The adaptive machine intelligence needed for advanced robots engaged in interplanetary and interstellar space exploration would include the following characteristics or capacities: learning, memory, and recognition. Learning is the capacity to form universal principles associated with information patterns present in the environment. This machine-intelligence capability subsumes a certain level of hypothesis formation and confirmation. In the process of learning, the smart robot may form new universal conclusions on a probationary basis. The smart robot would then adopt such new principles only after careful confirmation, such as may be achieved through reinforcement or rehearsal. Memory is the capacity to maintain universal conclusions indefinitely. The smart robot with a capacity for memory would exercise long-term recall aided by some recirculating or replicating process. Finally, recognition is the capacity to identify, or classify, information patterns present in the environment on the basis of pre-established universal principles.

A hypothetical scenario will now illustrate all of these points about the role and value of future space robots with advanced levels of machine intelligence. Imagine that in 2035, a very smart future robot rover is exploring the surface of Mars in an essentially free-range mode—that is, it is happily wandering across the surface with no direct human supervision and with a mission to find interesting or unusual sites related to the presence of water on ancient Mars. Suddenly, the smart rover encounters a very interesting geologic site, which it recognizes as being quite similar to a site detected, three decades earlier, and half-a-planet away, by its distant machine ancestor, the *Opportunity* Mars Exploration Rover. The very smart rover accesses its memory and pulls up all the available data from the *Opportunity* mission and the *Mars Science Laboratory (MSL)*, which also visited the other site in 2015. After quickly scanning gigabytes of stored data, the smart rover soon recognizes features common to both the new site and the distant site previously explored by the *Opportunity* and the *Mars Science Laboratory.*

The smart future rover then uses its adaptive machine intelligence to sort through all the information it has just collected about the new site. To reinforce hypothesis formation, it even initiates a few additional tests, using its automated laboratory. Once all the data are sorted, the smart rover begins to form a few new hypotheses and to draw several conclusions (that is, learn) about where the water was and went on ancient Mars. At that point, the smart future rover is ready to share the most recent discovery with its human controllers. So, it contacts the *Mars Telecommunications Orbiter* and sends an interplanetary Eureka message about these new data back to scientists on Earth or to any human explorers, who may be at the Mars surface site.

Does the automated exploration scenario appear a little far out? Perhaps. But mission planners at NASA's Jet Propulsion Laboratory are already considering an advanced space robot called the *Astrobiology Field Laboratory*—a sophisticated roving laboratory that will have the capacity to "scratch and sniff" the Red Planet for signs of past (or present) life at candidate sites after about 2015. These potentially life-bearing (extinct or extant) sites are those associated with the history of water on Mars.

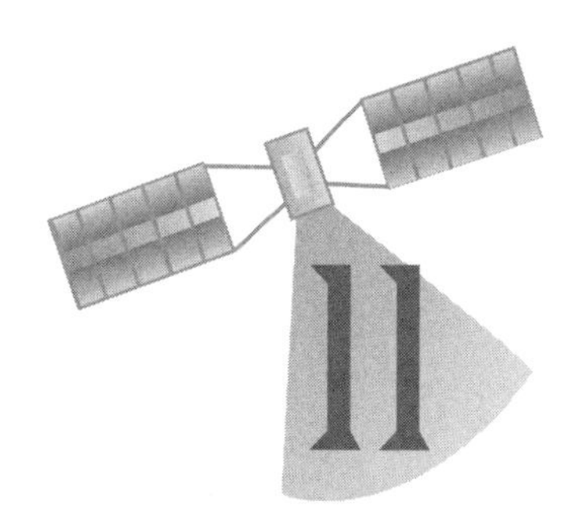

Self-Replicating Systems

Two of the most important products that will be manufactured in space later this century are robots and teleoperator systems. The ultimate goal of advanced space manufacturing systems cannot be achieved without a large expansion of the automation equipment initially provided from Earth. Eventually, space robots and teleoperators must be manufactured in space, drawing from the working experience gathered during the use of the first generation of space industrial robots. A teleoperator is an advanced space-robot system designed with special sensors to accommodate near-real-time operation by a human controller through virtual reality and telepresence technologies. Space industrial robots will support development and construction of the lunar base, assembly of large orbiting facilities and human settlements, and site preparation (including resource processing) for the first Mars surface base.

One lunar base construction scenario has the human workers located in a comfortable (virtual-reality environment) control facility on Earth, while teleoperated space robots do all the pushing, pulling, pounding, and lifting on the lunar surface. Similarly, an interesting human exploration of Mars mission scenario requires that automated resource-extraction facilities be sent ahead to the Red Planet to manufacture (extract) from Martian resources water, oxygen, and even the return-journey rocket propellant—well before the first human explorers even depart from Earth.

The second and third generation of space industrial robots must be far more versatile and fault-tolerant than the first generation devices created on Earth and shipped to extraterrestrial locations as seed or starter machines. The most critical technologies needed for the manufacture of second- and third-generation space industrial robots and teleoperator systems appear to be space-adaptive sensors and computer vision. Enhanced decision-making capabilities and self-preservation features must also be provided in space robots and teleoperators.

Once engineers develop the ability to make robots in space, another fascinating step in the evolution of space robots becomes possible—that of the self-replicating system (SRS). In fact, an SRS unit would appear to behave much like a biological organism.

✧ The Theory and Operation of Self-Replicating Systems

The brilliant Hungarian-American mathematician John von Neumann (1903–57) was the first person to seriously consider the problem of self-replicating systems. His book on the subject, *Theory of Self-reproducing Automata,* was edited by a colleague, Arthur W. Burks (b. 1915), and published posthumously in 1966—almost a decade after von Neumann's untimely death from cancer.

Von Neumann became interested in the study of automatic replication as part of his wide-ranging interests in complicated machines. His work during the World War II Manhattan project (the top secret American atomic bomb project) led him into automatic computing. Through this association, he became fascinated with the idea of large complex computing machines. In fact, he invented the scheme used today in the great majority of general-purpose digital computers—the von Neumann concept of the serial-processing stored program—which is also referred to as the von Neumann machine.

In 1945, von Neumann drafted a report in which he introduced the concept of the stored-program computer. He also recognized that the base 2 approach represented a considerable gain in computer design simplicity over the base 10 approach, which had been used in the world's first working electronic calculator, called the Electronic Numerical Integrator and Computer (ENIAC). ENIAC was the world's first digital computer. It was completed in 1946 and contained 18,000 vacuum tubes. While it was a major step forward in the evolution of "thinking machines," ENIAC stored and manipulated numbers in base 10. Von Neumann's suggestion of using the base 2 allowed circuits in the digital computer to assume only two states: on or off, or 0 or 1 (in binary notation).

Von Neumann is also credited as being one of the first to see the value of the digital computer as a device capable solving challenging real-world physical problems through applied mathematics. His work at the Los Alamos National Laboratory helped to develop a synergy between the capabilities of the new digital computers that were being developed at the time and the need for computational solutions to complex thermonuclear nuclear-weapon design problems. Many scientific historians cite von Neumann's contributions in this area as significantly accelerating the development of the American hydrogen bomb, which was first tested on October 31, 1952, in the Pacific Ocean.

Following his pioneering work in computer science, of which he is one of the founding fathers, von Neumann decided to tackle the larger problem of developing a self-replicating machine. The theory of automata provided him with a convenient synthesis of his early efforts in logic and proof theory and his more recent efforts (during and after World War II) on large-scale electronic computers. Von Neumann continued to work on the intriguing idea of a self-replicating machine and its implications until his death in 1957.

Von Neumann actually conceived of several types of self-replicating systems, which he called the kinetic machine, the cellular machine, the neuron-type machine, the continuous machine, and the probabilistic machine. Unfortunately, he was only able to develop a very informal description of the kinetic machine before his death in 1957.

The kinematic machine is the most often discussed of the von Neumann-type self-replicating systems. For this type of SRS, von Neumann envisioned a machine residing in a "sea of spare parts." The kinematic machine would have a memory tape that instructed the device to go through certain mechanical procedures. Using manipulator arms and its ability to move around, this type of SRS would gather and assemble parts. The stored computer program would instruct the machine to reach out and pick up a certain part, and then go through an identification and evaluation routine to determine whether the part selected was or was not called for by the master tape. (*Note:* In von Neumann's day, microprocessors, minicomputers, floppy disks, CD ROMs, and multi-gigabyte capacity hard drives did not exist.) If the component picked up by the manipulator arm did not meet the selection criteria, it was tossed back into the parts bin (that is, back into the "sea of parts.") The process would continue until the required part was found and then an assembly operation would be performed. In this way, von Neumann's kinematic SRS would eventually make a complete replica of itself—without, however, understanding what it was doing. When the duplicate was physically completed, the parent machine would make a copy of its own memory tape on the (initially) blank tape of its offspring. The last instruction on the parent's machine tape would be to activate the tape of its mechanical progeny. The offspring kinematic SRS could then start searching the "sea of parts" for components to build yet another generation of SRS units.

In dealing with his self-replicating system concepts, von Neumann concluded that these machines should include the following characteristics and capabilities: (1) logical universality, (2) construction capability, (3) constructional universality, and (4) self-replication. Logical universality is simply the device's ability to function as a general-purpose computer. To be able to make copies of itself, a machine must be capable of manipulating information, energy, and materials. This is what is meant by the term *construction capability.* The closely related term *constructional universality* is a characteristic which implies the machine's ability to manufacture any of the finite-sized machines that can be built from a finite number of different parts, which are available from an indefinitely large supply. The characteristic of self-reproduction means that the original machine, given a sufficient number of component parts (of which it is made) and sufficient instructions, can make additional replicas of itself.

One characteristic of SRS devices that von Neumann did not address, but that has been addressed by subsequent investigators, is the concept of evolution. In a long sequence of machines making machines like themselves, can successive robot generations learn how to make themselves better machines? Robot engineers and artificial intelligence experts are

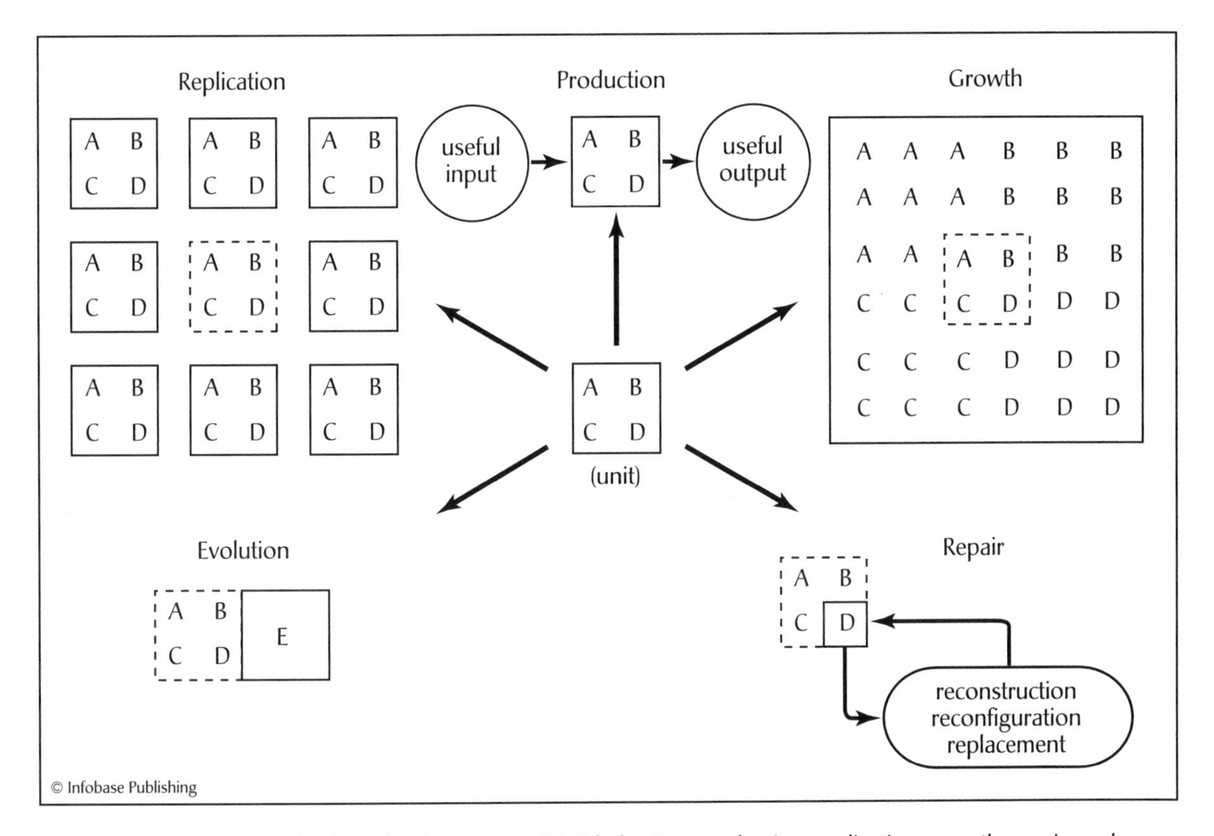

The five general classes of self-replicating system (SRS) behavior: production, replication, growth, repair, and evolution *(NASA)*

exploring this intriguing issue as part of the larger question of thinking machines that are self-aware. Can robots be made smart and alert enough to learn from the experiences encountered in daily operations and thus improve their performance? If so, will such improvements simply reflect a primitive level of machine learning? Or, will the smart machines somehow begin to develop an internal sense of "knowing" that they know? If and when this ever occurs, the smart robot will begin to mimic the consciousness of its human creators. Some AI researchers like to boldly speculate that an advanced "thinking" robot in the distant future could be capable of formulating famous philosophical postulate of René Descartes (1596–1650): *Cogito ergo sum* (I think, therefore I am). An SRS unit exhibiting the behavior of evolution might certainly be capable of achieving some form of machine self-awareness.

From von Neumann's work and the more recent work of other investigators, five broad classes of SRS behavior have been suggested:

1—*Production.* The generation of useful output from useful input. In the production process, the unit machine remains unchanged. Production is a simple behavior demonstrated by all working machines, including SRS devices.

2—*Replication.* The complete manufacture of a physical copy of the original machine unit by the machine unit itself.

3—*Growth.* An increase in the mass of the original machine unit by its own actions, while still retaining the integrity of its original design. For example, the machine might add an additional set of storage compartments in which to keep a larger supply of parts or constituent materials.

4—*Evolution.* An increase in the complexity of the unit machine's function or structure. This is accomplished by additions or deletions to existing subsystems, or by changing the characteristics of these subsystems.

5—*Repair.* Any operation performed by a unit machine on itself that helps reconstruct, reconfigure, or replace existing subsystems—but does not change the SRS unit population, the original unit mass, or its functional complexity.

In theory, replicating systems can be designed to exhibit any or all of these machine behaviors. When such machines are actually built, however, a particular SRS unit will most likely emphasize just one or several kinds of machine behavior, even if it is capable of exhibiting all of them. For example, the fully autonomous, general-purpose self-replicating lunar factory, proposed in 1980 by Georg von Tiesenhausen and Wesley A. Darbo of the Marshall Space Flight Center (MSFC), is an SRS design concept that is intended for unit replication. There are four major subsystems that make up this proposed SRS unit. First, a materials processing subsystem gathers raw materials from its extraterrestrial environment (the lunar surface) and prepares industrial feedstock. Next, a parts production subsystem uses this feedstock to manufacture other parts or entire machines.

At this point, the conceptual SRS unit has two basic outputs. Parts may flow to the universal constructor (UC) subsystem, where they are used to make a new SRS unit (this is replication); or else, parts may flow to a production facility subsystem, where they are made into commercially useful products. This self-replicating lunar factory has other secondary subsystems, such as a materials depot, parts depot, power supply, and command-and-control center.

The general components of a conceptual self-replicating lunar factory *(NASA)*

The universal constructor manufactures complete SRS units that are exact reproductions of the original SRS unit. Each replica can then make additional replicas of itself until a pre-selected SRS unit population is achieved. The universal constructor retains overall command and control (C&C) responsibilities for its own SRS unit as well as for its mechanical progeny—until, at least, the C&C functions themselves have been duplicated and transferred to the new units. To avoid cases of uncontrollable exponential growth of such SRS units in some planetary resource environment, the human masters of these devices may reserve the final step of the C&S transfer function to themselves, or so design the SRS units that the final C&C transfer function from machine to machine can be overridden by external human commands.

✧ Extraterrestrial Impact of Self-Replicating Systems

The issue of closure (total self-sufficiency) is one of the fundamental problems in designing self-replicating systems. In an arbitrary SRS unit there are three basic requirements necessary to achieve closure: (1) matter closure, (2) energy closure, and (3) information closure. In the case of matter closure engineers ask: Can the SRS unit manipulate matter in all the ways needed for complete self-construction? If not, the SRS unit has not achieved matter or material closure. Similarly, engineers ask whether the SRS unit can generate a sufficient amount of energy needed, and in the proper form, to power the processes needed for self-construction. Again, if the answer is no, then the SRS unit has not achieved energy closure. Finally, engineers must ask: Does the SRS unit successfully command and control all the processes necessary for complete self-construction? If not, information closure has not been achieved.

If the machine device is only partly self-replicating, then engineers say that only partial closure of the system has occurred. In this case, some essential energy or information must be provided from external sources, or the machine system will fail to reproduce itself.

Just what are the applications of self-replicating systems? The early development of SRS technology for use on Earth and in space should trigger an era of super-automation that will transform most terrestrial industries and lay the foundation for efficient space-based industries. One interesting machine is called the *Santa Claus machine*—originally suggested and named by the American physicist Theodore Taylor (1925–2004). In this particular version of an SRS unit, a fully automatic mining, refining and manufacturing facility gathers scoopfuls of terrestrial or extraterrestrial materials. It then processes these raw materials by means of a giant mass spectrograph that has huge super-conducting magnets. The material is converted into an ionized atomic beam and sorted into stockpiles of basic elements, atom by atom. To manufacture any item, the Santa Claus machine selects the necessary materials from its stockpile, vaporizes them, and injects them into a mold that changes the materials into the desired item. Instructions for manufacturing, including directions on adapting new processes and replication, are stored in a giant computer within the Santa Claus machine. If the product demands becomes excessive, the Santa Claus machine simply reproduces itself.

SRS units might be used in very large space construction projects (such as lunar mining operations) to facilitate and accelerate the exploitation of extraterrestrial resources and to make possible feats of planetary engineering. For example, mission planners could deploy a seed SRS unit on Mars as a prelude to permanent human habitation. This machine would use local Martian resources to automatically manufacture a large number of robot-explorer vehicles. This armada of vehicles would be dispersed over the surface of the Red Planet searching for the minerals and frozen volatiles needed in the establishment of a Martian civilization. In just a few years, a population of some 1,000 to 10,000 smart machines could scurry across the planet, completely exploring its entire surface and preparing the way for permanent human settlements.

Replicating systems would also make possible large-scale interplanetary mining operations. Extraterrestrial materials could be discovered, mapped, and mined, using teams of surface and subsurface prospector robots that were manufactured in large quantities in an SRS factory complex. Raw materials would be mined by hundreds of machines and then sent wherever they were needed in heliocentric space. Some of the raw materials might even be refined in transit, with the waste slag being used as the reaction mass for an advanced propulsion system.

Atmospheric mining stations could be set up at many interesting and profitable locations throughout the solar system. For example, Jupiter and Saturn could have their atmospheres mined for hydrogen, helium (including the very valuable isotope helium-3) and hydrocarbons, using aerostats. Cloud-enshrouded Venus might be mined for carbon dioxide, Europa for water, and Titan for hydrocarbons. Intercepting and mining comets with fleets of robot spacecraft might also yield large quantities of useful volatiles. Similar mechanized space armadas could mine water ice from Saturn's ring system. All of these smart space-robot devices would be mass-produced by seed SRS units. Extensive mining operations in the main asteroid belt would yield large quantities of heavy metals. Using extraterrestrial materials, these replicating machines could, in principle, manufacture huge mining or processing plants or even ground-to-orbit or interplanetary vehicles. This large-scale manipulation of the solar system's material resources would occur in a very short period of time, perhaps within one or two decades of the initial introduction of replicating machine technology.

From the viewpoint of a solar system civilization, perhaps the most exciting consequence of the self-replicating system is that it would provide a technological pathway for organizing potentially infinite quantities of matter. Large reservoirs of extraterrestrial matter might be gathered and organized to create an ever-widening human presence throughout heliocentric space. Self-replicating space stations, space settlements, and domed cities on certain alien worlds of the solar system would provide a diversity of environmental niches never before experienced in the history of the human race.

The SRS unit would provide such a large amplification of matter-manipulating capability that it is possible even now to start seriously considering planetary engineering (or terraforming) strategies for the Moon, Mars, Venus, and certain other alien worlds. In time, advanced self-replicating systems could be used in the 22nd century as part of humans' solar-system civilization to perform incredible feats of astroengineering. The harnessing of the total radiant energy output of the Sun, through the robot-assisted construction of a Dyson sphere, is an exciting example of the large-scale astroengineering projects that might be undertaken. The Dyson sphere is a huge artificial biosphere or habitable zone created around a parent star by an intelligent alien species.

The British-American theoretical physicist Freeman John Dyson (b. 1923) suggested this hypothesized structure as the upper limit of growth by an advanced civilization within a particular star system. The intelligent alien species would channel all the material resources of their star system into the construction of a Dyson sphere—a swarm of manufactured-space habitats capable of harvesting all the radiant energy output from the parent star. SRS units would support this type of grand-scale engineering project.

Advanced SRS technology also appears to be the key to human exploration and expansion beyond the very confines of the solar system. Although such interstellar missions may today appear highly speculative, and, indeed, they certainly require technologies that exceed contemporary or even projected levels in many areas, a consideration of possible interstellar applications is actually quite an exciting and useful mental exercise. It illustrates immediately the fantastic power and virtually limitless potential of the SRS concept.

It appears likely that before humans move out across the interstellar void, smart-robot probes will be sent ahead as scouts. (See chapter 12.) Interstellar distances are so large and search volumes so vast, that self-replicating probes (sometimes referred to as von Neumann probes) represent a highly desirable, if not totally essential, approach to performing detailed studies of a large number of other star systems, including the search for extraterrestrial life.

One speculative study on galactic exploration suggests that search patterns beyond the 100 nearest stars would most likely be optimized by the use of SRS probes. In fact, reproductive probes might permit the direct reconnaissance of the nearest 1 million stars in about 10,000 years and the entire Milky Way Galaxy in less than 1 million years—starting with a total investment by the human race in just one self-replicating interstellar robot spacecraft.

Of course, the problems of tracking, controlling, and assimilating all of the data sent back to the home-star system by an exponentially growing number of robot probes is simply staggering. Humans might avoid some of these problems by sending only very smart machines capable of greatly distilling the information gathered and transmitting only the most significant data, suitably abstracted, back to Earth. Robot engineers might also devise some type of command-and-control hierarchy, in which each robot probe only communicates with its parent. Thus, a chain of ancestral repeater stations could be used to control the flow of messages and exploration reports through interstellar space, as this bubble of machines pushes out into the galaxy.

Imagine the exciting chain reaction that might occur as one or two of the leading probes encountered an intelligent alien race. If the alien race proved hostile, an interstellar alarm would be issued, taking years to ripple across the interstellar void at the speed of light, repeater station by repeater station, until Earth received notification. Would future citizens of Earth respond by sending more sophisticated, possibly predator, robot probes to that area of the galaxy? Perhaps these human beings would decide instead to simply quarantine the belligerent species by positioning warning beacons all around the area, which would signal any other robot probes to swing clear of the hazardous alien encounter zone.

In time, as first hypothesized early in the 20th century by the American rocket expert, Robert Hutchings Goddard (1882–1945), giant space arks, representing an advanced level of synthesis between human crew and robot crew, will depart from the solar system and journey through the interstellar void. Upon reaching another star system that contained suitable material resources, the space ark itself could undergo replication. The human passengers (perhaps several generations of humans beyond the initial crew that departed the solar system) could then redistribute themselves between the parent space ark, offspring space arks, and any suitable extrasolar planets found orbiting that particular star. In a sense, the original space ark would serve as a self-replicating "Noah's Ark" for the human race and any terrestrial life-forms carried onboard the giant, mobile habitat. This dispersal of conscious intelligence (that is, intelligent human life) to a variety of ecological niches within other star systems would ensure that not even disaster on a cosmic scale, such as the death of the Sun, could threaten the complete destruction of the human species and all human accomplishments. (The death of the Sun will take place in about 5 billion years, when the Sun runs out of hydrogen for fusion in its core, leaves the main sequence, expands into a red giant, and then collapses into a white dwarf.) The self-replicating space ark would enable human beings to literally send a wave of consciousness and life (as known to humankind). From a millennial perspective, this is perhaps the grandest role for robotics in space. Sometimes referred to as the "greening of the galaxy," this propagating wave of human intelligence, in partnership with advanced machine intelligence, would promote a golden age of interstellar development—at least within a portion of the Milky Way Galaxy. How far this wave of conscious intelligence would propagate out into the galaxy is anyone's guess at this point. How far do the ripples spread on the surface, when a large fish jumps and makes a splash in the middle of the sea?

✧ Control of Self-Replicating Systems

Whenever engineers discuss the technology and role of self-replicating systems, their conversations inevitably turn to the interesting question: What happens if a self-replicating system (SRS) gets out of control? Before human beings seed the solar system or interstellar space with even a single SRS unit, engineers and mission planners should know how to pull an SRS unit's plug if things get out of control. Some engineers and scientists have already raised a very legitimate concern about SRS technology. Another question that robot engineers often encounter concerning SRS technology is whether smart machines represent a long-range threat to human life. In particular, will machines evolve with such advanced levels of artificial intelligence that they become the main resource competitors and adversaries of human beings—whether the ultra-smart machines can replicate or not? Even in the absence of advanced levels of machine intelligence that mimic human intelligence, the self-replicating system may represent a threat just through its potential for uncontrollable exponential growth.

These questions can no longer remain entirely in the realm of science fiction. Robot engineers must start examining the technical and social implications of developing advanced machine intelligences and self-replicating systems *before* they bring such systems into existence. Failure to engage in such prudent and reasonable forethought will avoid a future situation (now very popular in science fiction) in which human beings find themselves in a mortal conflict over planetary (or solar system) resources with their own intelligent machine creations.

Of course, human beings definitely need smart machines to improve life on Earth, to explore the solar system, to create a solar-system civilization, and to probe the neighboring stars. So robot engineers and scientists should proceed with the development of smart machines, but temper these efforts with safeguards to avoid the ultimate undesirable future situation, in which the machines turn against their human masters and eventually enslave or exterminate them. In 1942, the science fact/fiction writer Isaac Asimov (1920–92) suggested a set of rules for robot behavior in his story "Runaround," which appeared in *Astounding* magazine.

Over the years, Asimov's laws have become part of the cult and culture of modern robotics. They are: (Asimov's First Law of Robotics) "A robot may not injure a human being, or, through inaction, allow a human being to come to harm"; (Asimov's Second Law of Robotics) "A robot must obey the orders given it by human beings except where such orders would conflict with the first law"; and (Asimov's Third Law) "A robot must protect its own existence as long as such protection does not conflict with the first or second law." The message within these so-called laws represents a good starting point in developing benevolent, people-safe, smart machines.

Any machine sophisticated enough to survive and reproduce in largely unstructured environments, however, would probably also be capable of performing a certain degree of self-reprogramming, or automatic improvement (that is, to adopt the machine behavior of evolution). An intelligent SRS unit might eventually be able to program itself around any rules of behavior that were stored in its memory by its human creators. As it learned more about its environment, the smart SRS unit might decide to modify its behavior patterns to better suit its own needs. If this very smart SRS unit really "enjoys" being a machine and making (and perhaps improving) other machines, then when faced with a situation in which it must save a human master's life at the cost of its own, the smart machine may decide to simply shut down, instead of performing the life-saving task it was preprogrammed to do. Thus, while it did not harm the endangered human being, it also did not help the person out of danger either.

Science fiction contains many interesting stories about robots, androids, and even computers, turning on their human builders. The conflict between the human astronaut crew and the interplanetary spaceship's feisty computer, HAL, in Arthur C. Clarke and Stanley Kubrick's 1968 cinematic masterpiece *2001: A Space Odyssey* is an incomparable example. The purpose of this brief discussion is not to invoke a Luddite-type response against the development of very smart robots; only to suggest that such exciting research and engineering activities be tempered by some forethought concerning the potential technical and social impact of these developments.

Early in the Industrial Revolution, a group of British workers, ostensibly influenced by Ned Ludd, rioted and destroyed newly installed textile machinery that was taking their jobs away. The term *Luddite* now generally refers to a person who exhibits a very strong fear or hatred of technology—that is, a person who is an extreme technophobe. This term is often encountered during discussions about the social impact of robots here on Earth.

One or all of the following techniques might control an SRS population in space. First, the human builders could implant machine-genetic instructions (deeply embedded computer code) that contained a hidden or secret cutoff command. This cutoff command would be automatically activated after the SRS units had undergone a predetermined number of replications. For example, after each machine replica is made, one regeneration command could be deleted—until, at last, the entire replication process would be terminated with the construction of the last (predetermined) replica. A very simple example, which illustrates the principle behind a embedded reproduction-limit code, is that of a motion picture rented on a disposable DVD. After two or three plays, the disposable DVD disables (or erases) itself and the motion picture on the DVD can no longer be viewed.

Second, a special signal from Earth at some predetermined emergency frequency might be used to shut down individual, selected groups, or all SRS units at any time. This approach is like having an emergency stop button, which, when pressed by a human being, causes the affected SRS units to cease all activities and go immediately into a safe, hibernation posture. Many modern machines have either an emergency stop button, flow cutoff valve, heat-limit switch, or master circuit breaker. The signal-activated "all-stop" button on an SRS unit would just be a more sophisticated version of this engineered safety device.

For low-mass SRS units (perhaps in the 200-pound [100-kg] to 10,000-pound [4,500-kg] class) population control might prove more difficult because of the shorter replication times, when compared to much larger-mass SRS factory units. (Refer to the self-replicating lunar factory concept.) To keep these mechanical critters in line, robot engineers might decide to use a predator robot. The predator robot would be programmed to attack and destroy only SRS units whose populations were out of control because of some malfunction or other. Robot engineers have also considered SRS unit population control through the use of a universal destructor (UD). This machine would be capable of taking apart any other machine it encountered. The universal destructor would recover any information found in the prey robot's memory, prior to recycling the prey machine's parts. Wildlife managers use (biological) predator species on Earth today to keep animal populations in balance. Similarly, robot managers in the future could use a linear supply of non-replicating machine predators to control an exponentially growing population of misbehaving SRS units.

Robot engineers might also design the initial SRS units to be sensitive to population density. Whenever the smart robots sensed overcrowding or overpopulation, the machines could lose their ability to replicate (that is, become infertile), stop their operations and go into a hibernation state, or even (like lemmings on Earth) report to a central facility for disassembly. Unfortunately, SRS units might mimic the behavior patterns of their human creators too closely. So, without preprogrammed behavior safeguards, overcrowding could force such intelligent machines to compete among themselves for dwindling supplies of resources (terrestrial or extraterrestrial). Dueling, mechanical cannibalism, or even some highly organized form of robot-versus-robot conflict might result.

One hopes that future human engineers and scientists will create smart machines that only mimic the best characteristics of the human mind. For it is only in partnership with very smart and well-behaved self-replicating systems that the human race can some day hope to send a wave of life, conscious intelligence, and organization through the Milky Way Galaxy.

In the very long term, there appear to be two general pathways for the human species: either human beings are at a very important biological stage in the overall evolutionary scheme of the universe; or else they are at an evolutionary dead end. If the human race decides to limit itself to just one planet (Earth), a natural disaster or humankind's own foolhardiness will almost certainly terminate the species—perhaps in just a few centuries or few millennia from now. Even excluding such unpleasant natural or human-caused catastrophes, a planetary society without an extraterrestrial frontier will simply stagnate from isolation, while other alien civilizations (should such exist) flourish and populate the galaxy.

Replicating robot-system technology offers the human race very interesting options for continued evolution beyond the boundaries of Earth. Future generations of human beings might decide to create autonomous, interstellar self-replicating robot probes (von Neumann probes) and send these systems across the interstellar void on missions of exploration. Or future generations of human beings could elect to develop a closely knit (symbiotic) human-machine system—a highly automated interstellar ark—that is capable of crossing interstellar regions and then replicating itself when it encounters star systems with suitable planets and resources.

According to some scientists, any intelligent civilization that desires to explore a portion of the galaxy more than 100 light-years from its parent star would probably find it more efficient to use self-replicating robot probes. This galactic-exploration strategy would produce the largest amount of directly sampled data about other star systems for a given period of exploration. One estimate suggests that the entire galaxy could be explored in about one million years, assuming the replicating interstellar probes could achieve speeds of at least one-tenth the speed of light. If other alien civilizations (should such exist) follow this approach, then the most probable initial contact between extraterrestrial civilizations would involve a self-replicating robot probe from one civilization's encountering a self-replicating probe from another civilization.

If these encounters are friendly, the probes could exchange a wealth of information about their respective parent civilizations and any other civilizations previously encountered in their journeys through the galaxy. The closest terrestrial analogy would be a message placed in a very smart bottle that is then tossed into the ocean. If the smart bottle encounters another smart bottle, the two bump gently and provide each other a copy of their entire content of messages. One day, a beachcomber finds a smart bottle and discovers the entire collection of messages that has accumulated within.

If the interstellar probes have a hostile, belligerent encounter, they will most likely severely damage or destroy each other. In this case, the journey through the galaxy ceases for both probes and the wealth of information about alien civilizations, extant or extinct, vanishes. Returning to the simple message-in-smart-bottle analogy here on Earth, a hostile encounter damages both bottles, they sink to the bottom of the ocean, and their respective information contents are lost forever. No beachcomber will ever discover either bottle and so will never have the chance of reading the messages contained within.

One very distinct advantage of using interstellar robot probes in the search for other intelligent civilizations is the fact that these probes could also serve as a cosmic safety deposit box, carrying information about the technical, social, and cultural aspects of a particular civilization through the galaxy long after the parent civilization has vanished. The gold-anodized records NASA engineers included on the *Voyager 1* and *2* spacecraft and the special plaques they placed on the *Pioneer 10* and *11* spacecraft were humans' first attempts at achieving a tiny degree of cultural immortality in the cosmos. (Chapter 12 discusses these spacecraft and the special messages they carry.)

Starfaring self-replicating machines should be able to keep themselves running for a long time. One speculative estimate by exobiologists suggests that there may exist at present only 10 percent of all alien civilizations that ever arose in the Milky Way Galaxy—the other 90 percent have perished. If this estimate is correct then, on a simple statistical basis, nine out of every 10 robotic star probes within the galaxy could be the only surviving artifacts from long-dead civilizations. These self-replicating star probes would serve as emissaries across interstellar space and through eons of time. Here on Earth, the discovery and excavation of ancient tombs and other archaeological sites provides a similar contact through time with long-vanished peoples.

Perhaps later in this century, human space explorers and/or their machine surrogates will discover a derelict alien robot probe, or recover an artifact the origins of which are clearly not from Earth. If terrestrial scientists and cryptologists are able to decipher any language or message contained on the derelict probe (or recovered artifact), humans may eventually learn about at least one other ancient alien society. The discovery of a functioning or derelict robot probe from an extinct alien civilization might also lead human investigators to many other alien societies. In a sense, by encountering and successfully interrogating an alien robot star probe, the human team of investigators may actually be treated to a delightful edition of the proverbial *Encyclopedia Galactica*—a literal compendium of the technical, cultural, and social heritage of thousands of extraterrestrial civilizations within the galaxy (most of which are probably now extinct).

There are a number of interesting ethical questions concerning the use of interstellar self-replicating probes. Is it morally right, or even equitable, for a self-replicating machine to enter an alien star system and harvest a portion of that star system's mass and energy to satisfy its own mission objectives? Does an intelligent species legally "own" its parent star, home planet, and any material or energy resources residing on other celestial objects within its star system? Does it make a difference whether the star system is inhabited by intelligent beings? Or is there some lower threshold of galactic intelligence quotient (GIQ) below which starfaring races may ethically (on their own value scales) invade an alien star system and appropriate the resources needed to continue their mission through the galaxy? As an alien robot probe enters a star system to extract resources, by what criteria does the smart machine judge the intelligence level of an indigenous life-form—perhaps in an effort not to severely disturb existing life-bearing ecospheres? Further discussion about, and speculative responses to, such intriguing SRS-related questions extends far beyond the scope of this book. The brief line of inquiry introduced here cannot end, however, without at least mention of the most important question in cosmic ethics: Now that the human species has developed space technology, are humans and their solar system above (or below) any galactic appropriations threshold?

In summary, the self-replicating system is a very potent and exciting concept. If ever developed, the SRS unit would represent an extremely powerful robot tool, with ramifications on a cosmic scale. With properly developed and controlled SRS technologies, humans could set in motion a chain reaction that would spread organization, life, and conscious intelligence across the galaxy in an expanding wave-like bubble, limited in propagation velocity only by the speed of light itself.

Interstellar Probes

This artist's concept is an accurate representation of the view toward the Sun from NASA's *Pioneer 10* spacecraft on June 13, 1983. On that historic date, this far-traveling robot spacecraft crossed the orbit of Neptune, which at the time was the farthest major planet from the Sun because of the eccentric orbit of Pluto. With this passage, *Pioneer 10* became the first human-made object to travel beyond the planetary boundary of the solar system. From *Pioneer 10*'s vantage, a person looks across the solar system toward the center of the Milky Way Galaxy, which is the bright bulge in the background. The dust lanes of the galaxy's spiral arms are also apparent. *(NASA)*

Although often unrecognized as a particularly significant date, June 13, 1983, was a very special day in human history. On that date NASA's *Pioneer 10* robot spacecraft crossed the orbit of Neptune, which at the time was the planet farthest from the Sun. When it made this historic passage, *Pioneer 10* became the first human-made object to cross the planetary boundary of the solar system and begin a journey into the interstellar void.

During the last quarter of the 20th century, a total of four human-made objects, NASA's *Pioneer 10* and *11* spacecraft and *Voyager 1* and *2* spacecraft, successfully performed deep-space missions that eventually placed each of them on distinct escape trajectories from the solar system. The figure below shows the approximate path that each far-traveling robot spacecraft is now traveling as it begins a perpetual journey among the stars. Although none of these spacecraft was designed or intended to serve as an interstellar probe, NASA engineers had the foresight to install upon each space robot a special message, in the hope that millennia from now some intelligent alien species might find at least one of the spacecraft drifting among the

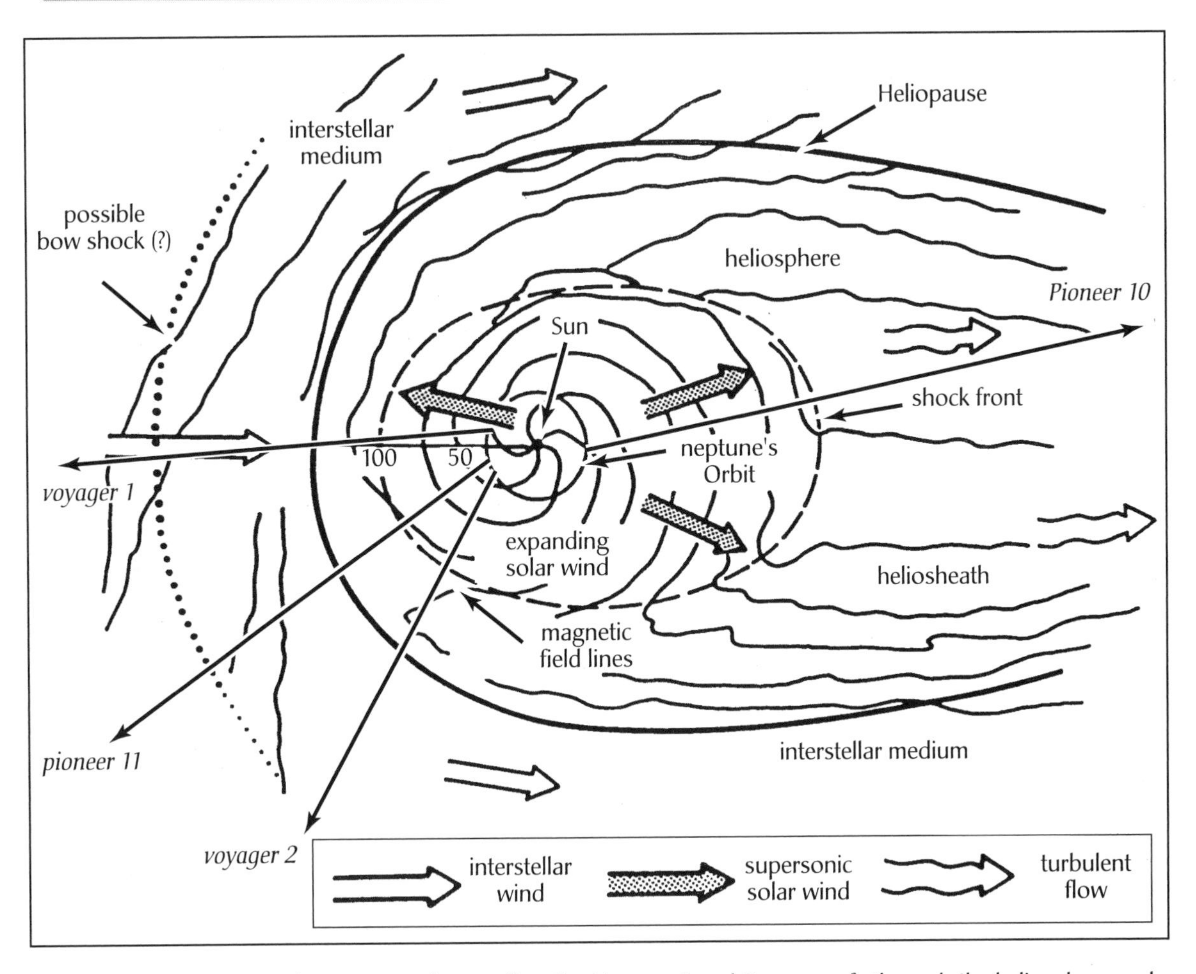

Paths of the *Pioneer 10* and *11* spacecraft, as well as the *Voyager 1* and *2* spacecraft, through the heliosphere and into the interstellar medium *(NASA)*

stars. The alien civilization would then decipher the spacecraft's message and learn about the people of Earth.

This chapter discusses each of these special robot spacecraft and the messages they carry from Earth to an alien civilization. The chapter also introduces several concepts that have been suggested for an early interstellar probe, which might be built and launched by the close of this century.

The *Thousand Astronomical Unit (TAU) Probe* is a conceptual robot spacecraft designed to travel into deep space for a distance of about 1,000 astronomical units from Earth. (One thousand astronomical units correspond to a distance of approximately 93 billion miles [150 billion km].) During its long journey, the *TAU Probe* would demonstrate the advanced robot spacecraft technologies needed to allow humans to launch a scientific robot probe to a nearby star system by about the year 2099.

Project Daedalus is the name given to a conceptual engineering design study sponsored by the British Interplanetary Society. This study represents a serious attempt to extend late 20th-century technology to the level of technology believed necessary to build and launch a successful robot interstellar probe to a nearby star in the latter portion of the 21st century.

Chapter 11 encourages each reader to take a giant intellectual leap and consider future robot technologies such as might be available to perform space exploration several centuries, or even a millennium, from now. This mental exercise includes some really *over-the-horizon technologies*, involving self-replicating systems that could wander through the entire galaxy. This chapter focuses on technologies that populate a technical horizon, which is a little closer in time. Specifically, the chapter begins with a discussion of the first robot spacecraft that have already left the solar system and concludes with some of the hypothesized robot spacecraft that could be developed by the late 21st century. With such horizon technologies, human beings would be able to construct robot spacecraft that would perform the first directed and highly focused exploration missions to nearby star systems. As it continues to evolve in the third millennium, the robot-human partnership in space exploration makes the universe both a destination and a destiny.

✧ Interstellar Journeys of the *Pioneer 10* and *11* Spacecraft

The *Pioneer 10* and *11* spacecraft, as their names imply, are true deep-space explorers—the first human-made objects to navigate the main asteroid belt, the first spacecraft to encounter Jupiter and its fierce radiation belts, the first to encounter Saturn, and the first spacecraft to leave the solar system. These spacecraft also investigated magnetic fields, cosmic rays, the solar wind, and interplanetary dust concentrations, as they flew through interplanetary space.

The *Pioneer 10* spacecraft was launched from Cape Canaveral Air Force Station, Florida, by an Atlas-Centaur rocket, on March 2, 1972. It became the first spacecraft to cross the main asteroid belt and the first to make close-range observations of the Jovian system. Sweeping past Jupiter on December 3, 1973 (its closest approach to the giant planet), it discovered no solid surface under the thick layer of clouds enveloping the giant planet—an indication that Jupiter is a liquid hydrogen planet. *Pioneer 10* also explored the giant Jovian magnetosphere, made close-up pictures of the intriguing Red Spot, and observed at relatively close range the Galilean satellites Io, Europa, Ganymede, and Callisto. When *Pioneer 10* flew past Jupiter, it acquired sufficient kinetic energy to carry it completely out of the solar system.

Departing Jupiter, *Pioneer 10* continued to map the *heliosphere* (the Sun's giant magnetic bubble, or field, drawn out from it by the action of the solar wind). Then, on June 13, 1983, *Pioneer 10* crossed the orbit of Neptune, which at the time was (and was until 1999) the major planet farthest out from the Sun. This unusual circumstance owed to the eccentricity in Pluto's orbit, which had taken the icy (ninth) planet inside the orbit of Neptune. The historic date marked the first passage of a human-made object beyond the known planetary boundary of the solar system. Beyond this solar system boundary, *Pioneer 10* measured the extent of the heliosphere as the spacecraft began its travels into interstellar space. Along with its sister ship (*Pioneer 11*), the *Pioneer 10* spacecraft helped scientists investigate the deep-space environment.

The *Pioneer 10* spacecraft is heading generally toward the red star Aldebaran. The robot spacecraft is more than 68 light-years away from Aldebaran, and the journey will require about 2 million years to complete. Budgetary constraints forced NASA to terminate routine tracking and project-data-processing operations for *Pioneer 10* on March 31, 1997. Occasional tracking of *Pioneer 10* continued beyond that date, however. The last successful data acquisitions from *Pioneer 10* by NASA's Deep Space Network (DSN) occurred in 2002 on March 3 (30 years after launch) and again on April 27. The spacecraft signal was last detected on January 23, 2003, after an uplink message was transmitted to turn off the remaining operational experiment, the *Geiger Tube Telescope.* No downlink data signal was achieved however; and by early February 2003, no signal at all was detected. NASA personnel concluded that the spacecraft's radioisotope thermoelectric generator (RTG) unit, which supplied electric power, had finally fallen below the level needed to operate the onboard transmitter. Consequently, no further attempts were made to communicate with *Pioneer 10.*

The *Pioneer 11* spacecraft was launched on April 5, 1973, and swept by Jupiter at an encounter distance of only 26,725 miles (43,000 km) on December 2, 1974. The spacecraft provided additional detailed data and pictures of Jupiter and its moons, including the first views of Jupiter's polar regions. Then, on September 1, 1979, *Pioneer 11* flew by Saturn, demonstrating a safe flight path through the rings for the more sophisticated *Voyager 1* and *2* spacecraft to follow. *Pioneer 11* (by then officially renamed *Pioneer Saturn*) provided the first close-up observations of Saturn, its rings, satellites, magnetic field, radiation belts, and atmosphere. The space robot found no solid surface on Saturn, but discovered at least one additional satellite and ring. After rushing past Saturn, *Pioneer 11* also headed out of the solar system toward the distant stars.

The *Pioneer 11* spacecraft has operated on a backup transmitter since launch. Instrument power-sharing began in February 1985 because of declining RTG power output. Science operations and daily telemetry ceased on September 30, 1995, when the RTG power level became insufficient to operate any of the spacecraft's instruments. All contact with *Pioneer 11* ceased at of the end of 1995. At that time, the spacecraft was 44.7 astronomical units (AU) away from the Sun and traveling through interstellar space at a speed of about 2.5 AU per year.

Both Pioneer spacecraft carry a special message (called the Pioneer plaque) for any intelligent alien civilization that might find them wandering through the interstellar void millions of years from now. This message is an illustration, engraved on an anodized aluminum plaque. The plaque depicts the location of Earth and the solar system, a man and a woman,

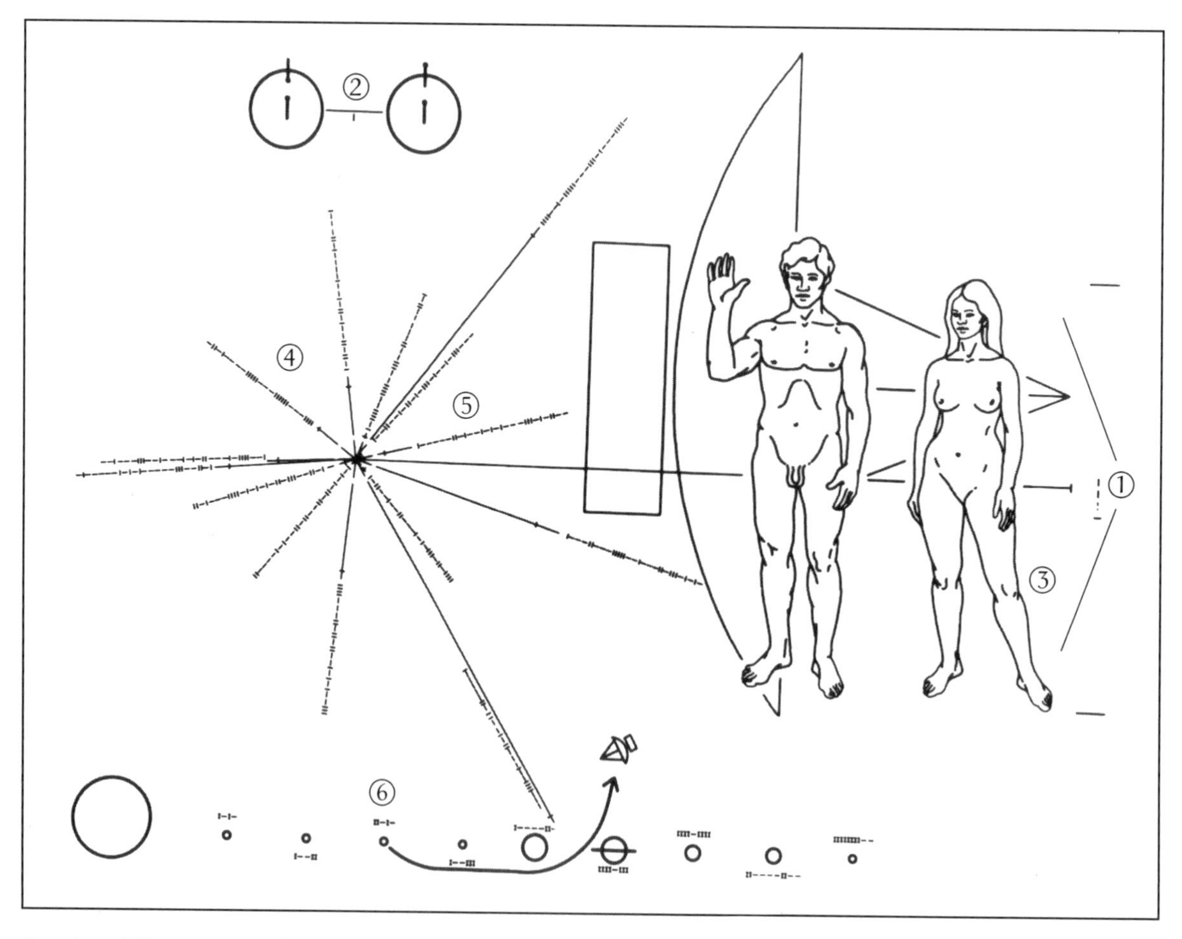

Annotated *Pioneer 10* (and *11*) plaque *(NASA)*

and other points of science and astrophysics that should be decipherable by a technically intelligent civilization.

The plaque is intended to show any intelligent alien civilization that might detect and intercept either Pioneer spacecraft millions of years from now when the spacecraft was launched, from where it was launched and by what type of intelligent beings it was built. The plaque's design is engraved into a gold-anodized aluminum plate, 6 inches (15.2 cm) by 9 inches (22.9 cm). The plate is approximately .05 inches (.127 cm) thick. Engineers attached the plaque to the Pioneer spacecraft's antenna support struts in a position that helps shield it from erosion by interstellar dust.

The figure on page 222 shows an annotated version of the Pioneer plaque. The numbers (1 to 6) have been intentionally superimposed on the plaque to assist in the discussion of its message. At the far right, the bracketing bars (1) show the height of the woman compared to the Pioneer spacecraft. The drawing at the top left of the plaque (2) is a schematic of the hyperfine transition of neutral atomic hydrogen, used here as a universal "yardstick" that provides a basic unit of both time and space (length) throughout the Milky Way Galaxy. This figure illustrates a reverse in the direction of the spin of the electron in a hydrogen atom. The transition depicted emits a characteristic radio wave with an approximately 8.3-inch (21-cm) wavelength. Therefore, by providing this drawing, people of Earth are telling any technically knowledgeable alien civilization finding it that they have chosen 8.3 inches (21 cm) as a basic length in the message. While extraterrestrial civilizations will certainly have different names and defining dimensions for their basic system of physical units, the wavelength size associated with the hydrogen radio-wave emission will still be the same throughout the galaxy. Science and commonly observable physical phenomena represent a general galactic language—at least for starters.

The horizontal and vertical ticks (3) represent the number 8 in binary form. It is hoped that the alien beings pondering over this plaque will eventually realize that the hydrogen wavelength (8.3-inch [21-cm]) multiplied by the binary number representing 8 (indicated alongside the woman's silhouette) describes her overall height—namely, 8×8.3 inches $= 66$ inches (8×21 centimeters $= 168$ cm), or approximately five and one-half feet tall. Both human figures are intended to represent the intelligent beings that built the Pioneer spacecraft. The man's hand is raised as a gesture of goodwill. These human silhouettes were carefully selected and drawn to maintain ethnic neutrality. Furthermore, no attempt was made to explain terrestrial "sex" to an alien culture—that is, the plaque makes no specific effort to explain the potentially mysterious differences between the man and woman depicted.

The radial pattern (4) should help alien scientists locate the solar system within the Milky Way Galaxy. The solid bars indicate distance, with the long horizontal bar (5) with no binary notation on it representing the distance from the Sun to the galactic center, while the shorter solid bars denote directions and distances to 14 pulsars from the Sun. The binary digits following these pulsar lines represent the periods of the pulsars. From the basic time unit established by the use of the hydrogen-atom transition, an intelligent alien civilization should be able to deduce that all times indicated are about 0.1 second—the typical period of pulsars. Since pulsar periods appear to be slowing down at well-defined rates, the pulsars serve as a form of galactic clock. Alien scientists should be able to search their astrophysical records and identify the star system from which the Pioneer spacecraft originated and approximately when it was launched, even if each spacecraft is not found for hundreds of millions of years. Consequently, through the use of this pulsar map, NASA's engineers and scientists have attempted to locate Earth, both in galactic space and in time.

As a further aid to identifying the *Pioneer*'s origin, a diagram of the solar system (6) is also included on the plaque. The binary digits accompanying each planet indicate the relative distance of that planet from the Sun. The *Pioneer*'s trajectory is shown starting from the third planet (Earth), which has been offset slightly above the others. As a final clue to the terrestrial origin of the Pioneer spacecraft, its antenna is depicted pointing back to Earth.

This message was designed for NASA by Frank Drake and the late Carl Sagan (1934–96), and Linda Salzman Sagan prepared the artwork.

✧ Voyager Interstellar Mission

As the influence of the Sun's magnetic field and solar wind grow weaker, both Voyager robot spacecraft eventually will pass out of the heliosphere and into the interstellar medium. Through NASA's Voyager Interstellar Mission (VIM), which began officially on January 1, 1990, the *Voyager 1* and *2* spacecraft will continue to be tracked on their outward journey. The two major objectives of the VIM are an investigation of the interplanetary and interstellar media and a characterization of the interaction between the two and a continuation of the successful Voyager program of ultraviolet astronomy. During the VIM, the spacecraft will search for the heliopause (the outermost extent of the solar wind, beyond which lies interstellar space). Scientists hope that at least one Voyager spacecraft will still be functioning when it penetrates the heliopause and will provide them with the first true sampling of the interstellar environment. Barring a catastrophic failure on board either Voyager spacecraft, their nuclear power systems should provide useful levels of electric power until at least 2015.

Each Voyager spacecraft has a mass of 1,815 pounds (825 kg) and carries a complement of scientific instruments to investigate the outer planets and their many moons and intriguing ring systems. These instruments, provided with electric power by a long-lived nuclear system called a radioisotope thermoelectric generator (RTG), recorded spectacular close-up images of the giant outer planets and their interesting moon systems, explored complex ring systems, and measured properties of the interplanetary medium.

Once every 176 years, the giant outer planets—Jupiter, Saturn, Uranus, and Neptune—align themselves in such a pattern that a spacecraft launched from Earth to Jupiter at just the right time might be able to visit the other three planets on the same mission, using a technique called gravity assist. NASA space scientists named this multiple giant planet encounter mission the Grand Tour and took advantage of a unique celestial alignment opportunity in 1977 by launching two sophisticated spacecraft, called *Voyager 1* and *2*.

The *Voyager 2* spacecraft lifted off from Cape Canaveral, Florida, on August 20, 1977, onboard a Titan-Centaur rocket. (NASA called the first Voyager spacecraft launched *Voyager 2,* because the second Voyager spacecraft to be launched eventually would overtake it and become *Voyager 1.*) *Voyager 1* was launched on September 5, 1977. This spacecraft followed the same trajectory as its twin (*Voyager 2*) and overtook its sister ship just after entering the asteroid belt in mid-December 1977.

Voyager 1 made its closest approach to Jupiter on March 5, 1979, and then used Jupiter's gravity to swing itself to Saturn. On November 12, 1980, *Voyager 1* successfully encountered the Saturnian system and then was flung up out of the ecliptic plane on an interstellar trajectory. The *Voyager 2* spacecraft encountered the Jovian system on July 9, 1979 (closest approach), and then used the gravity-assist technique to follow *Voyager 1* to Saturn. On August 25, 1981, *Voyager 2* encountered Saturn and then went on to successfully encounter both Uranus (January 24, 1986) and Neptune (August 25, 1989). Space scientists consider the end of *Voyager 2*'s encounter of the Neptunian system as the end of a truly extraordinary epoch in planetary exploration. In the first 12 years since they were launched from Cape Canaveral, these incredible robot spacecraft contributed more to the understanding of the giant outer planets of the solar system than had been accomplished in more than three millennia of Earth-based observations. Following its encounter with the Neptunian system, *Voyager 2* was also placed on an interstellar trajectory and, like its *Voyager 1* twin, now continues to travel outward from the Sun.

Since both Voyager spacecraft would eventually journey beyond the solar system, their designers placed a special interstellar message on each in the hope that, perhaps millions of years from now, some intelligent alien race will find either spacecraft drifting quietly through the interstellar void. If they are able to decipher the instructions for using this record,

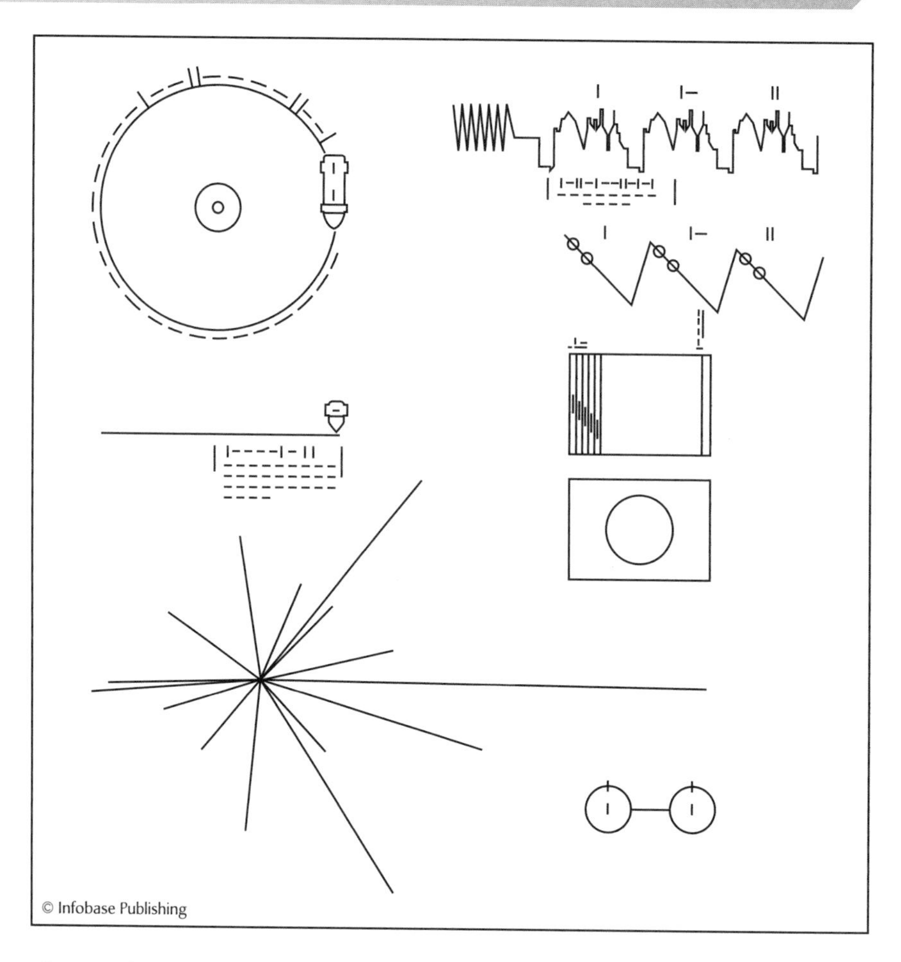

The set of instructions to any alien civilization that might find the *Voyager 1* or *2* spacecraft, explaining how to operate the Voyager record and where the robot spacecraft and message came from *(NASA)*

they will learn about the contemporary terrestrial civilization and the men and women who sent *Voyager* on its stellar journey.

The Voyager interstellar message is a phonograph record called "The Sounds of Earth." Electronically imprinted on it are words, photographs, music, and illustrations that will tell an extraterrestrial civilization about planet Earth. Included are greetings in more than 50 different languages, music from various cultures and periods and a variety of natural terrestrial sounds such as the wind, the surf, and different animals. The Voyager record also includes a special message from former president Jimmy Carter. The late Carl Sagan described in detail the full content of this phonograph message to the stars in his delightful book, *Murmurs of Earth.*

Each record is made of copper with gold plating and is encased in an aluminum shield that also carries instructions on how to play it. The figure on page 226 shows the set of instructions that accompany the Voyager record. In the upper left is a drawing of the phonograph record and the stylus carried with it. Written around it in binary notation is the correct time for one rotation of the record, 3.6 seconds. Here, the time units are 0.70 billionths of a second, the time period associated with a fundamental transition of the hydrogen atom. The drawing further indicates that the record should be played from the outside in. Below this drawing is a side view of the record and stylus, with a binary number giving the time needed to play one side of the record (approximately one hour).

The information provided in the upper-right portion of the instructions is intended to show how pictures (images) are to be constructed from the recorded signals. The upper-right drawing illustrates the typical waveform that occurs at the start of a picture. Picture lines 1, 2, and 3 are given in binary numbers and the duration of one of the picture "lines" is also noted (about eight milliseconds). The drawing immediately below shows how these lines are to be drawn vertically, with a staggered interlace to give the correct picture rendition. Immediately below this is a drawing of an entire picture raster, showing that there are 512 vertical lines in a complete picture. Then, immediately below this is a replica of the first picture on the record. This should allow extraterrestrial recipients to verify that they have properly decoded the terrestrial pictures. A circle was selected for this first picture to guarantee that any aliens who find the message use the correct aspect ratio in picture reconstruction.

Finally, the drawing at the bottom of the protective aluminum shield is that of the same pulsar map drawn on the *Pioneer 10* and *11* plaques. The map shows the location of the solar system with respect to 14 pulsars, whose precise periods are also given. The small drawing with two circles in the lower right-hand corner is a representation of the hydrogen atom in its two lowest states, with a connecting line and digit 1. This indicates that the time interval associated with the transition from one state to the other is to be used as the fundamental time scale, both for the times given on the protective aluminum shield and in the decoded pictures.

✧ Thousand Astronomical Unit (TAU) Probe Mission

The *Thousand Astronomical Unit (TAU) Probe* is a conceptual mid-century NASA space exploration mission involving an advanced-technology robot spacecraft that would travel on a 50-year journey into very deep space—out to a distance of about 1,000 astronomical units (some 93 billion miles [150 billion km]) away from Earth. The astronomical unit (AU) is a unit of distance in astronomy and space technology defined as the distance from the center of Earth to the center of the Sun. One AU is equally to approximately 93 million miles (149.6 million km), or 499 light-seconds.

The *Thousand Astronomical Unit* (TAU) *Probe* would feature an advanced multi-megawatt nuclear reactor, ion propulsion, and a laser (optical) communications system. Initially, the *TAU Probe* would be directed for an encounter with Pluto and its large moon Charon, followed by passage through the Kuiper belt and the heliopause, possibly reaching the innermost portion of Oort cloud. The heliopause is the theoretical end of the solar system—the region space where the Sun's influence ends and the solar wind smashes into the thin gas between the stars.

The advanced robot spacecraft would investigate low-energy cosmic rays, low-frequency radio waves, interstellar gases, and deep-space phenomena. It would also perform high-precision astrometry, the precise measurement of distances between stars. One key technology for this mission is an advanced space nuclear-reactor power system capable of autonomously and automatically providing a nominal 100 kilowatts of electric power for a period of at least 50 years. The power plant must be reliable and perform its task unattended by human controllers for five decades or more. With thrust provided by an advanced electric propulsion system, the *TAU Probe* would travel at a cruising speed of approximately 20 AU per year. This means that the journey would take roughly 50 years to achieve a total distance from Earth of 1,000 AU.

As a pre-interstellar probe mission, the *TAU Probe* would demonstrate sustained autonomous operations for more than 50 years. Spacecraft systems, subsystems, and components would require lifetimes and levels of reliability from one to two orders of magnitude greater than that currently available. The machine intelligence of this robot probe must be capable of autonomous assessment of the external environment and internal conditions onboard the probe. The smart robot probe must be able to make appropriate decisions and implement physical changes within itself, as circumstances warrant. In particular the *TAU Probe* must perform spacecraft health management, which involves the capability to predict, detect, and correct system performance. The robot spacecraft must be designed to practice fault management through repair, redundancy, and performance of workarounds—all without human guidance or assistance. Finally, the *TAU Probe* must be smart enough to perform unsupervised resource management, involving electric power usage and distribution, thermal control, use of consumables, the commitment of spare parts and emergency supplies, and data flow and data management.

✧ Designing an Interstellar Probe

An interstellar probe is a highly automated robot spacecraft sent from this solar system to explore another star systems. Most likely, this type of probe would make use of very smart machine systems capable of operating autonomously for decades or centuries.

Once the robot probe arrives at a new star system, it would begin a detailed exploration procedure. The target star system is scanned for possible life-bearing planets, and if any are detected, they become the object of more intense scientific investigations. Data collected by the mother spacecraft probe and any mini-probes (deployed to explore individual objects of interest within the new star system) are transmitted back to Earth. There, after light-years of travel, the signals are intercepted and analyzed by scientists, and interesting discoveries and information are used

This artist's concept shows the human race's first interstellar robot probe departing the solar system (ca. 2099) on an epic journey of scientific exploration. *(NASA)*

to enrich human knowledge and understanding about the galaxy and, by extrapolation, the universe.

The robot interstellar probe could also be designed to carry a payload of specially engineered microorganisms, spores, and bacteria. If the robot probe encounters ecologically suitable planets on which life has not yet evolved, then it could make the decision to "seed" such barren, but potentially fertile, worlds with primitive life-forms or, at least, life precursors. In that way, human beings (in partnership with their robot probes) would not only be exploring neighboring star systems but would be participating in the spreading of life itself through some portion of the Milky Way Galaxy.

NASA's long-range strategic planners have examined some of the engineering and operational requirements of the first interstellar probe, which might be launched at the end of this century to a nearby (10 light-years or less away) star system. Some of these challenging requirements (all of which exceed current levels of technology by one or two orders of magnitude) are briefly mentioned here. The interstellar probe must be capable of sustained, autonomous operation for more than 100 years. The robot spacecraft must be capable of managing its own health—that is, being able to anticipate or predict a potential problem, detect an emerging abnormality, and then prevent or correct the situation. For example, if a subsystem is about to overheat (but has not yet exceeded thermal-design limits), the smart robot probe would redirect operations and adjust the thermal-control system to avoid the potentially serious overheating condition.

The first interstellar robot probe must have a very high level of machine intelligence and be capable of exercising fault management through repair, redundancy, and workarounds without any human guidance or assistance. The smart robot must also be able to carefully manage its onboard resources, supervising the generation and distribution of electric power, allocating the use of consumables, deciding when and where to commit emergency reserves and a limited supply of spare parts and components. The main onboard computer (or machine brain) of the probe must exercise data-management skills and be capable of an inductive response to unknown or unanticipated environmental changes. When faced with unknown difficulties or opportunities, the robot probe must be able to modify the mission plan and generate new tasks.

For example, during the mission, long-range sensors onboard the probe might discover that a hot-Jupiter-type extrasolar planet within the target star system has a large (previously unknown) moon with an atmosphere and a liquid-water ocean. Instead of sending its last mini-probe ahead to investigate the hot Jupiter-type planet, the smart robot mother spacecraft makes a decision to release its last mini-probe to make close-up measurements of this interesting moon. Since the mother spacecraft is more than eight light-years from Earth when the (hypothesized) discovery is made, the decision to change the mission plan must be made exclusively by the robot spacecraft, which is only a few *light-days* away from the encounter. Sending a message back to Earth and asking for instructions would take more than 16 years (for round-trip communications), and by then the interstellar probe would have completely passed through the target star system and disappeared into the interstellar void.

Similarly, instruments onboard the interstellar probe (regarded here as the mother spacecraft) and its supporting cadre of mini-probes must be capable of deductive and inductive learning, so as to adjust how measurements are taken in response to unfolding opportunities, feedback, and unanticipated values (high and low). Some of the greatest scientific discoveries on Earth happened because of an accidental measurement or unanticipated reading.

For example, while studying the energy content of sunlight with the help of a thermometer and a prism, the German-born-British astronomer Sir (Frederick) William Herschel (1738–1822) slowly ran his thermometer across the visible portion of the solar spectrum. As he pushed the thermometer past red light into a black (to the eye) region, he was astonished to see a higher temperature reading. He had accidentally discovered the infrared portion of the electromagnetic spectrum, which is invisible to the human eye but certainly has a measurable energy content.

The instruments onboard the robot probe must be capable of exercising a similar level of curious inquiry and then be able to respond to unanticipated, but quite significant, new findings. The robot probe must have machine intelligence capable of knowing when new information is quite significant. This is a difficult task for human scientists, who often overlook the most significant pieces of data in an experiment or observation. To ask a robot's brain to respond "Eureka" (I've found it) at the moment of a great discovery is pushing machine intelligence well beyond the technical horizon projected for the next few decades. Yet, if the human race is going to make significant discoveries with robot interstellar probes, that is precisely what these advanced exploring machines must be capable of doing.

From a pure spacecraft-engineering perspective, the interstellar robot probe should consist of low-density, high-strength materials to minimize propulsion requirements. To make a mission to the nearby stars last 100 years or so, the robot spacecraft should be capable of cruising at about one-tenth the speed of light (or more). Any speed less than that would make star-probe mission to even the nearest stars last several centuries. The great, great, great grandchildren of the probe engineers would have to remain interested in receiving the signals from the (potentially long-forgotten) probe. So the first interstellar probe mission (using advanced but non-replicating technology) will, more than likely, last about 100 years.

The materials used on the outside of the robot probe must maintain their integrity for more than 100 years, even when they are subjected to deep-space conditions, especially ionizing radiation, cold, vacuum, and interstellar dust. The structure of the robot spacecraft should be capable of autonomous reconfiguration. The power system must be able to provide reliable base power (typically at a level of 100 kilowatts-electric up to possibly one megawatt-electric) on an autonomous and self-maintaining basis for more than 100 years. Finally, the star probe must be capable of autonomous data collection, assessment, storage, and communications (back to Earth) from a wide variety of scientific instruments and onboard spacecraft state-of-health sensors.

Some of the intriguing challenges in information technology include the proper calibration of instruments and collection of data over a period of years after decades of sensor dormancy. The robot probe must be able to transmit data back to Earth over distances ranging from 4.5 to 10.0 light-years. Finally, after decades of handling modest levels of data, the spacecraft's information systems must be capable of handling a gigantic burst of incoming data as the robot probe and its mini-probes encounter the target star system.

✧ Project Daedalus

Project Daedalus is the name given to an extensive study of interstellar space exploration conducted from 1973 to 1978 by a team of scientists and engineers under the auspices of the British Interplanetary Society. This hallmark effort examined the feasibility of performing a simple interstellar mission using only contemporary technologies and/or reasonable extrapolations of imaginable near-term capabilities.

In mythology, Daedulus was the grand architect of King Minos's labyrinth for the Minotaur on the island of Crete. But Daedalus also showed the Greek hero Theseus, who slew the Minotaur, how to escape from the labyrinth. An enraged King Minos imprisoned both Daedalus and his son Icarus. Undaunted, Daedalus (a brilliant engineer) fashioned two pairs of wings out of wax, wood, and leather. Before their aerial escape from a prison tower, Daedalus cautioned his son not to fly too high, so that the Sun would not melt the wax and cause the wings to disassemble. The two made good their escape from King Minos's Crete, but, while over the sea, Icarus, an impetuous teenager, ignored his father's warnings and soared high into the air. Daedalus (who reached Sicily safely) watched his young son, wings collapsed, tumble to his death in the sea below.

The proposed Daedalus spaceship structure, communications systems, and much of the payload were designed entirely within the parameters of 20th-century technology. Other components, such as the advanced machine intelligence flight controller and onboard computers for in-flight repair, required artificial-intelligence capabilities expected to be available in the mid-21st century. The propulsion system, perhaps the most challenging aspect of any interstellar mission, was designed as a nuclear-powered, pulsed-fusion rocket engine that burned an exotic thermonuclear fuel mixture of deuterium and helium-3 (a rare isotope of helium). This pulsed-fusion system was believed capable of propelling the robot interstellar probe to velocities in excess of 12 percent of the speed of light (that is, more than .12 c). The best source of helium-3 was considered to be the planet Jupiter, and one of the major technologies that had to be developed for Project Daedalus was an ability to mine the Jovian atmosphere for helium-3. This mining operation might be achieved by using "aerostat" extraction facilities (floating balloon-type factories).

The Project Daedalus team suggested that this ambitious interstellar flyby (one-way) mission might possibly be undertaken by the end of the 21st century—when the successful development of humankind's extraterrestrial civilization had generated the necessary wealth, technology base, and exploratory zeal. The target selected for this first interstellar probe was Barnard's star, a red dwarf (spectral type M) about 5.9 light-years away in the constellation Ophiuchus.

The Daedalus spaceship would be assembled in cislunar space (partially fueled with deuterium from Earth) and then ferried to an orbit around Jupiter, where it could be fully fueled with the helium-3 propellant that had been mined out of the Jovian atmosphere. These thermonuclear fuels would then be prepared as pellets, or "targets," for use in the ship's two-stage pulsed-fusion power plant. Once fueled and readied for its epic interstellar voyage, somewhere around the orbit of Callisto, the ship's mighty pulsed-fusion first-stage engine would come alive. This first-stage pulsed-fusion unit would continue to operate for about two years. At first-stage shutdown, the vessel would be traveling at about 7 percent of the speed of light (0.07 c).

The expended first-stage engine and fuel tanks would be jettisoned in interstellar space, and the second-stage pulsed-fusion engine would ignite. The second stage would also operate in the pulsed-fusion mode for about two years. Then, it, too, would fall silent, and the giant robot spacecraft, with its cargo of sophisticated remote sensing equipment and nuclear fission-powered probe ships, would be traveling at about 12 percent of the speed of light (0.12 c). It would take the Daedalus spaceship about 47 years of coasting (after second-stage shutdown) to encounter Barnard's star.

In this scenario, when the Daedalus interstellar probe was about 3 light-years away from its objective (about 25 years of mission-elapsed time), smart computers on board would initiate long-range optical and radio astronomy observations. A special effort would be made to locate and identify any extrasolar planets that might exist in the Barnardian system.

Of course, traveling at 12 percent of the speed of light, Daedalus would only have a very brief passage through the target star system. This would amount to a few days of "close-range" observation of Barnard's star itself and only "minutes" of observation of any planets or other interesting objects by the robot mother spacecraft.

Several years before the Daedalus mother spacecraft passed through the Barnardian system, however, it would launch its complement of nuclear-powered probes (also traveling at 12 percent of the speed of light initially). These probe ships, individually targeted to objects of potential interest by computers on board the robot mother spacecraft, would fly ahead and act as data-gathering scouts. A complement of 18 of these scout craft or small robotic probes was considered appropriate in the Project Daedalus study.

Then, as the main Daedalus spaceship flashed through the Barnardian system, it would gather data from its own onboard instruments as well as information telemetered to it by the numerous probes. Over the next day or so, it would transmit all these mission data back toward our solar system, where team scientists would patiently wait the approximately six years it takes for these information-laden electromagnetic waves, traveling at light speed, to cross the interstellar void.

Its mission completed, the Daedalus mother spaceship without its probes—would continue on a one-way journey into the darkness of the interstellar void, to be discovered perhaps millennia later by an advanced alien race, which might puzzle over humankind's first attempt at the direct exploration of another star system.

Today, the main conclusions that can be drawn from the Project Daedalus study might be summarized as follows: (1) exploration missions to other star systems are, in principle, technically feasible; (2) a great deal could be learned about the origin, extent, and physics of the Milky Way Galaxy, as well as the formation and evolution of stellar and planetary systems, by missions of this type; (3) the prerequisite interplanetary and initial interstellar space system technologies necessary to conduct this class of mission successfully also contribute significantly to humankind's search for extraterrestrial intelligence (for example, smart robot probes and interstellar communications); (4) a long-range societal commitment on the order of a century would be required to achieve such a project; and (5) the prospects for interstellar flight by human beings do not appear very promising using current or foreseeable technologies in this century.

The Project Daedalus study also identified three key technology advances that would be needed to make even a robot interstellar mission possible. These are (1) the development of controlled nuclear fusion, especially the use of the deuterium/helium-3 thermonuclear reaction; (2) advanced machine intelligence; and (3) the ability to extract helium-3 in large quantities from the Jovian atmosphere.

Although the choice of Barnard's star as the target for the first interstellar mission was somewhat arbitrary, if future human generations can build such an interstellar robot spaceship and successfully explore the Barnardian system, then with modest technology improvements, all star systems within 10 to 12 light-years of Earth become potential targets for a more ambitious program of (robotic) interstellar exploration.

Conclusion

Robot spacecraft are sophisticated exploring machines that have visited all the major worlds of the solar system, including tiny Pluto. At the dawn of the Space Age, scientists and engineers began using relatively unsophisticated space robots in their quest to explore the previously unreachable worlds and mysterious cosmic phenomena beyond Earth's atmosphere. Today, a little more than four decades later, incredibly complex robotic exploring machines allow scientists to conduct detailed, firsthand investigations of alien worlds throughout the solar system.

Emerging out of the space race of the cold war, modern robot spacecraft have dramatically changed what scientists know about the solar system and the universe. Even more exciting, perhaps, is the fact that during the last quarter of the 20th century four human-made objects, NASA's *Pioneer 10* and *11* spacecraft and *Voyager 1* and *2* spacecraft, have performed deep space missions, which eventually placed each of them on distinct escape trajectories from the solar system. Although none of these spacecraft was designed or intended to serve as interstellar probes, NASA engineers had the foresight to install upon each space robot a special message in the hope that millennia from now some intelligent alien species might find at least one of the spacecraft drifting among the stars and learn about Earth.

The robot-human partnership in space exploration makes the universe both a destination and a destiny. In its ultimate form, this partnership leads to the very exciting concept of the self-replicating system. If ever developed, the SRS unit would represent an extremely powerful tool for robotic space exploration with ramifications on a cosmic scale. Using properly developed and controlled SRS technologies, a future generation of humans could set in motion a chain reaction that would spread organization, life, and conscious intelligence across the galaxy in an expanding wave-like bubble, limited in propagation velocity only by the speed of light itself.

Chronology

✧ ca. 3000 B.C.E. (to perhaps 1000 B.C.E.)

Stonehenge erected on the Salisbury Plain of Southern England (possible use: ancient astronomical calendar for prediction of summer solstice)

✧ ca. 1300 B.C.E.

Egyptian astronomers recognize all the planets visible to the naked eye (Mercury, Venus, Mars, Jupiter, and Saturn), and they also identify over 40 star patterns or constellations

✧ ca. 500 B.C.E.

Babylonians devise zodiac, which is later adopted and embellished by Greeks and used by other early peoples

✧ ca. 375 B.C.E.

The early Greek mathematician and astronomer Eudoxus of Cnidos starts codifying the ancient constellations from tales of Greek mythology

✧ ca. 275 B.C.E.

The Greek astronomer Aristarchus of Samos suggests an astronomical model of the universe (solar system) that anticipates the modern heliocentric theory proposed by Nicolaus Copernicus. However, these ideas, which Aristarchus presents in his work *On the Size and Distances of the Sun and the Moon,* are essentially ignored in favor of the geocentric model of the universe proposed by Eudoxus of Cnidus and endorsed by Aristotle

✧ ca. 129 B.C.E.

The Greek astronomer Hipparchus of Nicaea completes a catalog of 850 stars that remains important until the 17th century

✧ ca. 60 C.E.

The Greek engineer and mathematician Hero of Alexandria creates the aeoliphile, a toylike device that demonstrates the action-reaction principle that is the basis of operation of all rocket engines

✧ ca. 150 C.E.

Greek astronomer Ptolemy writes *Syntaxis* (later called the *Almagest* by Arab astronomers and scholars)—an important book that summarizes all the astronomical knowledge of the ancient astronomers, including the geocentric model of the universe that dominates Western science for more than one and a half millennia

✧ 820

Arab astronomers and mathematicians establish a school of astronomy in Baghdad and translate Ptolemy's work into Arabic, after which it became known as *al-Majisti* (The great work), or the *Almagest,* by medieval scholars

✧ 850

The Chinese begin to use gunpowder for festive fireworks, including a rocketlike device

✧ 1232

The Chinese army uses fire arrows (crude gunpowder rockets on long sticks) to repel Mongol invaders at the battle of Kaifung-fu. This is the first reported use of the rocket in warfare

✧ 1280–90

The Arab historian al-Hasan al-Rammah writes *The Book of Fighting on Horseback and War Strategies,* in which he gives instructions for making both gunpowder and rockets

✧ 1379

Rockets appear in western Europe; they are used in the siege of Chioggia (near Venice), Italy

✧ 1420

The Italian military engineer Joanes de Fontana writes *Book of War Machines,* a speculative work that suggests military applications of gunpowder rockets, including a rocket-propelled battering ram and a rocket-propelled torpedo

✧ 1429

The French army uses gunpowder rockets to defend the city of Orléans. During this period, arsenals throughout Europe begin to test various types of gunpowder rockets as an alternative to early cannons

✧ ca. 1500

According to early rocketry lore, a Chinese official named Wan-Hu attempted to use an innovative rocket-propelled kite assembly to fly through the air. As he sat in the pilot's chair, his servants lit the assembly's 47 gunpowder (black powder) rockets. Unfortunately, this early rocket test pilot disappeared in a bright flash and explosion

✧ 1543

The Polish church official and astronomer Nicolaus Copernicus changes history and initiates the Scientific Revolution with his book *De Revolutionibus Orbium Coelestium* (On the revolutions of the heavenly spheres). This important book, published while Copernicus lay on his deathbed, proposed a Sun-centered (heliocentric) model of the universe in contrast to the longstanding Earth-centered (geocentric) model advocated by Ptolemy and many of the early Greek astronomers

✧ 1608

The Dutch optician Hans Lippershey develops a crude telescope

✧ 1609

The German astronomer Johannes Kepler publishes *New Astronomy,* in which he modifies Nicolaus Copernicus's model of the universe by announcing that the planets have elliptical orbits rather than circular ones. Kepler's laws of planetary motion help put an end to more than 2,000 years of geocentric Greek astronomy

✧ 1610

On January 7, 1610, Galileo Galilei uses his telescope to gaze at Jupiter and discovers the giant planet's four major moons (Callisto, Europa, Io, and Ganymede). He proclaims this and other astronomical observations in his book, *Sidereus Nuncius* (Starry messenger). Discovery of these four Jovian moons encourages Galileo to advocate the heliocentric theory of Nicolaus Copernicus and brings him into direct conflict with church authorities

✧ 1642

Galileo Galilei dies while under house arrest near Florence, Italy, for his clashes with church authorities concerning the heliocentric theory of Nicolaus Copernicus

✧ 1647

The Polish-German astronomer Johannes Hevelius publishes *Selenographia,* in which he provides a detailed description of features on the surface (near side) of the Moon

✧ 1680

Russian czar Peter the Great sets up a facility to manufacture rockets in Moscow. The facility later moves to St. Petersburg and provides the czarist army with a variety of gunpowder rockets for bombardment, signaling, and nocturnal battlefield illumination

✧ 1687

Financed and encouraged by Sir Edmond Halley, Sir Isaac Newton publishes his great work, *Philosophiae Naturalis Principia Mathematica* (Mathematical principles of natural philosophy). This book provides the mathematical foundations for understanding the motion of almost everything in the universe including the orbital motion of planets and the trajectories of rocket-propelled vehicles

✧ 1780s

The Indian ruler Hyder Ally (Ali) of Mysore creates a rocket corps within his army. Hyder's son, Tippo Sultan, successfully uses rockets against the British in a series of battles in India between 1782 and 1799

✧ 1804

Sir William Congreve writes *A Concise Account of the Origin and Progress of the Rocket System* and documents the British military's experience in India. He then starts the development of a series of British military (black-powder) rockets

✧ 1807

The British use about 25,000 of Sir William Congreve's improved military (black-powder) rockets to bombard Copenhagen, Denmark, during the Napoleonic Wars

✧ 1809

The brilliant German mathematician, astronomer, and physicist Carl Friedrich Gauss publishes a major work on celestial mechanics that revolutionizes the calculation of perturbations in planetary orbits. His work paves the way for other 19th-century astronomers to mathematically anticipate and then discover Neptune (in 1846), using perturbations in the orbit of Uranus

✧ 1812

British forces use Sir William Congreve's military rockets against American troops during the War of 1812. British rocket bombardment of Fort William McHenry inspires Francis Scott Key to add "the rocket's red glare" verse in the "Star Spangled Banner"

✧ 1865

The French science fiction writer Jules Verne publishes his famous story *De la terre a la lune* (From the Earth to the Moon). This story interests many people in the concept of space travel, including young readers who go on to become the founders of astronautics: Robert Hutchings Goddard, Hermann J. Oberth, and Konstantin Eduardovich Tsiolkovsky

✧ 1869

American clergyman and writer Edward Everett Hale publishes *The Brick Moon*—a story that is the first fictional account of a human-crewed space station

✧ 1877

While a staff member at the U.S. Naval Observatory in Washington, D.C., the American astronomer Asaph Hall discovers and names the two tiny Martian moons, Deimos and Phobos

✧ 1897

British author H. G. Wells writes the science fiction story *War of the Worlds*—the classic tale about extraterrestrial invaders from Mars

✧ 1903

The Russian technical visionary Konstantin Eduardovich Tsiolkovsky becomes the first person to link the rocket and space travel when he publishes *Exploration of Space with Reactive Devices*

✧ 1918

American physicist Robert Hutchings Goddard writes *The Ultimate Migration*—a far-reaching technology piece within which he postulates the use of an atomic-powered space ark to carry human beings away from a dying Sun. Fearing ridicule, however, Goddard hides the visionary manuscript and it remains unpublished until November 1972—many years after his death in 1945

✧ 1919

American rocket pioneer Robert Hutchings Goddard publishes the Smithsonian monograph *A Method of Reaching Extreme Altitudes.* This important work presents all the fundamental principles of modern rocketry. Unfortunately, members of the press completely miss the true significance of his technical contribution and decide to sensationalize his comments about possibly reaching the Moon with a small, rocket-propelled package. For such "wild fantasy," newspaper reporters dubbed Goddard with the unflattering title of "Moon man"

✧ 1923

Independent of Robert Hutchings Goddard and Konstantin Eduardovich Tsiolkovsky, the German space-travel visionary Hermann J. Oberth publishes the inspiring book *Die Rakete zu den Planetenräumen* (The rocket into planetary space)

✧ 1924

The German engineer Walter Hohmann writes *Die Erreichbarkeit der Himmelskörper* (The attainability of celestial bodies)—an important work that details the mathematical principles of rocket and spacecraft motion. He includes a description of the most efficient (that is, minimum energy) orbit transfer path between two coplanar orbits—a frequently used space operations maneuver now called the Hohmann transfer orbit

✧ 1926

On March 16 in a snow-covered farm field in Auburn, Massachusetts, American physicist Robert Hutchings Goddard makes space technology history by successfully firing the world's first liquid-propellant rocket. Although his primitive gasoline (fuel) and liquid oxygen (oxidizer) device burned for only two and one half seconds and landed about 60 meters away, it represents the technical ancestor of all modern liquid-propellant rocket engines.

In April, the first issue of *Amazing Stories* appears. The publication becomes the world's first magazine dedicated exclusively to science fiction. Through science fact and fiction, the modern rocket and space travel become firmly connected. As a result of this union, the visionary dream for many people in the 1930s (and beyond) becomes that of interplanetary travel

✧ 1929

German space-travel visionary Hermann J. Oberth writes the award-winning book *Wege zur Raumschiffahrt* (Roads to space travel) that helps popularize the notion of space travel among nontechnical audiences

✧ 1933

P. E. Cleator founds the British Interplanetary Society (BIS), which becomes one of the world's most respected space-travel advocacy organizations

✧ 1935

Konstantin Tsiolkovsky publishes his last book, *On the Moon,* in which he strongly advocates the spaceship as the means of lunar and interplanetary travel

✧ 1936

P. E. Cleator, founder of the British Interplanetary Society, writes *Rockets through Space,* the first serious treatment of astronautics in the United Kingdom. However, several established British scientific publications ridicule his book as the premature speculation of an unscientific imagination

✧ 1939–1945

Throughout World War II, nations use rockets and guided missiles of all sizes and shapes in combat. Of these, the most significant with respect to space exploration is the development of the liquid propellant V-2 rocket by the German army at Peenemünde under Wernher von Braun

✧ 1942

On October 3, the German A-4 rocket (later renamed Vengeance Weapon Two or V-2 Rocket) completes its first successful flight from the Peenemünde test site on the Baltic Sea. This is the birth date of the modern military ballistic missile

✧ 1944

In September, the German army begins a ballistic missile offensive by launching hundreds of unstoppable V-2 rockets (each carrying a one-ton high explosive warhead) against London and southern England

✧ 1945

Recognizing the war was lost, the German rocket scientist Wernher von Braun and key members of his staff surrender to American forces near Reutte, Germany in early May. Within months, U.S. intelligence teams, under Operation Paperclip, interrogate German rocket personnel and sort through carloads of captured documents and equipment. Many of these German scientists and engineers join von Braun in the United States to continue their rocket work. Hundreds of captured V-2 rockets are also disassembled and shipped back to the United States.

On May 5, the Soviet army captures the German rocket facility at Peenemünde and hauls away any remaining equipment and personnel. In the closing days of the war in Europe, captured German rocket technology and personnel helps set the stage for the great missile and space race of the cold war

On July 16, the United States explodes the world's first nuclear weapon. The test shot, code named Trinity, occurs in a remote portion of southern New Mexico and changes the face of warfare forever. As part of the cold-war confrontation between the United States and the former Soviet Union, the nuclear-armed ballistic missile will become the most powerful weapon ever developed by the human race.

In October, a then-obscure British engineer and writer, Arthur C. Clarke, suggests the use of satellites at geostationary orbit to support global communications. His article, in *Wireless World* "Extra-Terrestrial Relays," represents the birth of the communications satellite concept—an application of space technology that actively supports the information revolution

✧ 1946

On April 16, the U.S. Army launches the first American-adapted, captured German V-2 rocket from the White Sands Proving Ground in southern New Mexico.

Between July and August the Russian rocket engineer Sergei Korolev develops a stretched-out version of the German V-2 rocket. As part of his engineering improvements, Korolev increases the rocket engine's thrust and lengthens the vehicle's propellant tanks

✧ 1947

On October 30, Russian rocket engineers successfully launch a modified German V-2 rocket from a desert launch site near a place called Kapustin Yar. This rocket impacts about 320 kilometers downrange from the launch site

✧ 1948

The September issue of the *Journal of the British Interplanetary Society* publishes the first in a series of four technical papers by L. R. Shepherd and A. V. Cleaver that explores the feasibility of applying nuclear energy to space travel, including the concepts of nuclear-electric propulsion and the nuclear rocket

✧ 1949

On August 29, the Soviet Union detonates its first nuclear weapon at a secret test site in the Kazakh Desert. Code-named First Lightning (Pervaya Molniya), the successful test breaks the nuclear-weapon monopoly enjoyed by the United States. It plunges the world into a massive nuclear arms race that includes the accelerated development of strategic ballistic missiles capable of traveling thousands of kilometers. Because they are well behind the United States in nuclear weapons technology, the leaders of the former Soviet Union decide to develop powerful, high-thrust rockets to carry their heavier, more primitive-design nuclear weapons. That decision gives the Soviet Union a major launch vehicle advantage when both superpowers decide to race into outer space (starting in 1957) as part of a global demonstration of national power

✧ 1950

On July 24, the United States successfully launches a modified German V-2 rocket with an American-designed WAC Corporal second-stage rocket from the U.S. Air Force's newly established Long Range Proving Ground at Cape Canaveral, Florida. The hybrid, multistage rocket (called Bumper 8) inaugurates the incredible sequence of military missile and space vehicle launches to take place from Cape Canaveral—the world's most famous launch site.

In November, British technical visionary Arthur C. Clarke publishes "Electromagnetic Launching as a Major Contribution to Space-Flight." Clarke's article suggests mining the Moon and launching the mined-lunar material into outer space with an electromagnetic catapult

✧ 1951

Cinema audiences are shocked by the science fiction movie *The Day the Earth Stood Still.* This classic story involves the arrival of a powerful, humanlike extraterrestrial and his robot companion, who come to warn the governments of the world about the foolish nature of their nuclear arms race. It is the first major science fiction story to portray powerful space aliens as friendly, intelligent creatures who come to help Earth.

Dutch-American astronomer Gerard Peter Kuiper suggests the existence of a large population of small, icy planetesimals beyond the orbit of Pluto—a collection of frozen celestial bodies now known as the Kuiper belt

✧ 1952

Collier's magazine helps stimulate a surge of American interest in space travel by publishing a beautifully illustrated series of technical articles written by space experts such as Wernher von Braun and Willey Ley. The first of the famous eight-part series appears on March 22 and is boldly titled "Man Will Conquer Space Soon." The magazine also hires the most influential space artist Chesley Bonestell to provide stunning color illustrations. Subsequent articles in the series introduce millions of American readers to the concept of a space station, a mission to the Moon, and an expedition to Mars

Wernher von Braun publishes *Das Marsprojekt* (The Mars project), the first serious technical study regarding a human-crewed expedition to

Mars. His visionary proposal involves a convoy of 10 spaceships with a total combined crew of 70 astronauts to explore the Red Planet for about one year and then return to Earth

✧ 1953

In August, the Soviet Union detonates its first thermonuclear weapon (a hydrogen bomb). This is a technological feat that intensifies the superpower nuclear arms race and increases emphasis on the emerging role of strategic, nuclear-armed ballistic missiles.

In October, the U.S. Air Force forms a special panel of experts, headed by John von Neumann to evaluate the American strategic ballistic missile program. In 1954, this panel recommends a major reorganization of the American ballistic missile effort

✧ 1954

Following the recommendations of John von Neumann, President Dwight D. Eisenhower gives strategic ballistic missile development the highest national priority. The cold war missile race explodes on the world stage as the fear of a strategic ballistic missile gap sweeps through the American government. Cape Canaveral becomes the famous proving ground for such important ballistic missiles as the Thor, Atlas, Titan, Minuteman, and Polaris. Once developed, many of these powerful military ballistic missiles also serve the United States as space launch vehicles. U.S. Air Force General Bernard Schriever oversees the time-critical development of the Atlas ballistic missile—an astonishing feat of engineering and technical management

✧ 1955

Walt Disney (the American entertainment visionary) promotes space travel by producing an inspiring three-part television series that includes appearances by noted space experts like Wernher von Braun. The first episode, "Man in Space," airs on March 9 and popularizes the dream of space travel for millions of American television viewers. This show, along with its companion episodes, "Man and the Moon" and "Mars and Beyond," make von Braun and the term *rocket scientist* household words

✧ 1957

On October 4, Russian rocket scientist Sergei Korolev with permission from Soviet premier Nikita S. Khrushchev uses a powerful military rocket to successfully place *Sputnik 1* (the world's first artificial satellite) into orbit around Earth. News of the Soviet success sends a political and technical shockwave across the United States. The launch of *Sputnik 1* marks the beginning of the Space Age. It also is the start of the great space race of the cold war—a period when people measure national strength and global prestige by accomplishments (or failures) in outer space.

On November 3, the Soviet Union launches *Sputnik 2*—the world's second artificial satellite. It is a massive spacecraft (for the time) that carries a live dog named Laika, which is euthanized at the end of the mission.

The highly publicized attempt by the United States to launch its first satellite with a newly designed civilian rocket ends in complete disaster on December 6. The Vanguard rocket explodes after rising only a few inches above its launch pad at Cape Canaveral. Soviet successes with *Sputnik 1* and *Sputnik 2* and the dramatic failure of the Vanguard rocket heighten American anxiety. The exploration and use of outer space becomes a highly visible instrument of cold-war politics

✧ 1958

On January 31, the United States successfully launches *Explorer 1*—the first American satellite in orbit around Earth. A hastily formed team from the U.S. Army Ballistic Missile Agency (ABMA) and Caltech's Jet Propulsion Laboratory (JPL), led by Wernher von Braun, accomplishes what amounts to a national prestige rescue mission. The team uses a military ballistic missile as the launch vehicle. With instruments supplied by Dr. James Van Allen of the State University of Iowa, *Explorer 1* discovers Earth's trapped radiation belts—now called the Van Allen radiation belts in his honor.

The National Aeronautics and Space Administration (NASA) becomes the official civilian space agency for the United States government on October 1. On October 7, the newly created NASA announces the start of the Mercury Project—a pioneering program to put the first American astronauts into orbit around Earth.

In mid-December, an entire Atlas rocket lifts off from Cape Canaveral and goes into orbit around Earth. The missile's payload compartment carries Project Score (Signal Communications Orbit Relay Experiment)—a prerecorded Christmas season message from President Dwight D. Eisenhower. This is the first time the human voice is broadcast back to Earth from outer space

✧ 1959

On January 2, the Soviet Union sends a 790 pound-mass (360-kg) spacecraft, *Lunik 1,* toward the Moon. Although it misses hitting the Moon by between 3,125 and 4,375 miles (5,000 and 7,000 km), it is the first human-made object to escape Earth's gravity and go in orbit around the Sun.

In mid-September, the Soviet Union launches *Lunik 2.* The 860 pound-mass (390-kg) spacecraft successfully impacts on the Moon and becomes the first human-made object to (crash-) land on another world. *Lunik 2* carries Soviet emblems and banners to the lunar surface.

On October 4, the Soviet Union sends *Lunik 3* on a mission around the Moon. The spacecraft successfully circumnavigates the Moon and takes the first images of the lunar farside. Because of the synchronous rotation of the Moon around Earth, only the near side of the lunar surface is visible to observers on Earth

✧ 1960

The United States launches the *Pioneer 5* spacecraft on March 11 into orbit around the Sun. The modest-sized (92 pound-mass [42-kg]) spherical American space probe reports conditions in interplanetary space between Earth and Venus over a distance of about 23 million miles (37 million km).

On May 24, the U.S. Air Force launches a MIDAS (Missile Defense Alarm System) satellite from Cape Canaveral. This event inaugurates an important American program of special military surveillance satellites intended to detect enemy missile launches by observing the characteristic infrared (heat) signature of a rocket's exhaust plume. Essentially unknown to the general public for decades because of the classified nature of their mission, the emerging family of missile surveillance satellites provides U.S. government authorities with a reliable early warning system concerning a surprise enemy (Soviet) ICBM attack. Surveillance satellites help support the national policy of strategic nuclear deterrence throughout the cold war and prevent an accidental nuclear conflict.

The U.S. Air Force successfully launches the *Discoverer 13* spacecraft from Vandenberg Air Force Base on August 10. This spacecraft is actually part of a highly classified Air Force and Central Intelligence Agency (CIA) reconnaissance satellite program called Corona. Started under special executive order from President Dwight D. Eisenhower, the joint agency spy satellite program begins to provide important photographic images of denied areas of the world from outer space. On August 18, *Discoverer 14* (also called *Corona XIV*) provides the U.S. intelligence community its first satellite-acquired images of the former Soviet Union. The era of satellite reconnaissance is born. Data collected by the spy satellites of the National Reconnaissance Office (NRO) contribute significantly to U.S. national security and help preserve global stability during many politically troubled times.

On August 12, NASA successfully launches the *Echo 1* experimental spacecraft. This large (100 foot [30.5 m] in diameter) inflatable, metalized balloon becomes the world's first passive communications satellite. At the dawn of space-based telecommunications, engineers bounce radio signals off the large inflated satellite between the United States and the United Kingdom.

The former Soviet Union launches *Sputnik 5* into orbit around Earth. This large spacecraft is actually a test vehicle for the new *Vostok* spacecraft that will soon carry cosmonauts into outer space. *Sputnik 5* carries two dogs, Strelka and Belka. When the spacecraft's recovery capsule functions properly the next day, these two dogs become the first living creatures to return to Earth successfully from an orbital flight

✧ 1961

On January 31, NASA launches a Redstone rocket with a Mercury Project space capsule on a suborbital flight from Cape Canaveral. The passenger astrochimp Ham is safely recovered down range in the Atlantic Ocean after reaching an altitude of 155 miles (250 km). This successful primate space mission is a key step in sending American astronauts safely into outer space.

The Soviet Union achieves a major space exploration milestone by successfully launching the first human being into orbit around Earth. Cosmonaut Yuri Gagarin travels into outer space in the *Vostok 1* spacecraft and becomes the first person to observe Earth directly from an orbiting space vehicle.

On May 5, NASA uses a Redstone rocket to send astronaut Alan B. Shepard, Jr., on his historic 15-minute suborbital flight into outer space from Cape Canaveral. Riding inside the Mercury Project *Freedom 7* space capsule, Shepard reaches an altitude of 115 miles (186 km) and becomes the first American to travel in space.

President John F. Kennedy addresses a joint session of the U.S. Congress on May 25. In an inspiring speech touching on many urgent national needs, the newly elected president creates a major space challenge for the United States when he declares: "I believe that this nation should commit itself to achieving the goal, before this decade is out, of landing a man on the Moon and returning him safely to Earth." Because of his visionary leadership, when American astronauts Neil A. Armstrong and Edwin E. "Buzz" Aldrin, Jr., step onto the lunar surface for the first time on July 20, 1969, the United States is recognized around the world as the undisputed winner of the cold-war space race

✧ 1962

On February 20, astronaut John Herschel Glenn, Jr., becomes the first American to orbit Earth in a spacecraft. An Atlas rocket launches the NASA Mercury Project *Friendship 7* space capsule from Cape Canaveral. After completing three orbits, Glenn's capsule safely splashes down in the Atlantic Ocean.

In late August, NASA sends the *Mariner 2* spacecraft to Venus from Cape Canaveral. *Mariner 2* passes within 21,700 miles (35,000 km) of the planet on December 14, 1962—thereby becoming the world's first successful interplanetary space probe. The spacecraft observes very high surface temperatures (~800°F [430°C]). These data shatter pre–space age visions about Venus being a lush, tropical planetary twin of Earth.

During October, the placement of nuclear-armed Soviet offensive ballistic missiles in Fidel Castro's Cuba precipitates the Cuban Missile Crisis. This dangerous superpower confrontation brings the world perilously close to nuclear warfare. Fortunately, the crisis dissolves when Premier Nikita S. Khrushchev withdraws the Soviet ballistic missiles after much skillful political maneuvering by President John F. Kennedy and his national security advisers

✧ 1964

On November 28, NASA's *Mariner 4* spacecraft departs Cape Canaveral on its historic journey as the first spacecraft from Earth to visit Mars. It successfully encounters the Red Planet on July 14, 1965 at a flyby distance of about 6,100 miles (9,800 km). *Mariner 4*'s closeup images reveal a barren, desertlike world and quickly dispel any pre–space age notions about the existence of ancient Martian cities or a giant network of artificial canals

✧ 1965

A Titan II rocket carries astronauts Virgil "Gus" I. Grissom and John W. Young into orbit on March 23 from Cape Canaveral, inside a two-person Gemini Project spacecraft. NASA's *Gemini 3* flight is the first crewed mission for the new spacecraft and marks the beginning of more sophisticated space activities by American crews in preparation for the Apollo Project lunar missions

✧ 1966

The former Soviet Union sends the *Luna 9* spacecraft to the Moon on January 31. The 220 pound-mass (100-kg) spherical spacecraft soft lands in the Ocean of Storms region on February 3, rolls to a stop, opens four petal-like covers, and then transmits the first panoramic television images from the Moon's surface.

The former Soviet Union launches the *Luna 10* to the Moon on March 31. This massive (3,300 pound-mass [1,500-kg]) spacecraft becomes the first human-made object to achieve orbit around the Moon.

On May 30, NASA sends the *Surveyor 1* lander spacecraft to the Moon. The versatile robot spacecraft successfully makes a soft landing (June 1) in the Ocean of Storms. It then transmits over 10,000 images from the lunar surface and performs numerous soil mechanics experiments in preparation for the Apollo Project human landing missions.

In mid-August, NASA sends the *Lunar Orbiter 1* spacecraft to the Moon from Cape Canaveral. It is the first of five successful missions to collect detailed images of the Moon from lunar orbit. At the end of each mapping mission the orbiter spacecraft is intentionally crashed into the Moon to prevent interference with future orbital activities

✧ 1967

On January 27, disaster strikes NASA's Apollo Project. While inside their *Apollo 1* spacecraft during a training exercise on Launch Pad 34 at Cape Canaveral, astronauts Virgil "Gus" I. Grissom, Edward H. White, Jr., and Roger B. Chaffee are killed when a flash fire sweeps through their spacecraft. The Moon landing program was delayed by 18 months, while major design and safety changes are made in the Apollo Project spacecraft.

On April 23, tragedy also strikes the Russian space program when the Soviets launch cosmonaut Vladimir Komarov in the new *Soyuz* (union) spacecraft. Following an orbital mission plagued with difficulties, Komarov dies (on April 24) during reentry operations, when the spacecraft's parachute fails to deploy properly and the vehicle hits the ground at high speed

✧ 1968

On December 21, NASA's *Apollo 8* spacecraft (command and service modules only) departs Launch Complex 39 at the Kennedy Space Center during the first flight of mighty Saturn V launch vehicle with a human crew as part of the payload. Astronauts Frank Borman, James Arthur Lovell, Jr., and William A. Anders become the first people to leave Earth's gravitational influence. They go into orbit around the Moon and capture images of an incredibly beautiful Earth "rising" above the starkly barren lunar horizon—pictures that inspire millions and stimulate an emerging environmental movement. After 10 orbits around the Moon, the first lunar astronauts return safely to Earth on December 27

✧ 1969

The entire world watches as NASA's *Apollo 11* mission leaves for the Moon on July 16 from the Kennedy Space Center. Astronauts Neil A. Armstrong, Michael Collins, and Edwin E. "Buzz" Aldrin, Jr., make a long-held dream of humanity a reality. On July 20, American astronaut Neil Armstrong cautiously descends the steps of the lunar excursion module's ladder and steps on the lunar surface, stating, "One small step for a man, one giant leap for mankind!" He and Buzz Aldrin become the first two people to walk on another world. Many people regard the Apollo Project lunar landings as the greatest technical accomplishment in all of human history

✧ 1970

NASA's *Apollo 13* mission leaves for the Moon on April 11. Suddenly, on April 13, a life-threatening explosion occurs in the service module portion of the Apollo spacecraft. Astronauts James A. Lovell, Jr., John Leonard Swigert, and Fred Wallace Haise, Jr., must use their lunar excursion module (LEM) as a lifeboat. While an anxious world waits and listens, the crew skillfully maneuvers their disabled spacecraft around the Moon. With critical supplies running low, they limp back to Earth on a free-return trajectory. At just the right moment on April 17, they abandon the LEM *Aquarius* and board the Apollo Project spacecraft (command module) for a successful atmospheric reentry and recovery in the Pacific Ocean

✧ 1971

On April 19, the former Soviet Union launches the first space station (called *Salyut 1*). It remains initially uncrewed because the three-cosmonaut crew of the *Soyuz 10* mission (launched on April 22) attempts to dock with the station but cannot go on board

✧ 1972

In early January, President Richard M. Nixon approves NASA's space shuttle program. This decision shapes the major portion of NASA's program for the next three decades.

On March 2, an Atlas-Centaur launch vehicle successfully sends NASA's *Pioneer 10* spacecraft from Cape Canaveral on its historic mission. This far-traveling robot spacecraft becomes the first to transit the main-belt asteroids, the first to encounter Jupiter (December 3, 1973) and by crossing the orbit of Neptune on June 13, 1983 (which at the time was the farthest planet from the Sun) the first human-made object ever to leave the planetary boundaries of the solar system. On an interstellar trajectory, *Pioneer 10* (and its twin, *Pioneer 11*) carries a special plaque, greeting any intelligent alien civilization that might find it drifting through interstellar space millions of years from now.

On December 7, NASA's *Apollo 17* mission, the last expedition to the Moon in the 20th century, departs from the Kennedy Space Center, propelled by a mighty Saturn V rocket. While astronaut Ronald E. Evans remains in lunar orbit, fellow astronauts Eugene A. Cernan and Harrison H. Schmitt become the 11th and 12th members of the exclusive Moon walkers club. Using a lunar rover, they explore the Taurus-Littrow region. Their safe return to Earth on December 19 brings to a close one of the epic periods of human exploration

✧ 1973

In early April, while propelled by Atlas-Centaur rocket, NASA's *Pioneer 11* spacecraft departs on an interplanetary journey from Cape Canaveral. The spacecraft encounters Jupiter (December 2, 1974) and then uses a gravity assist maneuver to establish a flyby trajectory to Saturn. It is the first spacecraft to view Saturn at close range (closest encounter on September 1, 1979) and then follows a path into interstellar space.

On May 14, NASA launches *Skylab*—the first American space station. A giant Saturn V rocket is used to place the entire large facility into orbit in a single launch. The first crew of three American astronauts arrives on May 25 and makes the emergency repairs necessary to save the station, which suffered damage during the launch ascent. Astronauts Charles (Pete) Conrad, Jr., Paul J. Weitz, and Joseph P. Kerwin stay onboard for 28 days. They are replaced by astronauts Alan L. Bean, Jack R. Lousma, and Owen K. Garriott, who arrive on July 28 and live in space for about 59 days. The final *Skylab* crew (astronauts Gerald P. Carr, William R. Pogue, and Edward G. Gibson) arrive on November 11 and resided in the station until February 8, 1974—setting a space endurance record (for the time) of 84 days. NASA then abandons *Skylab.*

In early November, NASA launches the *Mariner 10* spacecraft from Cape Canaveral. It encounters Venus (February 5, 1974) and uses a gravity assist maneuver to become the first spacecraft to investigate Mercury at close range

✧ 1975

In late August and early September, NASA launches the twin *Viking 1* (August 20) and *Viking 2* (September 9) orbiter/lander combination spacecraft to the Red Planet from Cape Canaveral. Arriving at Mars in 1976, all Viking Project spacecraft (two landers and two orbiters) perform exceptionally well—but the detailed search for microscopic alien lifeforms on Mars remains inconclusive

✧ 1977

On August 20, NASA sends the *Voyager 2* spacecraft from Cape Canaveral on an epic grand tour mission during which it encounters all four giant planets and then departs the solar system on an interstellar trajectory. Using the gravity assist maneuver, *Voyager 2* visits Jupiter (July 9, 1979), Saturn (August 25, 1981), Uranus (January 24, 1986), and Neptune (August 25, 1989). The resilient, far-traveling robot spacecraft (and its twin *Voyager 1*) also carries a special interstellar message from Earth—a digital record entitled *The Sounds of Earth.*

On September 5, NASA sends the *Voyager 1* spacecraft from Cape Canaveral on its fast trajectory journey to Jupiter (March 5, 1979), Saturn (March 12, 1980), and beyond the solar system

✧ 1978

In May, the British Interplanetary Society releases its Project Daedalus report—a conceptual study about a one-way robot spacecraft mission to Barnard's star at the end of the 21st century

✧ 1979

On December 24, the European Space Agency successfully launches the first Ariane 1 rocket from the Guiana Space Center in Kourou, French Guiana

✧ 1980

India's Space Research Organization successfully places a modest 77 pound-mass (35 kg) test satellite (called *Rohini*) into low Earth orbit on July 1. The launch vehicle is a four-stage, solid propellant rocket manufactured in India. The SLV-3 (Standard Launch Vehicle-3) gives India independent national access to outer space

✧ 1981

On April 12, NASA launches the space shuttle *Columbia* on its maiden orbital flight from Complex 39-A at the Kennedy Space Center. Astronauts John W. Young and Robert L. Crippen thoroughly test the new aerospace vehicle. Upon reentry, it becomes the first spacecraft to return to Earth by gliding through the atmosphere and landing like an airplane. Unlike all previous onetime use space vehicles, *Columbia* is prepared for another mission in outer space

✧ 1986

On January 24, NASA's *Voyager 2* spacecraft encounters Uranus.

On January 28, the space shuttle *Challenger* lifts off from the NASA Kennedy Space Center on its final voyage. At just under 74 seconds into the STS 51-L mission, a deadly explosion occurs, killing the crew and destroying the vehicle. Led by President Ronald Reagan, the United States mourns seven astronauts lost in the *Challenger* accident

✧ 1988

On September 19, the State of Israel uses a Shavit (comet) three-stage rocket to place the country's first satellite (called *Ofeq 1*) into an unusual east-to-west orbit—one that is opposite to the direction of Earth's rotation but necessary because of launch safety restrictions.

As the *Discovery* successfully lifts off on September 29 for the STS-26 mission, NASA returns the space shuttle to service following a 32-month hiatus after the *Challenger* accident

✧ 1989

On August 25, the *Voyager 2* spacecraft encounters Neptune

✧ 1994

In late January, a joint Department of Defense and NASA advanced technology demonstration spacecraft, *Clementine,* lifts off for the Moon from Vandenberg Air Force Base. Some of the spacecraft's data suggest that the Moon may actually possess significant quantities of water ice in its permanently shadowed polar regions

✧ 1995

In February, during NASA's STS-63 mission, the space shuttle *Discovery* approaches (encounters) the Russian *Mir* space station as a prelude to the development of the *International Space Station.* Astronaut Eileen Marie Collins serves as the first female shuttle pilot.

On March 14, the Russians launch the *Soyuz TM-21* spacecraft to the *Mir* space station from the Baikanour Cosmodrome. The crew of three includes American astronaut Norman Thagard—the first American to travel into outer space on a Russian rocket and the first to stay on the *Mir* space station. The *Soyuz TM-21* cosmonauts also relieve the previous *Mir* crew, including cosmonaut Valeri Polyakov, who returns to Earth on March 22 after setting a world record for remaining in space for 438 days.

In late June, NASA's space shuttle *Atlantis* docks with the Russian *Mir* space station for the first time. During this shuttle mission (STS-71), *Atlantis* delivers the *Mir 19* crew (cosmonauts Anatoly Solovyev and Nikolai Budarin) to the Russian space station and then returns the *Mir 18* crew back to Earth—including American astronaut Norman Thagard, who has just spent 115 days in space onboard the *Mir.* The Shuttle-*Mir* docking program is the first phase of the *International Space Station.* A total of nine shuttle-*Mir* docking missions will occur between 1995 and 1998

✧ 1998

In early January, NASA sends the *Lunar Prospector* to the Moon from Cape Canaveral. Data from this orbiter spacecraft reinforces previous hints that the Moon's polar regions may contain large reserves of water ice in a mixture of frozen dust lying at the frigid bottom of some permanently shadowed craters.

In early December, the space shuttle *Endeavour* ascends from the NASA Kennedy Space Center on the first assembly mission of the *International Space Station.* During the STS-88 shuttle mission, *Endeavour* performs a rendezvous with the previously launched Russian-built *Zarya* (sunrise) module. An international crew connects this module with the American-built *Unity* module carried in the shuttle's cargo bay

✧ 1999

In July, astronaut Eileen Marie Collins serves as the first female space shuttle commander (STS-93 mission) as the *Columbia* carries NASA's *Chandra X-ray Observatory* into orbit

✧ 2001

NASA launches the *Mars Odyssey 2001* mission to the Red Planet in early April—the spacecraft successfully orbits the planet in October

✧ 2002

On May 4, NASA successfully launches its *Aqua* satellite from Vandenberg Air Force Base. This sophisticated Earth-observing spacecraft joins the *Terra* spacecraft in performing Earth system science studies.

On October 1, the United States Department of Defense forms the U.S. Strategic Command (USSTRATCOM) as the control center for all American strategic (nuclear) forces. USSTRATCOM also conducts military space operations, strategic warning and intelligence assessment, and global strategic planning

✧ 2003

On February 1, while gliding back to Earth after a successful 16-day scientific research mission (STS-107), the space shuttle *Columbia* experiences a catastrophic reentry accident at an altitude of about 63 km over the Western United States. Traveling at 18 times the speed of sound, the orbiter vehicle disintegrates, taking the lives of all seven crew members: six American astronauts (Rick Husband, William McCool, Michael Anderson, Kalpana Chawla, Laurel Clark, and David Brown) and the first Israeli astronaut (Ilan Ramon).

NASA's Mars Exploration Rover (MER) *Spirit* is launched by a Delta II rocket to the Red Planet on June 10. *Spirit,* also known as MER-A, arrives safely on Mars on January 3, 2004 and begins its teleoperated surface exploration mission under the supervision of mission controllers at the NASA Jet Propulsion Laboratory.

NASA launches the second Mars Exploration Rover, called *Opportunity,* using a Delta II rocket launch, which lifts off from Cape Canaveral Air Force Station on July 7, 2003. *Opportunity,* also called MER-B, successfully lands on Mars on January 24, 2004, and starts its teleoperated surface exploration mission under the supervision of mission controllers at the NASA Jet Propulsion Laboratory

✧ 2004

On July 1, NASA's *Cassini* spacecraft arrives at Saturn and begins its four-year mission of detailed scientific investigation.

In mid-October, the Expedition 10 crew, riding a Russian launch vehicle from Baikonur Cosmodrome, arrives at the *International Space Station* and the Expedition 9 crew returns safely to Earth.

On December 24, the 703 pound-mass (319-kg) *Huygens* probe successfully separates from the *Cassini* spacecraft and begins its journey to Saturn's moon, Titan

✧ 2005

On January 14, the *Huygens* probe enters the atmosphere of Titan and successfully reaches the surface some 147 minutes later. *Huygens* is the first spacecraft to land on a moon in the outer solar system.

On July 4, NASA's Deep Impact mission successfully encountered Comet Tempel 1.

NASA successfully launched the space shuttle *Discovery* on the STS-114 mission on July 26 from the Kennedy Space Center in Florida. After docking with the *International Space Station*, the *Discovery* returned to Earth and landed at Edwards AFB, California, on August 9.

On August 12, NASA launched the *Mars Reconnaissance Orbiter* from Cape Canaveral AFS, Florida.

On September 19, NASA announced plans for a new spacecraft designed to carry four astronauts to the Moon and to deliver crews and supplies to the *International Space Station*. NASA also introduced two new, shuttle-derived launch vehicles: a crew-carrying rocket and a cargo-carrying, heavy-lift rocket.

The Expedition 12 crew (Commander William McArthur and Flight Engineer Valery Tokarev) arrived at the *International Space Station* on October 3 and replaced the Expedition 11 crew.

The People's Republic of China successfully launched its second human spaceflight mission, called *Shenzhou 6*, on October 12. Two taikonauts, Fei Junlong and Nie Haisheng, traveled in space for almost five days and made 76 orbits of Earth before returning safely to Earth, making a soft, parachute-assisted landing in northern Inner Mongolia

✧ 2006

On January 15, the sample package from NASA's *Stardust* spacecraft, containing comet samples, successfully returned to Earth.

NASA launched the *New Horizons* spacecraft from Cape Canaveral on January 19 and successfully sent this robot probe on its long one-way mission to conduct a scientific encounter with the Pluto system (in 2015) and then to explore portions of the Kuiper belt that lie beyond.

Follow-up observations by NASA's *Hubble Space Telescope*, reported on February 22, have confirmed the presence of two new moons around the distant planet Pluto. The moons, tentatively called S/2005 P 1 and S/2005 P 2, were first discovered by *Hubble* in May 2005, but the science team wanted to further examine the Pluto system to characterize the orbits of the new moons and validate the discovery.

NASA scientists announced on March 9 that the *Cassini* spacecraft may have found evidence of liquid water reservoirs that erupt in Yellowstone Park–like geysers on Saturn's moon Enceladus.

On March 10, NASA's *Mars Reconnaissance Orbiter* successfully arrived at Mars and began a six-month-long process of adjusting and trimming the shape of its orbit around the Red Planet prior to performing its operational mapping mission.

The Expedition 13 crew (Commander Pavel Vinogradov and Flight Engineer Jeff Williams) arrived at the *International Space Station* on April 1 and replaced the Expedition 12 crew. Joining them for several days before returning back to Earth with the Expedition 12 crew was Brazil's first astronaut, Marcos Pontes

On August 24, members of the International Astronomical Union (IAU) met for the organization's 2006 General Assembly in Prague, Czech Republic. After much debate, the 2,500 assembled professional astronomers decided (by vote) to demote Pluto from its traditional status as one of the nine major planets and place the object into a new class, called a dwarf planet. The IAU decision now leaves the solar system with eight major planets and three dwarf planets: Pluto (which serves as the prototype dwarf planet), Ceres (the largest asteroid), and the large, distant Kuiper belt object identified as 2003 UB313 (nicknamed "Xena"). Astronomers anticipate the discovery of other dwarf planets in the distant parts of the solar system.

Glossary

absorption spectrum The collection of dark lines superimposed upon a continuous spectrum that occurs when radiation from a hot source passes through a cooler medium, allowing some of the radiant energy to get absorbed at selected wavelengths.

accelerometer An instrument that measures acceleration or gravitational forces capable of imparting acceleration. Frequently used on space vehicles to assist in guidance and navigation, and on planetary probes to support scientific data collection.

acquisition The process of locating the orbit of a satellite or the trajectory of a space probe so that mission-control personnel can track the object and collect its telemetry data.

acronym A word formed from the first letters of a name, such as *HST* which means the *Hubble Space Telescope,* or a word formed by combining the initial parts of a series of words, such as *lidar,* which means *li*ght *d*etecting *a*nd *r*anging. Acronyms are frequently used in space technology and astronomy.

active remote sensing A remote-sensing technique in which the sensor supplies its own source of electromagnetic radiation to illuminate a target. A synthetic aperture radar (SAR) system is an example.

aeroassist The use of the thin, upper regions of a planet's atmosphere to provide the lift or drag needed to maneuver a spacecraft. Near a planet with a sensible atmosphere, aeroassist allows a spacecraft to change direction or to slow down without expending propellant from the control rocket.

aerobraking The use of a specially designed spacecraft structure to deflect rarefied (very-low-density) airflow around a spacecraft, thereby supporting aeroassist maneuvers in the vicinity of a planet. Such maneuvers reduce the spacecraft's need to perform large propulsive burns when making orbital changes near a planet. In 1993, NASA's *Magellan* spacecraft became the first planetary-exploration system to use aerobraking as a means of changing its orbit around the target planet (Venus).

aerodynamic skip An atmospheric entry abort, caused by entering a planet's atmosphere at too shallow an angle. Much like a stone skipping across the surface of a pond, this condition results in a trajectory that sends a space vehicle back out into space rather than downward toward the planet's surface.

aerospace A term, derived from *aero*nautics and *space,* meaning of or pertaining to Earth's atmospheric envelope and outer space beyond it.

alphanumeric (*alpha*bet plus *numeric*) Including letters and numerical digits, as, for example, in the term *JEN75WX11.*

altimeter An instrument for measuring the height (altitude) above a planet's surface; generally reported relative to a common planetary reference point, such as sea level on Earth.

altitude In space-vehicle navigation, the height above the mean surface of the reference celestial body. Note that the *distance* of a spacecraft from the reference celestial body is taken as the distance from the center of the object.

Amor group A collection of near-Earth asteroids that cross the orbit of Mars but do not cross the orbit of Earth. This asteroid group acquired its name from the 0.6-mile- (1-km-) diameter Amor asteroid, discovered in 1932 by the Belgian astronomer Eugène-Joseph Delporte (1882–1955).

antenna A device used to detect, collect, or transmit radio waves. A radio telescope is a large receiving antenna, while many spacecraft have both a directional antenna and an omnidirectional antenna to transmit (downlink) telemetry and to receive (uplink) instructions.

aperture The opening in front of a telescope, camera, or other optical instrument through which light passes.

aphelion The point in an object's orbit around the Sun that is most distant from the Sun. Compare with PERIHELION.

Aphrodite Terra A large, fractured highland region near the equator of Venus.

apogee The point in the orbit of a satellite that is farthest from Earth. Term applies to both the orbit of the Moon and to the orbits of artificial satellites around Earth. At apogee, the orbital velocity of a satellite is at a minimum. Compare with PERIGEE.

Apollo group A collection of near-Earth asteroids that have perihelion distances of 1.017 astronomical units (AU) or less, taking them across the orbit of Earth around the Sun. This group acquired its name from the asteroid Apollo, the first to be discovered, by the German astronomer, Karl Reinmuth (1892–1979), in 1932.

apolune That point in an orbit around the Moon of a spacecraft launched from the lunar surface that is farthest from the Moon. Compare with PERILUNE.

artificial intelligence (AI) Information-processing functions (including thinking and perceiving) performed by machines that imitate (to some extent) the mental activities performed by the human brain. Anticipated advances in AI should allow "very smart" future robot spacecraft to explore distant alien worlds with minimal human supervision.

artificial satellite A human-made object, such as a spacecraft, placed in orbit around Earth or another celestial body. *Sputnik 1* was the first artificial satellite to be placed in orbit around Earth.

ascending node That point in the orbit of a celestial body at which it travels from south to north across a reference plane, such as the equatorial plane of the celestial sphere or the plane of the ecliptic. Also called the northbound node. Compare with DESCENDING NODE.

asteroid A small, solid rocky object that orbits the Sun but is independent of any major planet. Most asteroids (or minor planets) are found in the main asteroid belt. The largest asteroid is Ceres, which was discovered in 1801 by the Italian astronomer Giuseppe Piazzi (1746–1826). Earth-crossing asteroids, or near-Earth asteroids (NEAs), have orbits that take them near or across Earth's orbit around the Sun and are divided into the Aten, Apollo, and Amor groups.

astro- A prefix that means star or (by extension) outer space or celestial, as in, for example, the terms *astronaut, astronautics,* or *astrophysics.*

astronomical unit (AU) A convenient unit of distance defined as the semimajor axis of Earth's orbit around the Sun. One AU, the average distance between Earth and the Sun, is equal to approximately 93×10^6 miles (149.6×10^6 km), or 499.01 light-seconds.

astrophysics The branch of science that investigates the nature of stars and star systems.

Aten group A collection of near-Earth asteroids that cross the orbit of Earth, but whose average distances from the Sun lie inside Earth's orbit. This asteroid group acquired its name from the .55-mile- (.9-km-) diameter asteroid Aten, discovered in 1976 by the American astronomer Eleanor Kay Helin (née Francis).

atmosphere In general, the gravitationally bound gaseous envelope that forms an outer region around a planet or other celestial body.

atmospheric probe The special collection of scientific instruments (usually released by a mother spacecraft) for determining the pressure, composition, and temperature of a planet's atmosphere at different altitudes. An example is the probe released by NASA's *Galileo* spacecraft in December 1995. As it plunged into Jupiter's atmosphere, the probe successfully transmitted its scientific data to the *Galileo* spacecraft (the mother spacecraft) for about 58 minutes.

atomic clock A precise device for measuring or standardizing time that is based on periodic vibrations of certain atoms (cesium) or molecules (ammonia). Often used in robot exploration spacecraft.

attitude The position of an object as defined by the inclination of its axes with respect to a frame of reference. The orientation of a spacecraft that is either in motion or at rest, as established by the relationship between the vehicle's axes and a reference line or plane. Attitude is often expressed in terms of pitch, roll, and yaw.

attitude control system The onboard system of computers, low-thrust rockets (thrusters), and mechanical devices (such as a momentum wheel) used to keep a spacecraft stabilized during flight and to precisely point its instruments in some desired direction. Stabilization is achieved by spinning the spacecraft or by using a three-axis active approach that maintains the spacecraft in a fixed, reference attitude, by firing a selected combination of thrusters when necessary.

auxiliary power unit (APU) A power unit carried on a spacecraft that supplements the main source of electric power on the craft.

band A range of (radio wave) frequencies; or a closely spaced set of spectral lines that are associated with the electromagnetic radiation (EMR) characteristic of some particular atomic or molecular energy levels.

berthing The joining of two orbiting spacecraft, using a manipulator or other mechanical device to move one into contact (or very close proximity) with the other at a selected interface. *See also* DOCKING; RENDEZVOUS.

calibration The process of translating the signals collected by a measuring instrument (such as a telescope) into something that is scientifically useful. The calibration procedure generally removes most of the errors caused by instabilities in the instrument or in the environment through which the signal has traveled.

Caloris basin A very large, ringed impact basin (about 800 miles [1,300 km] across) on Mercury.

Cape Canaveral The region on Florida's east central coast from which the United States Air Force and NASA have launched more than 3,000 rockets since 1950. Cape Canaveral Air Force Station (CCAFS) is the major East Coast launch site for the Department of Defense, while the adjacent NASA Kennedy Space Center is the spaceport for the fleet of space shuttle vehicles.

Cassini mission The joint NASA–European Space Agency planetary exploration mission to Saturn launched from Cape Canaveral on October 15, 1997. Since July 2004, the *Cassini* spacecraft has performed detailed studies of Saturn, its rings, and moons. The *Cassini* mother spacecraft also delivered the *Huygens* probe, which successfully plunged into the nitrogen-rich atmosphere of Titan (Saturn's largest moon) on January 14, 2005. The mother spacecraft is named after the Italian-French astronomer Giovanni Cassini (1625–1712), the Titan probe after the Dutch astronomer, Christiaan Huygens (1629–95).

celestial body A heavenly body. Any aggregation of matter in outer space constituting a unit for study in astronomy, such as planets, moons, comets, asteroids, stars, nebulae, and galaxies.

Centaur group A group of unusual celestial objects (such as Chiron) that reside in the outer solar system and exhibit a dual asteroid/comet nature. Named after the centaurs in Greek mythology, which were half-human and half-horse.

Ceres The first and largest (580-mile- [940-km-] diameter) asteroid to be found. It was discovered on January 01,1801, by the Italian astronomer Giuseppe Piazzi (1746–1826).

***Chandra X-ray Observatory* (CXO)** One of NASA's major orbiting astronomical observatories, launched in July 1999 and named after the Indian-American astrophysicist, Subrahmanyan Chandrasekar (1910–95). NASA previously called the spacecraft the *Advanced X-ray Astrophysics Facility (AXAF).* This Earth-orbiting facility studies some of the most interesting and puzzling X-ray sources in the universe, including emissions from active galactic nuclei, exploding stars, neutron stars, and matter falling into black holes.

charge-coupled device (CCD) An electronic (solid state) device, containing a regular array of sensor elements that are sensitive to various types of electromagnetic radiation (e.g., light) and emit electrons when exposed to such radiation. The emitted electrons are collected and the resulting charge analyzed. CCDs are used as the light-detecting component in modern television cameras and telescopes.

Charon The large (about 745-mile- [1,200-km-] diameter) moon of Pluto discovered in 1978 by the American astronomer James Walter Christy.

chaser spacecraft The spacecraft that actively performs the key maneuvers during orbital rendezvous and docking/berthing operations. The other space vehicle serves as the target and remains essentially passive during the encounter.

chasma A canyon or deep linear feature on a planet's surface.

Chiron An unusual celestial body in the outer solar system with a chaotic orbit that lies almost entirely between the orbits of Saturn and Uranus. This massive asteroid-sized object has a diameter of about 125 miles (200 km) and was the first object placed in the Centaur group, because it also has a detectable coma—a feature characteristic of comets.

Chryse Planitia A large plain on Mars characterized by many ancient channels that could once have contained flowing surface water. Landing site for NASA's *Viking 1* lander (robot spacecraft) in July 1976.

cislunar Of or pertaining to phenomena, projects, or activities happening in the region of outer space between Earth and the Moon. From the Latin word *cis,* meaning "on this side" and lunar, which means "of or pertaining to the Moon." Therefore, it means "on this side of the Moon."

clean room A controlled work environment for robot spacecraft in which dust, temperature, and humidity are carefully controlled during the fabrication, assembly, and/or testing of critical components.

cold war The ideological conflict between the United States and the former Soviet Union from approximately 1946 to 1989, involving rivalry, mistrust, and hostility just short of overt military action. The tearing down of the Berlin Wall in November 1989 is generally considered to be the (symbolic) end of the cold war period.

collimator A device for focusing or confining a beam of particles or electromagnetic radiation, such as X-ray photons.

coma The gaseous envelope that surrounds the nucleus of a comet.

comet A dirty ice "rock" consisting of dust, frozen water, and gases that orbits the Sun. As a comet approaches the inner solar system from deep space, solar radiation causes its frozen materials to vaporize (sublime), creating a coma and a long tail of dust and ions. Scientists think these icy planetesimals are the remainders of the primordial material from which the outer planets were formed billions of years ago. *See also* KUIPER BELT and OORT CLOUD.

Comet Halley *(1P/Halley)* The most famous periodic comet. Named after the British astronomer Edmund Halley (1656–1742), who successfully predicted its 1758 return. With sightings reported since 240 B.C.E., this comet reaches perihelion approximately every 76 years. During its most recent inner-solar-system appearance, an international fleet of five different robot spacecraft, including the *Giotto* spacecraft, performed scientific investigations that supported the dirty ice-rock model of a comet's nucleus.

Compton Gamma Ray Observatory (CGRO) A major NASA orbiting astrophysical observatory dedicated to gamma-ray astronomy. The *CGRO* was placed in orbit around Earth in April 1991. At the end of its useful scientific mission, flight controllers intentionally commanded the massive (35,900 pound- [16,300-kg-] mass) spacecraft to perform a deorbit burn. This caused it to reenter and safely crash in June 2000 in a remote region of the Pacific Ocean. The spacecraft was named in honor of the American physicist Arthur Holly Compton (1892–1962).

cooperative target A three-axis stabilized orbiting object that has signaling devices to support rendezvous and docking/capture operations by a chaser spacecraft.

co-orbital Sharing the same or very similar orbit; for example, during a rendezvous operation the chaser spacecraft and its cooperative target are said to be co-orbital.

Copernicus Observatory A scientific robot spacecraft launched into orbit around Earth by NASA on August 21,1972. Named in honor of the Polish astronomer Nicolaus Copernicus (1473–1543), this space-based observatory examined the universe from 1972 to 1981 in the ultraviolet portion of the electromagnetic spectrum. Also called the *Orbiting Astronomical Observatory-3 (OAO-3).*

Cosmic Background Explorer (COBE) A NASA robot spacecraft placed in orbit around Earth in November 1989. It successfully measured the spectrum and intensity distribution of the cosmic-microwave background (CMB).

cosmic-microwave background (CMB) The background of microwave radiation that permeates the universe and has a blackbody temperature of about 2.7 K. Sometimes called the *primal glow,* scientists believe it represents the remains of the ancient fireball in which the universe was created.

cosmic rays Extremely energetic particles (usually bare atomic nuclei) that move through outer space at speeds just below the speed of light and bombard Earth from all directions. Their existence was discovered in about 1912 by the Austrian-American physicist Victor Francis Hess (1883–1964). Hydrogen nuclei (protons) make up the highest proportion of the cosmic-ray population (approximately 85 percent), but these particles range over the entire periodic table of elements. *Galactic cosmic rays* are samples of material from outside the solar system; *solar cosmic rays* are ejected from the Sun during solar-flare events.

Cosmos spacecraft The general name given to a large number of Soviet and later Russian spacecraft, ranging from military satellites to scientific platforms investigating near-Earth space. *Cosmos 1* was launched in March 1962; since then, well over 2,000 Cosmos satellites have been sent into outer space. Also called *Kosmos.*

cruise phase For a robot spacecraft on an interplanetary scientific mission, the part of the mission (usually months or even years in duration) following launch and prior to planetary (or celestial object) encounter.

Dactyl A tiny natural satellite of the asteroid Ida, discovered in February 1994 when NASA scientists were reviewing *Galileo* spacecraft data from the flyby encounter with the asteroid Ida on August 28, 1993.

decay (orbital) The gradual lessening of both the apogee and perigee of an orbiting object from its primary body. For example, the orbital-decay process for artificial satellites and debris often results in their ultimate fiery plunge back into the denser regions of the Earth's atmosphere.

Deep Space Network (DSN) NASA's global network of antennae that serve as the radio-wave communications link to distant interplanetary spacecraft and probes, transmitting instructions to them and receiving data from them. Large radio antennae of the DSN's three Deep Space Communications Complexes (DSCCs) are located in Goldstone, California, near Madrid, Spain, and near Canberra, Australia—providing almost continuous contact with a spacecraft in deep space as Earth rotates on its axis.

deep-space probe A robot spacecraft designed for exploring deep space, especially to the vicinity of the Moon and beyond. This includes lunar probes, Mars probes, outer-planet probes, solar probes, asteroid probes, comet probes, and so on.

degrees of freedom (DOF) A mode of motion, either angular or linear, motion with respect to a coordinate system, independent of any other mode. A body in motion has six possible degrees of freedom, three linear (sometimes called: *x*-, *y*-, and *z*- motion with reference to linear [axial] movements in the Cartesian coordinate system) and three angular movements (sometimes called pitch, yaw, and roll).

Deimos The tiny (about 7.5 miles [12 km] average diameter), irregularly shaped outer moon of Mars, discovered in 1877 by the American astronomer Asaph Hall (1829–1907). *See also* PHOBOS.

delta-V (symbol: Δ) Velocity change; a numerical index of the maneuverability of a spacecraft. This term often represents the maximum change in velocity that a space vehicle's propulsion system can provide—for example, it is the delta-V capability of an upper stage propulsion system that places an Earth-orbiting spacecraft on an interplanetary trajectory. Often described in terms of feet per second (m/s).

de-orbit burn A retrograde (opposite direction) rocket engine firing, by which a space vehicle's velocity is reduced to less than that required to remain in orbit around a celestial body.

descending node That point in the orbit of a celestial body at which it travels from north to south across a reference plane, such as the equatorial plane of the celestial sphere, or the plane of the ecliptic. Also called the southbound node. Compare with ASCENDING NODE.

direct conversion The conversion of thermal energy (heat) or other forms of energy (such as sunlight) directly into electrical energy, without intermediate conversion into mechanical work—that is, without the use of moving components such as those found in a conventional electric generator system. The main approaches for converting heat directly into electricity include thermoelectric conversion, thermionic conversion, and magnetohydrodynamic conversion. Solar energy is directly converted into electrical energy by means of solar cells (photovoltaic conversion). Batteries and fuel cells directly convert chemical energy into electrical energy. *See also* RADIOISOTOPE THERMOELECTRIC GENERATOR.

directional antenna An antenna that radiates or receives radio-frequency (RF) signals more efficiently in some directions that in others.

docking The act of physically joining two orbiting spacecraft. Usually accomplished by independently maneuvering one spacecraft (the chaser spacecraft) into contact with the other (the target spacecraft) at a chosen physical interface.

downlink The telemetry signal received at a ground station from a spacecraft or probe.

Earth-crossing asteroid (ECA) An inner solar-system asteroid whose orbital path takes it across Earth's orbit around the Sun.

Earth-observing satellite (EOS) A robot spacecraft in orbit around Earth that has a specialized collection of sensors capable of monitoring important environmental variables. Data from such satellites help support Earth-system science. Also called an environmental satellite or a green satellite.

eccentric orbit An orbit that deviates from a circle, thus forming an ellipse.

electromagnetic (EM) spectrum Comprises the entire range of wavelengths of electromagnetic radiation, from the most energetic, shortest-wavelength gamma rays to the longest-wavelength radio waves, and everything in between.

encounter The close flyby or rendezvous of a robot spacecraft with a target body. The target of an encounter can be a natural celestial body (such as a planet, asteroid, or comet) or a human-made object (such as another spacecraft).

escape velocity (common symbol: V_e) The minimum velocity that an object must acquire to overcome the gravitational attraction of a celestial body. The escape velocity for an object launched from the surface of Earth is approximately 7 miles per second (11.2 km/s), while the escape velocity from the surface of Mars is about three miles per second (5.0 km/s).

Europa The smooth, ice-covered moon of Jupiter, discovered in 1610 by the Italian astronomer Galileo Galilei (1564–1642), currently thought to have a possibly life-bearing liquid-water ocean beneath its frozen surface.

European Space Agency (ESA) An international organization that promotes the peaceful use of outer space and cooperation among the European member states in space research and applications.

exoatmospheric Occurring outside Earth's atmosphere; events and actions that take place at altitudes above about 62 miles (100 km).

Explorer 1 The first U.S. Earth-orbiting satellite, which was launched successfully from Cape Canaveral on January 31, 1958, by a Juno I four-stage configuration of the Jupiter C launch vehicle.

Explorer spacecraft NASA has used the name *Explorer* to designate members of a large family of scientific robot spacecraft and satellites intended to "explore the unknown." Since 1958, Explorer spacecraft have studied Earth's atmosphere and ionosphere; the planet's precise shape and geophysical features; the planet's magnetosphere and interplanetary space; and various astronomical and astrophysical phenomena.

extraterrestrial (ET) Occurring, located, or originating beyond planet Earth and its atmosphere.

extraterrestrial contamination The contamination of one world by life-forms, especially microorganisms, from another world. Taking Earth's biosphere as the reference, planetary-contamination is called *forward-contamination* when an alien world is contaminated by contact with terrestrial organisms and *back-contamination* when alien organisms are released into Earth's biosphere.

extraterrestrial life Life-forms that may have evolved independently of, and now exist beyond, the terrestrial biossphere.

Extreme Ultraviolet Explorer (EUVE) The 70th NASA Explorer spacecraft. After being successfully launched from Cape Canaveral in June 1992, this scientific robot spacecraft went into orbit around Earth and provided astronomers with a survey of the until-then relatively unexplored extreme-ultraviolet portion of the electromagnetic spectrum.

extremophile A hardy (terrestrial) microorganism that can exist under extreme environmental conditions, such as in frigid polar regions or boiling hot springs. Astrobiologists speculate that similar (extraterrestrial) microorganisms might exist elsewhere in this solar system, perhaps within subsurface biological niches on Mars, or in a suspected liquid-water ocean beneath the frozen surface of Europa.

field of view (FOV) The area or solid angle than can be viewed through, or scanned by, a remote sensing (optical) instrument.

flyby An interplanetary or deep-space mission in which the *flyby spacecraft* passes close to its target celestial body (e.g., a distant planet, moon, asteroid, or comet), but does not impact the target or go into orbit around it.

free-flying spacecraft (free-flyer) Any spacecraft or payload that can be detached from NASA's space shuttle or the *International Space Station* and then operate independently in orbit around Earth or in interplanetary space.

frequency (usual symbol: f or ν) The rate of repetition of a recurring or regular event; the number of cycles of a wave per second. For electromagnetic radiation, the frequency (ν) is equal to the speed of light (c) divided by the wavelength (λ). *See also* HERTZ.

fuel cell A direct-conversion device that transforms chemical energy directly into electrical energy by reacting with continuously supplied chemicals. In a modern fuel cell, an electrochemical catalyst (like platinum) promotes a noncombustible reaction between a fuel (such as hydrogen) and an oxidant (such as oxygen).

Galilean satellites The four largest and brightest moons of Jupiter, discovered in 1610 by the Italian astronomer Galileo Galilei (1564–1642). They are Io, Europa, Ganymede, and Callisto.

Galileo Project NASA's highly successful scientific mission to Jupiter, launched in October 1989. With electricity supplied by two radioisotope-thermoelectric generator (RTG) units, the *Galileo* spacecraft extensively studied the Jovian system from December 1995 until February 2003. Upon arrival, it also released a probe into the upper portions of Jupiter's atmosphere. On February 28, 2003, the NASA flight team terminated its operation of the *Galileo* spacecraft and commanded the robot craft to plunge into Jupiter's atmosphere. This mission-ending plunge took place in late September 2003.

gamma ray (symbol: γ) Very-short-wavelength, high-frequency packets (or quanta) of electromagnetic radiation. Gamma-ray photons are similar to X-rays, except that they originate within the atomic nucleus and have energies between 10,000 electron volts (10 keV) and 10 million electron volts (10 MeV).

giant planets In this solar system, the large, gaseous outer planets: Jupiter, Saturn, Uranus, and Neptune. Any detected or suspected extrasolar planets as large or larger than Jupiter.

***Giotto* spacecraft** Scientific robot spacecraft launched by the European Space Agency (ESA) in July 1985 that successfully encountered the nucleus of Comet Halley in mid-March 1986, at a (closest-approach) distance of about 370 miles (600 km).

gravity-assist The change in a robot spacecraft's direction and speed achieved by a carefully calculated flyby through a planet's gravitational field. This change in spacecraft velocity occurs without the use of supplementary propulsive energy.

Great Dark Spot (GDS) A large, dark, oval-shaped feature in the clouds of Neptune, discovered in 1989 by NASA's *Voyager 2* spacecraft.

Greenwich mean time (GMT) Mean solar time at the meridian of Greenwich, England, used as the basis for standard time throughout the world. Normally expressed in four numerals, 0001 to 2400. Also called *universal time* (UT).

guidance system A system that evaluates flight information; correlates it with destination data; determines the desired flight path of the spacecraft; and communicates the necessary commands to the craft's flight-control system.

halo orbit A circular or elliptical orbit, in which a spacecraft remains in the vicinity of a Lagrangian libration point.

Halley's comet *See* COMET HALLEY.

hard landing The relatively high-velocity impact of a lander spacecraft or probe on a solid planetary surface. The impact usually destroys all equipment, except perhaps for a very rugged instrument package or payload container.

heliocentric With the Sun as a center.

heliopause The boundary of the heliosphere. It is thought to occur about 100 astronomical units from the Sun, and marks the edge of the Sun's influence and the beginning of interstellar space.

hertz (symbol: Hz) The SI unit of frequency. One hertz is equal to one cycle per second. Named in honor of the German physicist Heinrich Rudolf Hertz (1857–94), who produced and detected radio waves for the first time in 1888.

high-Earth orbit (HEO) An orbit around Earth at an altitude greater than 3,475 miles (5,600 km).

High Energy Astronomy Observatory (HEAO) A series of three NASA robot spacecraft placed in Earth orbit (*HEAO-1* launched in August 1977; *HEAO-2* in November 1978; and *HEAO-3* in September 1979) to support x-ray astronomy and gamma-ray astronomy. After launch, NASA renamed *HEAO-2* the *Einstein Observatory* to honor of the famous German-Swiss-American physicist Albert Einstein (1879–1955).

highlands Oldest exposed areas on the surface of the Moon; extensively cratered and chemically distinct from the *maria.*

Hohmann transfer orbit The most efficient orbit-transfer path between two coplanar circular orbits. The maneuver consists of two impulsive high thrust burns (or firings) of a spacecraft's propulsion system. The technique was suggested in 1925 by the German engineer Walter Hohmann (1880–1945).

"housekeeping" (spacecraft) The collection of routine tasks that must be performed to keep a spacecraft functioning properly during an orbital flight or interplanetary mission.

Hubble Space Telescope (HST) A cooperative European Space Agency (ESA) and NASA program to operate a long-lived, space-based, optical observatory. Launched on April 25, 1990, by NASA's space shuttle *Discovery* (STS-31 mission), subsequent on-orbit repair and refurbishment missions have allowed this powerful Earth-orbiting optical observatory to revolutionize knowledge of the size, structure, and makeup of the universe. Named in honor of the American astronomer Edwin Powell Hubble (1889–1953).

***Huygens* probe** A scientific probe sponsored by the European Space Agency (ESA) and named after the Dutch astronomer, Christiaan Huygens (1629–95). The *Cassini* mother spacecraft delivered *Huygens* to Saturn and the probe successfully plunged into the nitrogen-rich atmosphere of Titan (Saturn's largest moon) on January 14, 2005.

hyperbolic orbit An orbit in the shape of a hyperbola; all interplanetary, flyby spacecraft follow hyperbolic orbits, both for Earth-departure and again upon arrival at the target planet.

Ida A heavily cratered, irregularly shaped asteroid about 35 × 15 × 13 miles (56 × 24 × 21 km) in size that has its own tiny natural satellite, Dactyl.

image The representation of a physical object or scene formed by a mirror, lens, or electro-optical recording device.

Imbrium basin Large (about 810 miles [1,300 km] across), ancient impact crater on the Moon.

impact The event or moment when a high-speed object (such as an asteroid, comet, meteoroid, or human-made space probe) strikes the surface of a planetary body.

impact crater The crater or basin formed on the surface of a planetary body as a result of the high-speed impact of a meteoroid, asteroid, comet, or human-made space probe.

inclination (symbol: *i*) One of the six Keplerian (orbital) elements; inclination describes the angle of an object's orbital plane with respect to the central body's equator. For Earth-orbiting objects, the orbital plane always goes through the center of Earth, but it can tilt at any angle relative to the equator. By general agreement, inclination is the angle between Earth's equatorial plane and the object's orbital plane measured counterclockwise at the ascending node.

inferior planet(s) Mercury and Venus—the two planets that have orbits which lie inside Earth's orbit around the Sun. Compare with SUPERIOR PLANET(S).

in-flight phase The flight of a robot spacecraft from launch to the time of planetary flyby, encounter and orbit, or impact.

infrared radiation (IR) That portion of the electromagnetic (EM) spectrum between the optical (visible) and radio wavelengths. The infrared region extends from about one micrometer (μm) to 1,000 μm wavelength.

inner planets The terrestrial planets—Mercury, Venus, Earth, and Mars—all of which have orbits around the Sun that lie inside the main asteroid belt. Compare with OUTER PLANETS.

insertion The process of putting an artificial satellite or spacecraft into orbit.

interferometer An instrument that achieves high angular resolution by combining signals from at least two widely separated telescopes (optical interferometer) or a widely separated antenna array (radio interferometer).

interplanetary Between the planets; within the solar system.

interplanetary dust (IPD) Tiny particles of matter (generally less than 100 micrometers [μm] in diameter) that exist in outer space within the confines of this solar system.

interstellar Between or among the stars.

interstellar medium (ISM) The gas and tiny dust particles that are found between the stars in the Milky Way Galaxy. More than 100 different types of molecules have been discovered in interstellar space, including many organic molecules.

interstellar probe A conceptual, highly automated, robot spacecraft launched by human beings in this solar system (or perhaps by intelligent alien beings in some other solar system) to explore nearby star systems.

ionosphere That portion of Earth's upper atmosphere, extending from an altitude of about 30 to 620 miles (50 to 1,000 km), in which ions and free electrons exist in sufficient quantity to reflect radio waves.

Ishtar Terra A very large highland plateau in the northern hemisphere of Venus, about 3,100 miles (5,000 km) long and 370 miles (600 km) wide.

jettison To discard or toss away.

Jovian planet A large (Jupiter-like) planet characterized by a great total mass, low average density, mostly liquid interior, and an abundance of the lighter elements (especially hydrogen and helium). In this solar system, the Jovian planets are Jupiter, Saturn, Uranus, and Neptune.

Kennedy Space Center (KSC) Sprawling NASA spaceport on the east central coast of Florida adjacent to Cape Canaveral Air Force Station. Launch site (Complex 39) and primary landing/recovery site for the space shuttle.

Keplerian elements The six parameters that uniquely specify the position and path of a satellite (natural or human-made) in its orbit around the primary body as a function of time.

Kepler's laws The three empirical laws describing the motion of a satellite (natural or human-made) in orbit around its primary body, formulated in the early 17th century by the German astronomer Johannes Kepler (1571–1630).

Kuiper belt A region in the outer solar system beyond Neptune, and extending out to perhaps 1,000 astronomical units, that contains millions of icy planetesimals. These icy objects range in size from tiny particles to Pluto-sized planetary bodies. The Dutch-American astronomer Gerard Peter Kuiper (1905–73) first suggested the existence of this disk-shaped reservoir of icy objects in 1951. *See also* OORT CLOUD.

Lagrangian libration point The five points in outer space (called L_1, L_2, L_3, L_4, *and* L_5) where a small object can experience a stable orbit in spite of the force of gravity exerted by two much more massive celestial bodies when they orbit about a common center of mass. Joseph Louis Lagrange calculated the existence and location of these points in 1772.

lander (spacecraft) A spacecraft designed to safely reach the surface of a planet or moon and survive long enough on the planetary body to collect useful scientific data that it sends back to Earth by telemetry.

launch window An interval of time during which a launch may be made to satisfy some mission objective. Sometimes just a short period each day for a certain number of days.

light-year (symbol: ly) The distance light (or other forms of electromagnetic radiation) travels in one year. One light-year equals a distance of approximately 5.87×10^{12} miles (9.46×10^{12} km) or 63,240 astronomical units (AU).

line of apsides The line connecting the two points of an orbit that are nearest and farthest from the center of attraction, such as the perigee and apogee of a satellite in orbit around Earth.

line of sight (LOS) The straight line between a sensor or the eye of an observer and the object or point being observed. Sometimes called the *optical path.*

long period comet A comet with an orbital period around the Sun greater than 200 years. Compare with SHORT PERIOD COMET.

low Earth orbit (LEO) A circular orbit just above Earth's sensible atmosphere at an altitude of between 185 and 250 miles (300 and 400 km).

Luna A series of robot spacecraft sent to explore the Moon in the 1960s and 1970s by the former Soviet Union.

lunar Of or pertaining to Earth's natural satellite, the Moon.

lunar highlands The light-colored, heavily cratered mountainous part of the Moon's surface.

lunar orbiter A spacecraft placed in orbit around the Moon; specifically, the series of five *Lunar Orbiter* robot spacecraft NASA used from 1966 to 1967 to photograph the Moon's surface precisely, in support of the Apollo Project.

lunar probe A planetary probe for exploring and reporting conditions on or about the Moon.

Lunar Prospector A NASA orbiter spacecraft that circled the Moon from 1998 to 1999, searching for mineral resources. Data collected by this robot spacecraft suggest the possible presence of water-ice deposits in the Moon's permanently shadowed polar regions.

lunar rover Crewed or automated (robot) rover vehicles used to explore the Moon's surface. NASA's lunar rover vehicle (LRV) served as a Moon car for Apollo Project astronauts during the *Apollo 15, 16,* and *17* expeditions. Russian *Lunokhod 1* and *2* robot rovers were operated on Moon from Earth between 1970 and 1973.

Lunokhod A Russian eight-wheeled robot vehicle, controlled by radio wave signals from Earth and used to perform lunar surface exploration during the *Luna 17* (1970) and *Luna 21* (1973) missions to the Moon.

Magellan mission The planetary orbiter spacecraft that used its powerful radar-imaging system to make detailed surface maps of cloud-covered Venus from 1990 to 1994. NASA named this robot spacecraft *Magellan* after the Portuguese explorer Ferdinand Magellan (1480–1521).

magnetometer An instrument for measuring the strength and sometimes the direction of a magnetic field.

magnetosphere The region around a planet in which charged atomic particles are influenced (and often trapped) by the planet's own magnetic field, rather than the magnetic field of the Sun as projected by the solar wind.

main-belt asteroid One located in the asteroid belt between Mars and Jupiter.

manipulator The part of a robot capable of grasping or handling; a mechanical device on a robot spacecraft designed to handle objects.

maria (singular: mare) Latin word for "seas." Originally used by the Italian astronomer Galileo Galilei (1564–1642) to describe the large, dark ancient lava flows on the lunar surface, since he and other 17th century astronomers thought these features were bodies of water on the Moon's surface. Following tradition, this term is still used by modern astronomers.

Mariner A series of NASA planetary exploration robot spacecraft that performed important flyby and orbital missions to Mercury, Mars, and Venus in the 1960s and 1970s.

***Mars Exploration Rover (MER) 2003* mission** In 2003, NASA launched identical twin Mars rovers designed to operate on the surface of the Red Planet. *Spirit (MER-A)* was launched from Cape Canaveral on June 10, 2003, and successfully landed on Mars on January 4, 2004. *Opportunity (MER-B)* was launched from Cape Canaveral on July 7, 2003, and successfully landed on Mars on January 25, 2004. Both soft landings used the airbag bounce-and-roll arrival demonstrated during the *Mars Pathfinder* mission. *Spirit* landed in Gusev Crater and *Opportunity* landed at Terra Meridiania. As of July 1, 2006, both rovers were still functioning.

Mars Global Surveyor (MGS) A NASA orbiter spacecraft launched in November 1996 that has performed detailed studies of the Martian surface and atmosphere since March 1999.

Mars Odyssey Launched from Cape Canaveral by NASA in April 2001, the *2001 Mars Odyssey* is an orbiter spacecraft designed to conduct a detailed exploration of Mars, with emphasis being given to the search for geological features that would indicate the presence of water—flowing on the surface in the past, or frozen currently in subsurface reservoirs.

Mars Pathfinder An innovative NASA mission that successfully landed a Mars surface rover—a small robot called *Sojourner*—in the Ares Vallis region of the Red Planet in July 1997. For more than 80 days, human beings on Earth used teleoperation and telepresence to cautiously drive the six-wheeled mini-rover to interesting locations on the Martian surface.

Maxwell Montes A prominent mountain range on Venus located in Ishtar Terra, containing the highest peak (6.8-mile- [11-km-] altitude) on the planet. The mountain range was named after the Scottish theoretical physicist James Clerk Maxwell (1831–79).

minor planet *See* ASTEROID.

modulation The process of modifying a radio-frequency (RF) signal by shifting its phase, frequency, or amplitude to carry information. The respective processes are called phase modulation (PM), frequency modulation (FM), and amplitude modulation (AM).

moon A small natural celestial body that orbits a larger one; a natural satellite.

Moon Earth's only natural satellite and closest celestial neighbor. It has an equatorial diameter of 2,160 miles (3,476 km), keeps the same side (nearside) toward Earth, and orbits at an average distance (center-to-center) of 238,910 miles (384,488 km).

mother spacecraft A exploration spacecraft that carries and deploys one or several atmospheric probes, lander spacecraft, and/or lander and rover spacecraft combinations when it arrives at a target planet. The mother spacecraft then relays data back to Earth and may also orbit the planet in order to perform its own scientific mission. NASA's *Galileo* spacecraft to Jupiter and *Cassini* spacecraft to Saturn are examples.

multispectral sensing The remote-sensing method of simultaneously collecting several different bands (wavelength regions) of electromagnetic radiation (such as the visible, the near-infrared, and the thermal-infrared bands), when observing an object or region of interest.

nadir The direction from a spacecraft directly down toward the center of a planet. It is the opposite of the ZENITH.

NASA The National Aeronautics and Space Administration, the civilian space agency of the United States. Created in 1958 by an act of Congress, NASA's overall mission is to plan, direct, and conduct civilian (including scientific) aeronautical and space activities for peaceful purposes.

near-Earth asteroid (NEA) An inner solar-system asteroid whose orbit around the Sun brings it close to Earth, perhaps even posing a collision threat in the future.

nearside The side of the Moon that always faces Earth.

New Horizons Pluto-Kuiper Belt Flyby A reconnaissance-type exploration mission that will help scientists understand the icy worlds at the outer edge of the solar system. Launched on January 19, 2006, the *New Horizons* spacecraft could perform a flyby of Pluto and its moon (Charon) as early as 2015, depending on the exact gravity-assist trajectory. The robot spacecraft would then continue beyond Pluto and visit one or more Kuiper belt objects (of opportunity) by 2026. This robot spacecraft's long journey will help resolve some basic questions about the surface features and properties of these distant icy bodies as well as their geology, interior makeup, and atmospheres.

nuclear-electric propulsion (NEP) A space-deployed propulsion system that uses a space-qualified, compact nuclear reactor to produce the electricity needed to operate a space vehicle's electric propulsion engine(s).

Olympus Mons A huge mountain about 405 miles (650 km) wide and rising 16 miles (26 km) above the surrounding plains—the largest known volcano on Mars.

Oort cloud The large number (about 10^{12}) or cloud of comets postulated in 1950 by the Dutch astronomer, Jan Hendrik Oort (1900–92) to orbit the Sun at an enormous distance—ranging from some 50,000 and 80,000 astronomical units.

orbit The path followed by body in space, generally under the influence of gravity—as, for example, a satellite around a planet.

orbital injection The process of providing a spacecraft with sufficient velocity to establish an orbit.

orbital period The interval between successive passages of a spacecraft through the same point in its orbit. Often called PERIOD.

orbiter A robot spacecraft especially designed to travel through interplanetary space, achieve a stable orbit around the target planet (or other celestial body), and then conduct a program of detailed scientific investigation.

outer planets The planets in the solar system with orbits greater than the orbit of Mars, including Jupiter, Saturn, Uranus, Neptune, and Pluto.

outer space Any region beyond Earth's atmospheric envelope—usually considered to begin at 62–124 miles (100–200 km) altitude.

parking orbit A temporary (but stable) orbit of a spacecraft around a celestial body. This type of orbit is used for assembly and/or transfer of equipment, or to wait for conditions favorable for departure from that orbit.

peri- A prefix meaning near.

perigee The point at which a satellite's orbit is the closest to Earth; the minimum altitude attained by an Earth-orbiting object. Compare with APOGEE.

perihelion The point in an elliptical orbit around the Sun that is nearest to the center of the Sun. Compare with APHELION.

perilune The point in an elliptical orbit around the Moon that is nearest to the lunar surface. Compare with APOLUNE.

period (orbital) The time taken by a satellite to travel once around its orbit.

periodic comet A comet with a period of less than 200 years. Also called a *short period comet.*

Phobos The larger, innermost of the two small moons of Mars, discovered in 1877 by the American astronomer, Asaph Hall (1829–1907). *See also* DEIMOS.

photometer An instrument that measures light intensity and the brightness of celestial objects, such as stars.

***Pioneer 10, 11* spacecraft** NASA's twin robot spacecraft that were the first to navigate the main asteroid belt, the first to visit Jupiter (1973 and 1974), the first to visit Saturn (*Pioneer 11* in 1979), and the first human-made objects to leave the solar system (*Pioneer 10* in 1983). Each spacecraft is now on a different trajectory to the stars, carrying a special message (the Pioneer plaque) for any intelligent alien civilization that might find it millions of years from now.

Pioneer Venus mission Two spacecraft launched by NASA to Venus in 1978. *Pioneer 12* was an orbiter spacecraft that gathered data from 1978–92. The *Pioneer Venus Multiprobe* served as a mother-spacecraft, launching one large and three identical small planetary probes into Venus' atmosphere (December 1978).

pitch The rotation of a spacecraft about its lateral axis. *See also* ROLL; YAW.

pixel Contraction for picture element; the smallest unit of information on a screen or in an image.

planet A nonluminous celestial body that orbits around the Sun or some other star. The name "planet" comes from the ancient Greek *planetes* ("wanderers")—since early astronomers identified the planets as the wandering points of light relative to the fixed stars. There are nine major planets in the solar system and numerous minor planets (or asteroids). The distinction between a planet and a large satellite is not always precise. The Moon is nearly the size of Mercury and is very large in comparison to Earth—suggesting the Earth-Moon system might easily be treated as a double-planet system.

planetary probe An instrument-containing robot spacecraft deployed in the atmosphere or on the surface of a planetary body, in order to obtain environmental information.

planetesimals Small rock and rock/ice celestial objects found in the solar system, ranging from 0.06 mile (0.1 km) to about 62 miles (100 km) diameter. *See also* CENTAUR GROUP.

planet fall The act of landing of a spacecraft or space vehicle on a planet or moon.

plutino ***(little Pluto)*** Any of the numerous, small (~62-mile [100-km] diameter), icy celestial bodies that occupy the inner portions of the Kuiper belt, and whose orbital motion resonance with Neptune resembles that of Pluto—namely, that each object completes two orbits around the Sun in the time it takes Neptune to complete three orbits. *See also* TRANS-NEPTUNIAN OBJECT (TNO).

polar orbit An orbit around a planet (or primary body) that passes over or near its poles; an orbit with an inclination of about 90 degrees.

primary body The celestial body around which a satellite, moon, or other object orbits, or from which it is escaping or toward which it is falling.

prograde orbit An orbit having an inclination of between 0 and 90 degrees.

Quaoar Large, icy world with a diameter of about 780 miles (1,250 km) located in the Kuiper belt about 1 million miles (1.6 million km) beyond Pluto, first observed in June 2004.

radiation belt The region(s) in a planet's magnetosphere where there is a high density of trapped atomic particles from the solar wind.

radio frequency (RF) The portion of the electromagnetic spectrum useful for telecommunications with a frequency range between 10,000 and 3×10^{11} hertz.

radioisotope thermoelectric generator (RTG) Compact space nuclear-power system that uses direct conversion (based on the thermoelectric principle) to transform the thermal energy from a radioisotope source (generally plutonium-238) into electricity. All NASA spacecraft that have explored the outer regions of the solar system have used RTGs for their electric power.

Ranger Project The first NASA robot spacecraft sent to the Moon in the 1960s. These hard-impact planetary probes were designed to take a series of television images of the lunar surface before crash landing.

Red Planet The planet Mars—so named because of its distinctive reddish soil.

regolith (lunar) The unconsolidated mass of surface debris that overlies the Moon's bedrock. This blanket of pulverized lunar dust and soil was created by millions of years of meteoric and comentary impacts.

remote manipulator system (RMS) The dextrous, Canadian-built, 50-foot- (15.2-m-) long articulated arm that is remotely controlled by astronauts from the aft flight deck of NASA's space shuttle.

remote sensing The sensing of an object or phenomenon, using different portions of the electromagnetic spectrum, without having the sensor in direct contact with the object being studied.

rendezvous The close approach of two or more spacecraft in the same orbit, so that docking can take place. Orbiting objects meet at a preplanned location and time, and slowly come together with essentially zero relative velocity.

resolution The smallest detail (measurement) that can be distinguished by a sensor system under specific conditions, such as its spatial resolution or spectral resolution.

retrograde motion Motion in a reverse or backward direction.

ring (planetary) A disk of matter that encircles a planet. Such rings usually contain ice and dust particles, ranging in size from microscopic fragments up to chunks that are tens of meters in diameter.

robot spacecraft A semi-automated or fully automated spacecraft, capable of executing its primary exploration mission with minimal or no human supervision.

roll The rotational or oscillatory movement of a spacecraft about its longitudinal (lengthwise) axis. *See also* PITCH; YAW.

rover A human-crewed or robot space vehicle used to travel across and explore the surface of a planet or moon. The Mars Exploration Rovers *Spirit* and *Opportunity* are examples.

satellite A smaller (secondary) body in orbit around a larger primary body. For example, Earth is a natural satellite of the Sun, while the Moon is a natural satellite of Earth. A human-made spacecraft placed in orbit around Earth or another planet is called an artificial satellite—or more commonly just a satellite.

science payload The collection of scientific instruments on a robot spacecraft.

self-replicating system (SRS) An advanced robot system that was first postulated by the Hungarian-born, German-American mathematician John von Neumann (1903–57). Space-age versions of the SRS would be very smart machines capable of gathering materials, performing self-maintenance, manufacturing desired products, and even making copies of themselves (self-replication).

sensible atmosphere That portion of a planet's atmosphere that offers resistance to a body passing through it.

sensor The portion of a scientific instrument that detects and/or measures some physical phenomenon.

short period comet A comet with an orbital period of fewer than 200 years.

SI units The international system of units (the metric system) that uses the meter (m), kilogram (kg), second (s) as its basic units of length, mass, and time, respectively.

soft landing The act of landing on the surface of a planet or moon without damaging any portion of a spacecraft or its payload, except possibly an expendable landing-gear structure. Compare with HARD LANDING.

sol A Martian day (about 24 hours, 37 minutes, and 23 seconds in duration); seven sols is equal to about 7.2 Earth-days.

solar cell A direct-conversion device that transforms incoming sunlight (solar energy) directly into electricity. It is used extensively (in combination with rechargeable storage batteries) as the prime source of electric power for spacecraft orbiting Earth or on missions within the inner solar system. Also called *photovoltaic cell.*

solar-electric propulsion (SEP) A low-thrust propulsion system that uses solar cells to provide the electricity for a spacecraft's electric propulsion rocket engines.

solar flare A highly concentrated, explosive release of electromagnetic radiation and nuclear particles within the Sun's atmosphere near an active sunspot.

solar panel The winglike assembly of solar cells used by a spacecraft to convert sunlight (solar energy) directly into electrical energy. Also called a *solar array.*

solar system (1) Any star and its gravitationally bound collection of nonluminous objects, such as planets, asteroids, and comets. (2) Humans' home solar system, consisting of the Sun and all of the objects bound to it by gravitation. This includes nine major planets with more than 60 known moons, more than 2,000 minor planets, and a very large number of comets. Except for the comets, all the other celestial objects travel around the Sun in the same direction.

solar wind The variable stream of plasma (i.e., electrons, protons, alpha particles, and other atomic nuclei) that flows continuously outward from the Sun into interplanetary space.

space-based astronomy The use of astronomical instruments on spacecraft in orbit around Earth and in other locations throughout the solar system, to view the universe from above Earth's atmosphere. Major breakthroughs in astronomy, astrophysics, and cosmology have occurred because of the unhampered viewing advantages provided by space platforms.

spacecraft A platform that can function, move, and operate in outer space or on a planetary surface. Spacecraft can be human-occupied or uncrewed (robot) platforms. They can operate in orbit around Earth, or while on an interplanetary trajectory to another celestial body. Some spacecraft travel through space and orbit another planet, while others descend to a planet's surface, making a hard landing (collision impact) or a (survivable) soft landing. Exploration spacecraft are often categorized as flyby, orbiter, atmospheric probe, lander, or rover spacecraft.

spacecraft clock The time-keeping component within a spacecraft's command and data-handling system. It meters the passing time during a mission and regulates nearly all activity within the spacecraft.

space vehicle The general term describing a crewed or robot vehicle capable of traveling through outer space. An aerospace vehicle can operate both in outer space and in Earth's atmosphere.

spectrometer An optical instrument that splits incoming visible light (or other electromagnetic radiation) from a celestial object into a spectrum by diffraction and then measures the relative amplitudes of the different wavelengths. Infrared and ultraviolet spectrometers are often carried on scientific spacecraft.

speed of light (symbol: c) The speed at which electromagnetic radiation (including light) moves through a vacuum; a universal constant equal to approximately 186,000 miles per second (300,000 km/s).

spin stabilization Directional stability of a spacecraft, obtained as a result of spinning the moving body around its axis of symmetry.

Sputnik 1 Launched by the former Soviet Union on October 4, 1957, it was the first satellite to orbit Earth. *Sputnik* means "fellow traveler." This simple, spherically shaped, 184-pound- (84-kg-mass) Russian spacecraft inaugurated the space age.

star probe A conceptual NASA robot spacecraft, capable of approaching within 620,000 miles (1 million km) of the Sun's surface (photosphere) and providing the first in situ measurements of its corona (outer atmosphere).

station keeping The sequence of maneuvers that maintains a spacecraft in a predetermined orbit or on a desired trajectory.

sun-synchronous orbit A very useful polar orbit that allows a satellite's sensors to maintain a fixed relation to the Sun during each local data collection—an important feature for Earth-observing spacecraft or scientific-orbiter spacecraft conducting extended studies of other planets.

superior planet A planet that has an orbit around the Sun outside Earth's orbit—Mars, Jupiter, Saturn, Uranus, Neptune, or Pluto.

Surveyor Project The NASA Moon exploration effort in which five lander spacecraft softly touched down onto the lunar surface between 1966 and 1968—the robot precursor to the Apollo Project human expeditions.

synchronous orbit An orbit around a planet (or primary body) in which a satellite (secondary body) moves around the planet in the same amount of time it takes the planet to rotate on its axis.

synthetic aperture radar A space-based radar system that computer-correlates the echoes of signals emitted at different points along a satellite's orbit, thereby mimicking the performance of a radar-antenna system many times larger than the one actually being used.

tail (cometary) The long, wispy portion of some comets, containing the gas (plasma tail) and dust (dust tail) streaming out of the comet's head (coma) as it approaches the Sun. The plasma tail interacts with the solar wind and points straight back from the Sun, while the dust tail can be curved and fan-shaped.

telecommunications The transmission of information over great distances using radio waves or other portions of the electromagnetic spectrum.

telemetry The process of taking measurements at one point and transmitting the information via radio waves over some distance to another location for evaluation and use. Telemetered data on a robot spacecraft's communications downlink often includes scientific data, as well as spacecraft state-of-health data.

teleoperation The technique by which a human controller operates a versatile robot system that is at a distant, often hazardous, location. High-resolution vision and tactile sensors on the robot, reliable telecommunications links, and computer-generated virtual reality displays enable the human worker to experience telepresence.

telepresence The process, supported by an information-rich control station environment, that enables a human controller to manipulate a distant robot through teleoperation and almost feel physically present in the robot's remote location.

telescope An instrument that collects electromagnetic radiation from a distant object so as to form an image of the object or to permit the radiation signal to be analyzed. Optical (astronomical) telescopes are divided into two general classes: refracting telescopes and reflecting telescopes. Earth-based astronomers also use large radio telescopes, while orbiting observatories use optical, infrared radiation, ultraviolet radiation, x-ray, and gamma-ray telescopes to study the universe.

terrestrial planets In addition to Earth, the planets Mercury, Venus, and Mars—all of which are relatively small, high-density planetary bodies composed of metals and silicates with shallow or no atmospheres, in comparison to the Jovian planets.

Titan The largest moon of Saturn, discovered in 1655 by the Dutch astronomer, Christiaan Huygens (1629–95). It is the only moon in the solar system with a significant atmosphere.

trajectory The three-dimensional path traced by any object moving because of an externally applied force; the flight path of a space vehicle.

transfer orbit An elliptical interplanetary trajectory tangent to the orbits of both the departure planet and target planet (or moon). *See also* HOHMANN TRANSFER ORBIT.

Trans-Neptunian object (TNO) Any of the numerous small, icy celestial bodies that lie in the outer fringes of the solar system beyond Neptune. TNOs include plutinos and Kuiper belt objects.

ultraviolet astronomy The branch of astronomy, conducted primarily from space-based observatories, that uses the ultraviolet portion of the electromagnetic spectrum to study unusual interstellar and intergalactic phenomena.

ultraviolet radiation The region of the electromagnetic spectrum between visible (violet) light and X-rays, with wavelengths from 400 nanometers (just past violet light) down to about 10 nanometers (the extreme ultraviolet cutoff).

universal time (UT) The worldwide civil time standard, equivalent to Greenwich mean time.

uplink The telemetry signal sent from a ground station to a spacecraft or planetary probe.

Valles Marineris An extensive canyon system on Mars near the planet's equator, discovered in 1971 by NASA's *Mariner 9* spacecraft.

Vandenberg Air Force Base (VAFB) Located on the central California coast north of Santa Barbara, this U.S. Air Force facility is the launch site of military, NASA, and commercial space launches that require high inclination, especially polar orbits.

Venera The family of robot spacecraft (flybys, orbiters, probes, and landers) from the former Soviet Union that successfully explored Venus, including its inferno-like surface, between 1961 and 1984.

Viking Project NASA's highly successful Mars exploration effort in the 1970s in which two orbiter and two lander robot spacecraft conducted the first detailed study of the Martian environment and the first (albeit inconclusive) scientific search for life on the Red Planet.

Voyager NASA's twin robot spacecraft that explored the outer regions of the solar system, visiting all the Jovian planets. *Voyager 1* encountered Jupiter (1979) and Saturn (1980) before departing on an interstellar trajectory. *Voyager 2* performed the historic grand-tour mission by visiting Jupiter (1979), Saturn (1981), Uranus (1986), and Neptune (1989). Both RTG-powered spacecraft are now involved in the Voyager Interstellar Mission (VIM) and each carries a special recording ("Sounds of Earth")—a digital message for any intelligent species that finds them drifting between the stars millennia from now.

X-ray A penetrating form of electromagnetic radiation of very short wavelength (approximately 0.01 to 10 nanometers) and high photon energy (approximately 100 electron volts to some 100 kiloelectron volts).

X-ray astronomy The branch of astronomy, primarily space-based, that uses characteristic X-ray emissions to study very energetic and violent processes throughout universe. X-ray emissions carry information about the temperature, density, age, and other physical conditions of celestial objects that produced them—including supernova remnants, pulsars, active galaxies, and energetic solar flares.

yaw The rotation or oscillation of a spacecraft about its vertical axis so as to cause the longitudinal axis of the craft to deviate from the flight path in its horizontal plane. *See also* PITCH; ROLL.

Yohkoh A Japanese solar X-ray observation satellite launched in 1991.

zenith The point on the celestial sphere vertically overhead. Compare with NADIR.

Zond A family of robot spacecraft from the former Soviet Union that explored the Moon, Mars, Venus, and interplanetary space in the 1960s.

Further Reading

RECOMMENDED BOOKS

Angelo, Joseph A., Jr. *The Dictionary of Space Technology.* Rev. ed. New York: Facts On File, Inc., 2004.

———. *Encyclopedia of Space Exploration.* New York: Facts On File, Inc., 2000.

———, and Irving W. Ginsberg, eds. *Earth Observations and Global Change Decision Making, 1989: A National Partnership.* Malabar, Fla.: Krieger Publishing, 1990.

Brown, Robert A., ed. *Endeavour Views the Earth.* New York: Cambridge University Press, 1996.

Burrows, William E., and Walter Cronkite. *The Infinite Journey: Eyewitness Accounts of NASA and the Age of Space.* Discovery Books, 2000.

Chaisson, Eric, and Steve McMillian. *Astronomy Today.* 5th ed. Upper Saddle River, N.J.: Pearson Prentice Hall, 2005.

Cole, Michael D. *International Space Station. A Space Mission.* Springfield, N.J.: Enslow Publishers, 1999.

Collins, Michael. *Carrying the Fire.* New York: Cooper Square Publishers, 2001.

Consolmagno, Guy J., et al. *Turn Left at Orion: A Hundred Night Objects to See in a Small Telescope—And How to Find Them.* New York: Cambridge University Press, 2000.

Damon, Thomas D. *Introduction to Space: The Science of Spaceflight.* 3d ed. Malabar, Fla.: Krieger Publishing Co., 2000.

Dickinson, Terence. *The Universe and Beyond.* 3d ed. Willowdater, Ont.: Firefly Books Ltd., 1999.

Heppenheimer, Thomas A. *Countdown: A History of Space Flight.* New York: John Wiley and Sons, 1997.

Kluger, Jeffrey. *Journey beyond Selene: Remarkable Expeditions Past Our Moon and to the Ends of the Solar System.* New York: Simon & Schuster, 1999.

Kraemer, Robert S. *Beyond the Moon: A Golden Age of Planetary Exploration, 1971–1978.* Smithsonian History of Aviation and Spaceflight Series. Washington, D.C.: Smithsonian Institution Press, 2000.

Lewis, John S. *Rain of Iron and Ice: The Very Real Threat of Comet and Asteroid Bombardment.* Reading, Mass.: Addison-Wesley, 1996.

Logsdon, John M. *Together in Orbit: The Origins of International Participation in the Space Station.* NASA History Division, Monographs in Aerospace History 11, Washington, D.C.: Office of Policy and Plans, November 1998.

Matloff, Gregory L. *The Urban Astronomer: A Practical Guide for Observers in Cities and Suburbs.* New York: John Wiley and Sons, 1991.

Neal, Valerie, Cathleen S. Lewis, and Frank H. Winter. *Spaceflight: A Smithsonian Guide.* New York: Macmillan, 1995.

Pebbles, Curtis L. *The Corona Project: America's First Spy Satellites.* Annapolis, Md.: Naval Institute Press, 1997.

Seeds, Michael A. Horizons: *Exploring the Universe.* 6th ed. Pacific Grove, Calif.: Brooks/Cole Publishing, 1999.

Sutton, George Paul. *Rocket Propulsion Elements.* 7th ed. New York: John Wiley & Sons, 2000.

Todd, Deborah, and Joseph A. Angelo, Jr. *A to Z of Scientists in Space and Astronomy.* New York: Facts On File, Inc., 2005.

EXPLORING CYBERSPACE

In recent years, numerous Web sites dealing with astronomy, astrophysics, cosmology, space exploration, and the search for life beyond Earth have appeared on the Internet. Visits to such sites can provide information about the status of ongoing missions, such as NASA's *Cassini* spacecraft as it explores the Saturn system. This book can serve as an important companion, as you explore a new Web site and encounter a person, technology phrase, or physical concept unfamiliar to you and not fully discussed within the particular site. To help enrich the content of this book and to make your astronomy and/or space technology–related travels in cyberspace more enjoyable and productive, the following is a selected list of Web sites that are recommended for your viewing. From these sites you will be able to link to many other astronomy or space-related locations on the Internet. Please note that this is obviously just a partial list of the many astronomy and space-related Web sites now available. Every effort has been made at the time of publication to ensure the accuracy of the information provided. However, due to the dynamic nature of the Internet, URL changes do occur and any inconvenience you might experience is regretted.

Selected Organizational Home Pages

European Space Agency (ESA) is an international organization whose task is to provide for and promote, exclusively for peaceful purposes, cooperation among European states in space research and technology and their applications. URL: http://www.esrin.esa.it. Accessed on April 12, 2005.

National Aeronautics and Space Administration (NASA) is the civilian space agency of the United States government and was created in 1958 by an act of Congress. NASA's overall mission is to plan, direct, and conduct American civilian (including scientific) aeronautical and space activities for peaceful purposes. URL: http://www.nasa.gov. Accessed on April 12, 2005.

National Oceanic and Atmospheric Administration (NOAA) was established in 1970 as an agency within the U.S. Department of Commerce to ensure the safety of the general public from atmospheric phenomena and to provide the public with an understanding of Earth's environment and resources. URL: http://www.noaa.gov. Accessed on April 12, 2005.

National Reconnaissance Office (NRO) is the organization within the Department of Defense that designs, builds, and operates U.S. reconnaissance satellites. URL: http://www.nro.gov. Accessed on April 12, 2005.

United States Air Force (USAF) serves as the primary agent for the space defense needs of the United States. All military satellites are launched from Cape Canaveral Air Force Station, Florida or Vandenberg Air Force Base, California. URL: http://www.af.mil. Accessed on April 14, 2005.

United States Strategic Command (USSTRATCOM) is the strategic forces organization within the Department of Defense, which commands and controls U.S. nuclear forces and military space operations. URL: http://www.stratcom.mil. Accessed on April 14, 2005.

Selected NASA Centers

Ames Research Center (ARC) in Mountain View, California, is NASA's primary center for exobiology, information technology, and aeronautics. URL: http://www.arc.nasa.gov. Accessed on April 12, 2005.

Dryden Flight Research Center (DFRC) in Edwards, California, is NASA's center for atmospheric flight operations and aeronautical flight research. URL: http://www.dfrc.nasa.gov. Accessed on April 12, 2005.

Glenn Research Center (GRC) in Cleveland, Ohio, develops aerospace propulsion, power, and communications technology for NASA. URL: http://www.grc.nasa.gov. Accessed on April 12, 2005.

Goddard Space Flight Center (GSFC) in Greenbelt, Maryland, has a diverse range of responsibilities within NASA, including Earth system science, astrophysics, and operation of the *Hubble Space Telescope* and other Earth-orbiting spacecraft. URL: http://www.nasa.gov/goddard. Accessed on April 14, 2005.

Jet Propulsion Laboratory (JPL) in Pasadena, California, is a government-owned facility operated for NASA by Caltech. JPL manages and operates NASA's deep-space scientific missions, as well as the NASA's Deep Space Network, which communicates with solar system exploration spacecraft. URL: http://www.jpl.nasa.gov. Accessed on April 12, 2005.

Johnson Space Center (JSC) in Houston, Texas, is NASA's primary center for design, development, and testing of spacecraft and associated systems for human space flight, including astronaut selection and training. URL: http://www.jsc.nasa.gov. Accessed on April 12, 2005.

Kennedy Space Center (KSC) in Florida is the NASA center responsible for ground turnaround and support operations, prelaunch checkout, and launch of the space shuttle. This center is also responsible for NASA launch facilities at Vandenberg Air Force Base, California. URL: http://www.ksc.nasa.gov. Accessed on April 12, 2005.

Langley Research Center (LaRC) in Hampton, Virginia, is NASA's center for structures and materials, as well as hypersonic flight research and aircraft safety. URL: http://www.larc.nasa.gov. Accessed on April 15, 2005.

Marshall Space Flight Center (MSFC) in Huntsville, Alabama, serves as NASA's main research center for space propulsion, including contemporary rocket engine development as well as advanced space transportation system concepts. URL: http://www.msfc.nasa.gov. Accessed on April 12, 2005.

Stennis Space Center (SSC) in Mississippi is the main NASA center for large rocket engine testing, including space shuttle engines as well as future generations of space launch vehicles. URL: http://www.ssc.nasa.gov. Accessed on April 14, 2005.

Wallops Flight Facility (WFF) in Wallops Island, Virginia, manages NASA's suborbital sounding rocket program and scientific balloon flights to Earth's upper atmosphere. URL: http://www.wff.nasa.gov. Accessed on April 14, 2005.

White Sands Test Facility (WSTF) in White Sands, New Mexico, supports the space shuttle and space station programs by performing tests on and evaluating potentially hazardous materials, space flight components, and rocket propulsion systems. URL: http://www.wstf.nasa.gov. Accessed on April 12, 2005.

Selected Space Missions

Cassini Mission is an ongoing scientific exploration of the planet Saturn. URL: http://saturn.jpl.nasa.gov. Accessed on April 14, 2005.

Chandra X-ray Observatory (CXO) is a space-based astronomical observatory that is part of NASA's Great Observatories Program. *CXO* observes the universe in the X-ray portion of the electromagnetic spectrum. URL: http://www.chandra.harvard.edu. Accessed on April 14, 2005.

Exploration of Mars is the focus of this Web site, which features the results of numerous contemporary and previous flyby, orbiter, and lander robotic spacecraft. URL: http://mars.jpl.nasa.gov. Accessed on April 14, 2005.

National Space Science Data Center (NSSDC) provides a worldwide compilation of space missions and scientific spacecraft. URL: http://nssdc.gsfc.nasa.gov/planetary. Accessed on April 14, 2005.

Voyager (Deep Space/Interstellar) updates the status of NASA's *Voyager 1* and *2* spacecraft as they travel beyond the solar system. URL: http://voyager.jpl.nasa.gov. Accessed on April 14, 2005.

Other Interesting Astronomy and Space Sites

Arecibo Observatory in the tropical jungle of Puerto Rico is the world's largest radio/radar telescope. URL: http://www.naic.edu. Accessed on April 14, 2005.

Astrogeology (USGS) describes the USGS Astrogeology Research Program, which has a rich history of participation in space exploration efforts and planetary mapping. URL: http://planetarynames.wr.usgs.gov. Accessed on April 14, 2005.

Hubble Space Telescope **(HST)** is an orbiting NASA Great Observatory that is studying the universe primarily in the visible portions of the electromagnetic spectrum. URL: http://hubblesite.org. Accessed on April 14, 2005.

NASA's Deep Space Network (DSN) is a global network of antennas that provide telecommunications support to distant interplanetary spacecraft and probes. URL: http://deepspace.jpl.nasa.gov/dsn. Accessed on April 14, 2005.

NASA's Space Science News provides contemporary information about ongoing space science activities. URL: http://science.nasa.gov. Accessed on April 14, 2005.

National Air and Space Museum (NASM) of the Smithsonian Institution in Washington, D.C., maintains the largest collection of historic aircraft and spacecraft in the world. URL: http://www.nasm.si.edu. Accessed on April 14, 2005.

Planetary Photojournal is a NASA/JPL– sponsored Web site that provides an extensive collection of images of celestial objects within and beyond the solar system, historic and contemporary spacecraft used in space exploration, and advanced aerospace technologies URL: http://photojournal.jpl.nasa.gov. Accessed on April 14, 2005.

Planetary Society is the nonprofit organization founded in 1980 by Carl Sagan and other scientists that encourages all spacefaring nations to explore other worlds. URL: http://planetary.org. Accessed on April 14, 2005.

Search for Extraterrestrial Intelligence (SETI) Projects at UC Berkeley is a Web site that involves contemporary activities in the search for extraterrestrial intelligence (SETI), especially a radio SETI project that lets anyone with a computer and an Internet connection participate. URL: http://www.setiathome.ssl.berkeley.edu. Accessed on April 14, 2005.

Solar System Exploration is a NASA-sponsored and -maintained Web site that presents the last events, discoveries and missions involving the exploration of the solar system. URL: http://solarsystem.nasa.gov. Accessed on April 14, 2005.

Space Flight History is a gateway Web site sponsored and maintained by the NASA Johnson Space Center. It provides access to a wide variety of interesting data and historic reports dealing with (primarily U.S.) human space flight. URL: http://www11.jsc.nasa.gov/history. Accessed on April 14, 2005.

Space Flight Information (NASA) is a NASA-maintained and -sponsored gateway Web site that provides the latest information about human spaceflight activities, including the *International Space Station* and the space shuttle. URL: http://spaceflight.nasa.gov Accessed on April 14, 2005.

Index

Italic page numbers indicate illustrations.

A

G

H

I